Contemporary Sport, Leisure and Ergonomics

Understanding the 'human operator' is a central concern of both ergonomists and sport and exercise scientists. This cutting-edge collection of international research papers explores the interface between physical, cognitive and occupational ergonomics and sport and exercise science, illuminating our understanding of 'human factors' at work and at play. Drawing on a wide diversity of disciplines, including:

- applied anatomy
- biomechanics
- physiology
- engineering
- psychology
- design

The book explores themes of central importance within contemporary ergonomics and sport and exercise science, such as performance, health, environment, technology and special populations. *Contemporary Sport, Leisure and Ergonomics* establishes important methodological connections between the disciplines, advancing the research agenda within each. It is essential reading for all serious ergonomists and human scientists.

Thomas Reilly is Professor of Sports Science and Director of the Research Institute for Sport and Exercise Sciences at Liverpool John Moores University, and President of the World Commission of Science and Sports.

Greg Atkinson is Professor of Chronobiology in the School of Sport and Exercise Sciences, Liverpool John Moores University.

Contemporary Sport, Leisure and Ergonomics

Edited by
Thomas Reilly and Greg Atkinson

Routledge
Taylor & Francis Group

LONDON AND NEW YORK

First published 2009 by Routledge
2 Park Square, Milton Park, Abingdon, Oxon OX14 4RN

Simultaneously published in the USA and Canada
by Routledge
270 Madison Avenue, New York, NY 10016

Routledge is an imprint of the Taylor & Francis Group, an informa business

British Library Cataloguing in Publication Data
A catalogue record for this book is available from the British Library

Library of Congress Cataloging-in-Publication Data
Contemporary sport, leisure, and ergonomics / edited by Thomas Reilly and
Greg Atkinson.
    p. cm.
    1. Sports—Physiological aspects. I. Reilly, Thomas, 1941-II. Atkinson, G.
(Greg)
  RC1235.C66 2009
  613.7′1—dc22
                                     2008055400

ISBN13: 978-0-415-47272-2 hbk
ISBN13: 978-0-203-89245-9 ebk

ISBN10: 0-415-47272-5 hbk
ISBN10: 0-203-89245-3 ebk

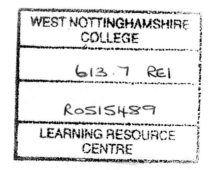

Contents

Preface

This book constitutes the proceedings of the Sixth International Conference on Sport, Leisure and Ergonomics held at Burton Manor, Cheshire in November 2007. The event was sponsored by the Ergonomics Society (now the Institute for Ergonomics and Human Factors) and co-sponsored by the World Commission of Science and Sports. The event is held every four years, following from the inaugural meeting in 1987.

The contents of the book demonstrate the common areas of interest between those applying the human sciences in sport and leisure and in occupational ergonomics contexts. The broad aim of the conference is to bring together specialists from these distinct areas so that knowledge between them may be shared. The book serves as a facility for ergonomists and sport and exercise scientists to communicate across their respective domains.

Greg Atkinson and Thomas Reilly, December 2008

Introduction

The areas in which ergonomics techniques may be applied are wide-ranging and cover sport, leisure and recreation as well as occupational and industrial contexts. Those practitioners working in the former areas are most likely to have had their academic training within the sports sciences whereas ergonomists operating in the latter area typically acquire their expertise in individual disciplines of the human sciences, in engineering or in an interdisciplinary ergonomics programme. The formal occasions in which knowledge is shared across these two domains are few and so the present collection of research reports provides a unique opportunity to view some common ground between them.

A common feature of all ergonomics work is the central focus on the individual, the so-called human operator. The concern is how the individual is related to the immediate activity, equipment used or machinery operated, the postures adopted or enforced, the working environment – its layout, organisation and prevailing conditions. In an ideal occupational context the task is altered to suit the characteristics of the individual, environmental stress is reduced and the individual is protected from harm. In a recreational context protection is still a priority but experience of heavy physical loading may be a part of mission objectives. Competitive sport is most uncompromising and, in the absence of a comfort zone, individual characteristics may need to be altered by training or acclimatization to meet the arduous physical demands entailed in successful performance.

The themes that stimulate projects within the ergonomics community have endured since the Ergonomics Research Society (now the Institute for Ergonomics and Human Factors) was established in 1949. They include the concepts of fatigue and stress, their reduction and elimination. There is a persisting interest in relative loading, its quantification and interpretation. Progress is marked by a continual search for improvement in methods of investigation and developing research models to suit prevailing problems. There may be a need to choose between laboratory and field projects, simulations or real-life observations, quantitative or qualitative approaches, or an appropriate mixture of methods may suit in addressing the research question. Compared to their predecessors, contemporary ergonomists have superior analytical facilities for data capture and analysis, better opportunities for simulations using virtual reality environments and better modelling possibilities using artificial neural networks.

The themes that stimulate projects within the ergonomics community have endured since the Ergonomics Research Society (now the Institute for Ergonomics and Human Factors) was established in 1949. They include the concepts of fatigue and stress, their reduction and elimination. There is a persisting interest in relative loading, its quantification and interpretation. Progress is marked by a continual search for improvement in methods of investigation and developing research models to suit prevailing problems. There may be a need to choose between laboratory and field projects, simulations or real-life observations, quantitative or qualitative approaches, or an appropriate mixture of methods may suit in addressing the research question. Compared to their predecessors, contemporary ergonomists have superior analytical facilities for data capture and analysis, better opportunities for simulations using virtual reality environments and better modelling possibilities using artificial neural networks.

Yet, human factors issues continue to arise whether the concern is with traditional industry, new technology or contemporary sport and leisure. The world of work not only generates recurring problems but also raises new questions for ergonomists to solve. The landscape of professional sport has changed profoundly over the last two decades: the systematised preparation and training programmes that participants commit themselves to bear little resemblance to the self-imposed stresses that those in leisure activities endure. Nevertheless, the risk of harm in leisure and recreational contexts cannot be dismissed easily – whether on water, land or air, indoors or outdoors. Highlighting of risks and difficulties in these domains raises awareness among participants and allows them to meet these encounters with nature with feelings of satisfaction.

The content of the current volume represents a selection of the material presented at the Sixth International Conference on Sport, Leisure and Ergonomics held at Burton Manor College in November 2007. The event has been held every four years since the inaugural meeting in 1987. The Proceedings of the meetings have been published either as edited books following peer-review of manuscripts (Atkinson and Reilly, 1995; Reilly and Greeves, 2002) or as special issues of the journal Ergonomics. Over this 20-year period, 180 scientific papers have been published arising from the conference.

A breakdown of the subject-area in those proceedings is shown in Table 1. The most frequent attention has been given to analysis of technique followed by equipment design and corporate health and fitness. Measurement methods, musculoskeletal loading and environmental stress are among other areas that have generated projects. An alternative configuration of the material might have highlighted other areas, for example special populations. What is clear is the diversity of topics that are covered and the variety of approaches used to investigate them.

The book is organised into five separate sections. They include musculoskeletal loading, occupational ergonomics, competitive sport, diet and fitness assessment, and expertise. The allocation of contributions to sections is in some instances arbitrary due to the overlap between topics and the multidisciplinary nature of some of the contributions.

In view of the range of topics covered in this volume, the book should be of interest to a range of specialisms. They include both researchers and practitioners in the sports sciences and in occupational ergonomics. The content is of relevance also to students working their way towards qualifications in the sports sciences, in ergonomics and in related disciplines. It should be helpful in guiding those hoping to communicate the outcomes of their current work when the next conference is convened in 2011.

Table 1. Breakdown of content from the proceedings of the six series of international conferences on sport, leisure and ergonomics.

Topic		Number of publications
Technique analysis		23
Corporate health and fitness		18
Equipment design		
Clothing	3	
Protective devices	5	
Machines	5	
Shoes/Orthotics	5	18
Musculoskeletal loading		16
Measurement methods		14
Disability		13
Environmental stress		13
Body composition/Functional anatomy		12
Training responses		9
Circadian factors		7
Drugs/Ergogenic aids		7
Fitness assessment		7
Paediatric ergonomics		5
Sports coaching		5
Ageing		4
Computers in sport		3
Injury		3
Psychological stress		3
		180

References

Atkinson, G. and Reilly, T., 1995, *Sport, Leisure and Ergonomics,* (London: E. and F.N. Spon).
Reilly, T. and Greeves, J., 2002, *Advances in Sport, Leisure and Ergonomics,* (London: Routledge).

Greg Atkinson and Thomas Reilly

Part I
Musculoskeletal loading

CHAPTER ONE

Manifestations of shoulder fatigue in prolonged activities involving low-force contractions

Michiel de Looze[1,2], Tim Bosch[1] and Jaap van Dieën[2]

[1]TNO, Research Center Body@Work, Hoofddorp, The Netherlands
[2]Research Institute Move, Faculty of Human Movement Sciences,
VU University, Amsterdam, The Netherlands

1. INTRODUCTION

Monotonous activities involving repetitive hand and finger movements at low force levels are increasingly common in our daily life, both in terms of numbers of people exposed to these activities and in terms of individual exposure times. This shift is mainly due to the widespread automation of work processes promoting activities like computer work in offices and other settings, joy-stick steering and control in construction, distribution, transport, refuse collecting, and light pick-and-place tasks in manufacturing. The frequent use of computers at home, when surfing on the Internet, e- mailing, chatting, and gaming, further contributes to this phenomenon.

Despite the low force levels involved, these types of activity are not without musculoskeletal health risks, particularly in relation to the shoulder girdle region. For many years it has been assumed that muscular contraction levels at 15%MVC (maximum voluntary contraction) are acceptable in the work place (Rohmert, 1973), but this assumption has changed over the years. It has been shown that musculoskeletal disorders in the shoulder region are frequent even in jobs with force levels of 2-5%MVC (Jonsson, 1988). Others reported that shoulder problems may relate to muscle loads as low as 0.5 to 1% (Veiersted *et al.*, 1990; Jensen *et al.*, 1993). Most recently, statements have been brought forward that muscle loads are not acceptable at all if sustained over longer periods of time (Sjøgaard and Jensen, 2006).

Shoulder disorders that are related to low-level activities are generally assumed to be muscular in nature. They result from several underlying mechanisms like blood flow impairment, accumulation of Ca^{2+}, muscle damage due to internal shear forces, and selective motor unit recruitment (Visser and van Dieën, 2006). The exact nature of these mechanisms, the plausible interactions and any dose relationships are as yet not clear.

Many hypothesize that shoulder muscle fatigue is a precursor of shoulder complaints (e.g. Herberts and Kadefors, 1976; Rempel *et al.*, 1992; Takala, 2002). If so, muscle fatigue when measured during work would be a relevant biomarker for cumulative, exposure to repetitive work and a surrogate indicator of risk, and as such may help to prevent health problems (Nussbaum, 2001; Dennerlein *et al.,* 2003). This suggestion assumes that in low-force activities not only muscle fatigue develops, but also that this fatigue development can be measured.

Compared to activities at high-level forces, the occurrence of fatigue due to low-force activity, is poorly understood. In high-level activity, major intramuscular changes related to blood flow, water fluxes, temperature and metabolite concentrations occur, which result in the development of fatigue, directly demonstrated by a decrease in the maximal muscle strength (Asmussen, 1993). Moreover, manifestations of muscle fatigue can be observed in the electromyographic (EMG) signals in the fatigued muscles. Extensive evidence specifically shows that the EMG signal amplitude increases while the frequency spectrum shifts towards lower frequencies (Basmajian and DeLuca, 1985).

In low-level activity however, only subtle intra-muscular changes occur and the maximal muscle strength or other measures of performance may not detectably decrease. Using a strict definition of fatigue as a demonstrable loss in performance (e.g. Bigland-Ritchie and Woods, 1984), one might state that fatigue in low-level activity does not even occur. However, fatigue is not only defined in terms of performance changes but also in the perceptual and physiological domain (Åhsberg, 1998). In this respect, it is interesting to note that various authors reported increases in subjective fatigue or discomfort due to low-level activities (Sjøgaard *et al.,* 1986; Byström and Kilbom, 1990). But are these subjective findings supported by any measurable physiological changes? It might well be that in stages where performance is not (yet) affected, the contractile capabilities of muscles are hampered and homeostatic disturbances may occur that in the long run may induce chronic morphological changes (Blangsted *et al.,* 2005).

In the present report, studies on objectively measurable fatigue related changes in time in low-level force activities were reviewed. The review was limited to the repetitive low-level force activities that are typical for the type of occupational tasks that were mentioned before. Inclusion was limited to studies on the shoulder region; only studies investigating shoulder fatigue were included. Our research question was: Is there evidence of objective signs of fatigue in the shoulder region during realistic low-force work tasks?

2. METHODS

This review was based on an electronic literature search in Medline (1950-August 2007) and NIOSHTIC2, CISDOC, HSELINE, MHIDAS, and OSHLINE (1985)-August 2007). The keywords used were: muscle fatigue, low frequency fatigue, discomfort, low intensity, light manual work, office, computer work, assembly work, static, dynamic, Visual Display Unit or VDU, electromyography, blood flow, mechanomyography, Near Infrared Spectroscopy or NIRS and position sense.

The references retrieved by this search were first screened on the basis of their abstracts. In cases where the abstracts did not provide sufficient information, the full paper text was screened. The papers apparently fulfilling our inclusion criteria (see below) were selected for further study. The literature retrieved in this way was supplemented with studies cited in the retrieved papers. Finally, personal databases of the authors were searched for relevant papers.

To be included studies had to be published in peer-reviewed journals in the English language and should consider the development of fatigue in the shoulder region. The other inclusion criteria were related to the type, duration and intensity of the task under investigation, and the method to obtain objective information about fatigue.

As we were interested in the typical repetitive, long-lasting and low-force occupational activities that are common in daily occupational life, studies on fatigue due to purely isometric muscle contractions were not considered in this review. This review did include studies on occupational tasks simulated in the laboratory as well as studies on occupational tasks performed in real life.

Next, studies on activities at high force levels were excluded from this review. In studies where the intensity of the work load was expressed as a percentage of the maximal electromyographic activity of the trapezius muscle (pars descendens), we used a percentage of 20% as a cut-off point. Additionally, we included studies on computer work, light manual work and assembly work, as for these type of activities the intensity levels in terms of median EMG amplitudes reported in the literature are generally below 20%MVC (Westgaard *et al.,* 1999; Thorn, 2007). Another inclusion criterion related to the task was a total duration of the task of one hour or more. Both studies on continuous tasks and studies on tasks that were interrupted by rest breaks were included.

Finally, all studies had to apply an objective measuring method to assess the development of fatigue over time. Among the potential measuring methodologies and parameters were electromyography (EMG), mechanomyography (MMG), blood oxygen saturation, and position sense. With respect to the EMG and MMG studies, we only included studies that considered both the power spectrum and the signal amplitude, as it is typically the combination of a power frequency decrease and amplitude increase that indicates the development of fatigue (Basmajian and DeLuca, 1985).

3. RESULTS

The search strategy resulted in 137 hits. The first screening of the abstracts resulted in the exclusion of 41 papers for various reasons (e.g. no realistic task, not a task involving the shoulder muscles, no study of fatigue development over time). After application of the selection criteria while considering the full text, another 83 papers were excluded, mainly because of another body region of interest than the shoulder (30 studies), too short a task duration (25 studies), the study of purely isometric muscle contractions (11 studies) and a level of intensity too high (6 studies). In a few papers, the issue of fatigue was addressed, but not its temporal development. Four EMG studies fulfilling all other selection criteria, either

Table 1. Overview of results*

study	subj.	task	duration	intensity expressed by the mean or median initial trapezius activation in %MVC	objective and subjective measurements	subjective fatigue development in time	objective manifestations of fatigue development in time	objective evidence for fatigue
Bosch et al. 2007	16F 4M	assembly of catheters (A) and picking and placing of covers of shavers (B)	8,5-10 hrs (A) 9,5 hrs (B) including pauses	not specified	EMG-test LPD (10)	shoulder discomfort increase from 0,5 to 2	• ARV increase by 27% (A) and 19% (B) • decrease in higher frequency bands by 4% (A) and 11% (B) and MPF decrease by 3% (in B only)	+
Christensen 1986	16F 9M	picking and assembling printed circuits (females); soldering and assembling chasses (males)	7 hrs 30 min including pauses	14.9%	EMG-task&test HR RPE (9)	RPE increase from 1 to 3-4 (females); from 2 to 3 (males),	• RMS (task): no change • MPF (test): no change	-
Hostens & Ramon 2005	22M	driving on a private asphalt road	1 hr	not specified	EMG-task		• RMS decrease • instantaneous MPF decreases in active muscle parts	0
Jensen et al. 1993	29F	sewing machine operation	8 hrs including pauses	15%	EMG-test muscle strength RPE (10)	RPE increase from 1-3	• RMS increase by 3% • MPF decrease by 4% • maximal muscle strength: no change	+
Kimura et al. 2007	6M	computer work: writing by keyboard + 1KG load on wrists	1 hr 55 min including three pauses (5 min)	not specified	EMG-test RPE (4)	RPE increase from 0 to 1-3	• RMS increase by 14.8% • MPF decrease by 10.6% • CV decrease by 13.6%	+
Kleine et al. 1999	9F	computer work: writing by keyboard	3 hrs 30 min including two pauses (15 min)	not specified	EMG-task		• RMS increase by 17, 18, 22% (in 1st, 2nd and 3rd hour) • MPF no changes	0
Mathiassen & Winkel 1996	8F	laboratory simulation of assembly of starters for power saws	6 hrs including breaks	not specified (30%MVC was exceeded during 5% of working time)	EMG-test fatigue PPT (10)	fatigue from 1 to 5 PPT decrease by 10-15% in initial hours	• RMS increase by 11%, • zero-crossing rate decrease by about 2.5%	+
McLean et al. 2000	15F 3M	computer terminal work	1 hr 20 min with and without microbreaks	not specified	EMG-task		• RMS: no change • MPF: no change	-
Nakata et al. 1992	13F	laboratory simulation of assembly work: picking nails and screws and inserting them into holes in boards	2 hrs	3-6%	EMG-test LPD (10)	LPD increase from 0.5 to 2.2 in right shoulder	• RMS increase by 27% in 'high discomfort subgroup' only • MPF: no significant decrease	0
Nussbaum 2001	8F 8M	laboratory simulation of overhead assembly: tapping motions between two targets above shoulder	3 hrs or until exhaustion	16%	EMG-task&test		• RMS decrease by 9.2% (test) and 12.9% (task) per hour; 57% of subjects show amplitude increase (in task) • MPF decrease by 0.9 (test) and 1.3% (task) per hour. 55% show MPF decrease (in task)	+
Rolander et al. 2005	17F 10M	dentist work	about 4 hrs including breaks	5-9%	EMG-task		• ARV increase by 14%; • MPF: no change;	0
Seghers et al. 2003	8F 8M	computer work: keyboard and mouse	1 hr 29 min	not specified	EMG-task local discomfort (10)	no statistical results provided	• ARV increase by 5% • MPF: no change	0
Sundelin 1993	12F	repetitive grasping a small cylinder and releasing it through a hole	2 hrs, one hour without pauses and one hour with pauses	not specified (median value of peak load: 31% median value of static load: 4.4%)	EMG-task RPE(20)	RPE from 0 to 12-14	in eight subjects a combination of • RMS increase 18-40% per hour • MPF decrease 2-15% per hour	+

* Only results significant at P<0.05 are presented (in 7th and 8th column). In the final column, +
means that an amplitude increase and frequency decrease was found, '-' means that none of these
changes were observed, and 'o' refers to the situation where only one of these changes was
found. Explanation of used terms and abbreviations: M and F indicate the male and female
gender; EMG-task and EMG-test refer to the cases where EMG is obtained during the task itself
and during test contractions, respectively; EMG amplitude is expressed by the averaged rectified
value (ARV) or the root mean square (RMS). LPD = local perceived discomfort, RPE = rating of
perceived exertion, and PPT = pressure pain threshold; number between brackets indicates the
applied number of scale units.

considered the power spectrum or the signal amplitude, but not the combination of
the two. The remaining 13 papers are presented in Table 1.

Out of these 13, there were eight studies on tasks performed in the laboratory,
and five studies on occupational tasks performed in real life. The tasks concerned
computer work in two studies, assembly work or simulated assembly work in five
studies, and other tasks like driving, sewing, drilling, and dentistry. There appeared
to be a large variation in task duration across the various studies, ranging from 1 to
9.5 hours. It is noticeable that the size of the samples in some of the studies is
fairly small.

The level of intensity if expressed as a percentage of the EMG amplitude in a
maximal voluntary contraction (in five studies) ranged from 3-6 up to 16%. Other
studies, which did not specify the load intensity, were included because the
intensity levels were assumed to remain below the 20% cut-off point. However,
one might question here the inclusion of several papers. The study of Sundelin
(1993) was included on the basis of the median amplitude of the peak load of 31%.
The studies of Hostens and Ramon (2005) and Kimura *et al.* (2007) were also
given the benefit of doubt, not knowing the load intensity of car driving and
keyboard typing with the wrists loaded by 1 kg, respectively.

Surprisingly, none of the included studies applied a measuring method other
than EMG. In all these EMG studies, the activation of the descending part of the
trapezius was analyzed. In some, the activation of the infraspinatus or deltoid
muscles was additionally studied, but fatigue manifestations were always more
prominent in the trapezius, and thus these are reported in the table.

Considering these results, six studies found objective evidence for
manifestations of muscle fatigue within the muscle, as deduced from a combination
of a frequency decrease and an amplitude increase. In two studies, however, no
such changes in power frequency and amplitude were observed. In the remaining
five studies only one of these changes occurred, either an increase in amplitude (in
four studies) or a decrease in the frequency content (in one study).

4. DISCUSSION

4.1 Objective signs of fatigue

Shoulder fatigue is of particular importance in ergonomics. Shoulder disorders are
among the most prevalent work-related musculoskeletal disorders (Walker-Bone
and Cooper, 2005), while local muscle fatigue is assumed to be a main contributor,

particularly when it is cumulative and adequate recovery is not possible (Sommerich *et al.* 1993). Moreover, a detrimental effect of fatigue on task performance is plausible in many working areas. Prevention of fatigue should thus be aimed for and any objective assessment of fatigue may help to define and organize the work such that fatigue remains acceptably low. However, the potential of measuring fatigue in realistic low-force work tasks has not been systematically addressed and clarified. This review shows that electromyographic manifestations of muscle fatigue in the trapezius muscle do appear in low-force activities: in six out of fourteen studies a significant decrease in the EMG power frequency was accompanied by an amplitude increase; the other seven studies did show none or only one of these changes.

The lack of a positive result in about half of the studies could be due to a lack of statistical power. However, comparing the number of subjects between the studies finding objective signs of fatigue vs. the studies not finding this result, we did not observe any differences. The same holds for the duration of the task under investigation: the duration was not different between the studies showing manifestations of fatigue and not doing so (see figure 1A and 1B). Another suggestion could be that fatigue-related changes in muscle activation become less predictable under less constraining conditions due to more confounding factors. When considering the way the included studies were structured and designed, we did not find any evidence supporting this suggestion. These considerations lead us to the conclusion that objective signs of fatigue in the shoulder region are present in at least some of the realistic low-force work tasks.

The question thus arises what the determinants of fatigue development in these tasks could be. A potential determinant could be gender of the subject groups. In overhead assembly work, females were observed to exhibit longer endurance times, delayed reports of discomfort, and slower declines in strength compared to males (Nussbaum *et al.*, 2001). However, comparing the studies with and without fatigue signs, we did not find any evidence for a gender effect.

Another possible determinant could be the intensity of the work task. Figure 1C presents the median amplitude levels of the trapezius activation expressed as a percentage of the maximum for the various studies. For studies not reporting this value, an estimated value is shown. Based on Westgaard *et al.* (1999) and Thorn (2007), we applied a value of 7% MVC for typing on a keyboard (for the study of Kleine *et al.* 1999) and 8% MVC for computer work using keyboard and mouse (McLean *et al.* 2000; Seghers *et al.* 2003) and a value of 16% MVC for assembly work (Mathiassen and Winkel, 1996; Bosch *et al.* 2007). The median intensity level in the study of Sundelin (1993) was estimated exactly in between the reported median and peak values. The studies of Hostens and Ramon (2005) and Kimura *et al.* (2007) with unpredictable intensity levels were omitted. From Figure 1C it appears that the studies with a positive result with regard to the existence of signs of fatigue have a rather high intensity specifically above the level of 15% MVC. The intensity level in studies showing no objective signs of fatigue are generally lower. Among these latter studies are all three included studies on computer work.

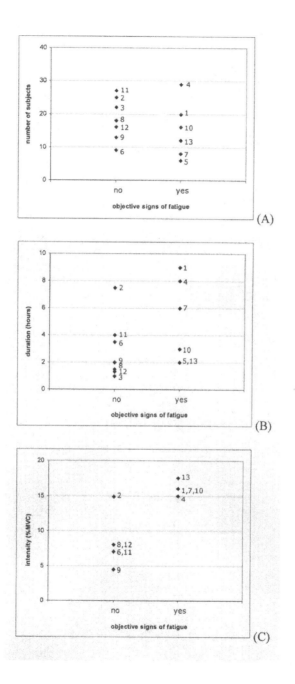

Figure 1. Number of subjects (a), task duration (b) and intensity level (c) in the studies included in this review. Studies showing objective signs of fatigue are separated from the other ones. (See text for further explanations).

4.2 Subjective experiences of fatigue

In all studies that report on subjective experiences, the subjective ratings were found to increase. No constant or decreasing patterns were reported.

In two out of the seven studies finding no objective sign of fatigue, the perceived exertion and discomfort was observed to increase from 1 to 2-4 (10-point scale, Christensen, 1986) and from 0.5 to 2.2 (9-points, Nakata *et al.,* 1992), respectively. In the studies demonstrating objective signs of fatigue, the reported increases in perceived fatigue, discomfort or exertion show a large range. Bosch *et al.* (2007) reported an increase in shoulder discomfort from 0.5 to 2 (on a 10-points scale), while Sundelin (1993) reported an increase of the general perceived exertion from 0 to14 (20-points scale). Apparently, subjective feelings of fatigue are more consistently found compared to objective fatigue manifestations, and thus, subjective fatigue is not in all cases supported by objective findings. In our view, there might be many psychological influences when performing a given task with potential effect on the perception of fatigue; only part of these might be objectively measurable.

4.3 Changes in muscle activation

The changes in the time and frequency domain of the EMG signal during a fatiguing contraction have been extensively studied. In particular for prolonged static efforts, it is accepted that a simultaneous downward shift of the frequency spectrum and an increase in the signal's amplitude can be considered as an objective sign of fatigue.

Even though the interpretation of the EMG changes is subject to caution (Dimitrova and Dimitrov, 2003), it is believed that the frequency decrease represent a decreased conduction velocity along the muscle fibres (e.g. DeLuca, 1984) and a larger amount of synchronization of motor unit firing in the fatigued muscle (Lippold, 1981). A recent study on sustained low-level activity, has revealed indications of increased interstitial potassium concentrations in the descending part of the trapezius muscle, possibly underlying the decreased conduction velocity (Rosendal *et al.,* 2004). The increase in amplitude when the muscle gets fatigued can be understood from the additional recruitment of motor units. These motor units might be required to compensate for the loss of force producing capacity of the already active units (Moritani *et al.,* 1986). It might also be that the increasing amplitude is a direct result from the decreasing frequency content, because of the low-pass filtering characteristics of skin tissue. At a lower frequency content, a larger proportion of the signal would be able to pass, resulting in the measurement of a higher amplitude (Lindstrom *et al.,* 1977).

The fatigue-related changes in the EMG signal can be confounded by various factors. The decrease in frequency content could be masked by the additional recruitment of motor units. Since the motor units being additionally recruited are of a larger size (according to the size-principle), an opposite (upward) shift in the frequency spectrum will be expected. Another potential confounder is the temperature. An increasing temperature in the working muscle would lead to a decrease in the EMG amplitude (Petrofsky and Lind, 1980), which may mask the increase in amplitude during fatigue. In non-isometric conditions, extra cautions should be raised: changes in muscle forces, muscle length and muscle contraction velocity may affect the fatigue-related spectral and amplitude changes (e.g. Gerdle *et al.*, 1991). It might be that in some studies subjects changed their working posture or working technique over time possibly due to fatigue, potentially affecting the load sharing within the measured muscle or between different muscles. The above confounders may have played a role in the various studies that were included in this review, but their effect in terms of direction and size remains unknown.

Of special interest are the studies where only one of the expected changes was found. Hostens and Ramon (2005) did find some indication of (temporal) decreased MPF during driving, which were not accompanied by a simultaneous increase in amplitude. Instead, the amplitude was generally found to decrease. One of their explanations for the decreased amplitude was the potentiation of muscle fibres, where for a given stimulus muscles produce more force, making the muscle more efficient per unit impulse (Garner *et al.*, 1989). Kleine *et al.* (1999), Nakata *et al.* (1992), Rolander *et al.* (2005) and Seghers *et al.* (2003) found an increase in EMG amplitude without a decrease of the frequency content. Generally, the combination of an increase in amplitude with no frequency shift indicates an increase in force generation without fatigue. Kleine *et al.* (1999) suggested that the intensity level in the studied computer task was too low for muscle fatigue to develop, whereas the increase in amplitude of the trapezius muscle was associated with a relative elevation of the shoulders during the task. Similarly, Seghers *et al.* (2003) assigned the amplitude increase of the right trapezius EMG during a computer game to an (unobserved) elevation of the right shoulder. Nakata *et al.* (1992) also concluded that there was no fatigue development in their study and speculated that the increased amplitude might have resulted from a change (which would mean a decrease) in muscle temperature.

4.4 Other measures

This review shows that no manifestations of muscle fatigue during realistic working tasks have been reported thus far, at trapezius activation levels below 15% MVC such as computer work. It cannot be answered whether muscle fatigue simply does not occur at these low intensities or any manifestations of muscle fatigue are not detectable given the unfavorable signal-to-noise ratios.

It might be that EMG measurement strategies other than the ones applied can increase the discriminative power. For instance, the conduction velocity of the electrical signal along the muscle fibres, which can be deduced when using two-dimensional electrode arrays, seems to be a more direct measure of muscle fatigue than the frequency content and thus temporal fatigue-related changes may be detected more easily. Kimura *et al.* (2007) did find a decrease in the conduction velocity of 13% in 1 hour and 55 minutes in computer work, while the wrist was loaded by an extra 1 kg. The question remains unanswered whether the conduction velocity would change in normal computer work.

For potential measures to assess fatigue other than the traditional EMG-based parameters, it is also still questionable whether these would be affected at very low intensities. We used the obvious search terms to find studies investigating shoulder fatigue development on the basis of mechanomyography (MMG), blood oxygen saturation, and position sense. However, none of the studies fulfilling our criteria involved one of these methods. Effects of fatigue on MMG, blood oxygen saturation and position sense were studied before by for example Björklund *et al.* (2000), Søgaard *et al.* (2003) and Heiden *et al.* (2005), but not during prolonged, low-force activities.

5. CONCLUSION

From the studies included in this review, it can be concluded that electromyographic manifestations of fatigue in the trapezius muscle do appear in low-force activities like light manual work and assembly when the intensity level is about 15-20%MVC.

In the studies finding these objective manifestations of fatigue the amplitude increases (group averages) ranged across the various studies from 3% to 27%, while the MPF decreases ranged (group averages) from 0.9 to 11%.

One can only speculate about the relevance of these findings from the perspective of health. Anyway, the individual fatigue-related changes rather than the group averages, may be of more relevance. Several authors mention that the inter-individual variation is large. Sundelin (1993) presented individual figures, showing amplitude and frequency changes ranging from 18-40% and 2-15%, respectively within the eight out of twelve subjects showing evidence of fatigue.

Another reason for ergonomists to be interested in fatigue is its potentially negative effect on performance. A negative effect of fatigue on performance has been demonstrated in low-force tasks, but only in isolated, unnatural and constrained tasks (e.g. Huysmans *et al.*, 2007). For more complex realistic low-force activities, the relationship between fatigue-related changes and meaningful performance losses however, has to our knowledge not been addressed previously. Without knowing this relationship, it is hard to define the relevance of these review results from the perspective of performance.

Local muscle fatigue seems to occur in some light manual activities, and could be considered a risk indicator. However, before having merit as a design and evaluation tool, more information is needed specifically related to the exact temporal pattern of fatigue development and the relationships with shoulder disorder risks and practical measures of performance. This may require fatigue measurement strategies other than the traditional ones applied in the included studies.

References

Åhsberg, E., 1998, *Perceived Fatigue Related to Work*, (National Institute for Working Life).

Asmussen, E., 1993, Muscle fatigue: Retrospection. *Medicine and Science in Sports and Exercise*, **25**, pp. 412-420.

Basmajian, J.V. and DeLuca, C.J., 1985, *Muscles Alive: their functions revealed by electromyography*, 5th ed., (Baltimore: Lippincott, Williams and Wilkins).

Bigland-Ritchie, B. and Woods, J.J., 1984, Changes in muscle contractile properties and neural control during human muscular fatigue. *Muscle and Nerve*, **7**, pp. 691-9.

Björklund, M., *et al.*, 2000, Position sense acuity is diminished following repetitive low-intensity work to fatigue in a simulated occupational setting. *European Journal of Applied Physiology*, **81**, pp. 361-7.

Blangsted, A.K., *et al.*, 2005, Voluntary low-force contraction elicits prolonged low-frequency fatigue and changes in surface electromyography and mechanomyography. *Journal of Electromyography and Kinesiology*, **15**, pp. 138-48.

Bosch, T., de Looze, M.P. and van Dieen, J.H., 2007, Development of fatigue and discomfort in the upper trapezius muscle during light manual work. *Ergonomics*, **50**, pp. 161-77.

Byström, S.E. and Kilbom, A., 1990, Physiological response in the forearm during and after isometric intermittent handgrip. *European Journal of Applied Physiology: Occupational Physiology*, **60**, pp. 457-66.

Christensen, H., 1986, Muscle activity and fatigue in the shoulder muscles of assembly-plant employees. *Scandinavian Journal of Work, Environment and Health*, **12**, pp. 582-7.

De Luca, C.J., 1984, Myoelectrical manifestations of localized muscular fatigue in humans. *Critical Reviews in Biomedical Engineering*, **11**, pp. 251-79.

Dennerlein, J.T., *et al.*, 2003, Fatigue in the forearm resulting from low-level repetitive ulnar deviation. *American Industrial Hygiene Association Journal*, **64**, pp. 799-805.

Dimitrova, N.A. and Dimitrov, G.V., 2003, Interpretation of EMG changes with fatigue: facts, pitfalls, and fallacies. *Journal of Electromyography and Kinesiology*, **13**, pp. 13-36.

Garner, S.H., Hicks, A.L. and McComas, A.J., 1989, Prolongation of twitch potentiating mechanism throughout muscle fatigue and recovery. *Experimental Neurology*, **103**, pp. 277-81.

Gerdle, B., *et al.*, 1991, Dependence of the mean power frequency of the electromyogram on muscle force and fibre type. *Acta Physiologica Scandinavica*, **142**, pp. 457-65

Heiden, M., *et al.*, 2005, Effects of time pressure and precision demands during computer mouse work on muscle oxygenation and position sense. *European Journal of Applied Physiology*, **94**, pp. 97-106.

Herberts, P. and Kadefors, R., 1976, A study of painful shoulder in welders. *Acta*

Orthopaedica Scandinavica, **47**, pp. 381-7.

Hostens, I. and Ramon, H., 2005, Assessment of muscle fatigue in low level monotonous task performance during car driving. *Journal of Electromyography and Kinesiology*, **15**, pp. 266-74.

Huysmans, M.A., *et al.*, 2007, Fatigue effects on tracking performance and muscle activity. *Journal of Electromyography and Kinesiology* (Epub ahead of print).

Jensen, B.R., *et al.*, 1993, Shoulder muscle load and muscle fatigue among industrial sewing-machine operators. *European Journal of Applied Physiology and Occupational Physiology*, **67**, pp. 467-75.

Jonsson, B., 1988, The static load component in muscle work. *European Journal of Applied Physiology and Occupational Physiology*, **57**, pp. 305-10.

Kimura, M., *et al.*, 2007, Electromyogram and perceived fatigue changes in the trapezius muscle during typewriting and recovery. *European Journal of Applied Physiology*, **100**, pp. 89-96.

Kleine, B.U., *et al.*, 1999, Surface EMG of shoulder and back muscles and posture analysis in secretaries typing at visual display units. *International Archives of Occupational and Environmental Health*, **72**, pp. 387-94.

Lindstrom, L., Kadefors, R. and Petersen, I., 1977, An electromyographic index for localized muscle fatigue. *Journal of Applied Physiology*, **43**, pp. 750-4.

Lippold, O., 1981, The tremor in fatigue. *Ciba Foundation Symposium*, **82**, pp. 234-48.

Mathiassen, S.E. and Winkel, J., 1996, Physiological comparison of three interventions in light assembly work: reduced work pace, increased break allowance and shortened working days. *International Archives of Occupational and Environmental Health*, **68**, pp. 94-108.

McLean, L., *et al.*, 2000, Myoelectric signal measurement during prolonged computer terminal work. *Journal of Electromyography and Kinesiology*, **10**, pp. 33-45.

Moritani, T., Muro, M. and Nagata, A., 1986, Intramuscular and surface electromyogram changes during muscle fatigue. *Journal of Applied Physiology*, **60**, pp. 1179-85.

Nakata, M., Hagner, I-M and Jonsson, B., 1992, Perceived musculoskeletal discomfort and electromyography during repetitive light work. *Journal of Electromyography and Kinesiology*, **2**, pp. 103-11.

Nussbaum, M.A., et al., 2001, Static and dynamic myoelectric measures of shoulder muscle fatigue during intermittent dynamic exertions of low to moderate intensity. *European Journal of Applied Physiology*, **85**, pp. 299-309.

Petrofsky, J.S. and Lind, A.R., 1980, The influence of temperature on the amplitude and frequency components of the EMG during brief and sustained isometric contractions. *European Journal of Applied Physiology and Occupational Physiology*, **44**, pp. 189-200.

Rempel, D.M., Harrison, R.J. and Barnhart, S., 1992, Work-related cumulative trauma disorders of the upper extremity. *Journal of the American Medical Association*, 12, **267**, pp. 838-42.

Rohmert, W., 1973, Problems in determining rest allowances Part 1: use of modern methods to evaluate stress and strain in static muscular work. *Applied*

Ergonomics, **4**, pp. 91-5.

Rolander, B., *et al.*, 2005, Evaluation of muscular activity, local muscular fatigue, and muscular rest patterns among dentists. *Acta Odontologica Scandinavica*, **63**, pp. 189-95.

Rosendal, L., *et al.*, 2004, Interstitial muscle lactate, pyruvate and potassium dynamics in the trapezius muscle during repetitive low-force arm movements, measured with microdialysis. *Acta Physiologica Scandinavica*, **182**, pp. 379-88.

Seghers, J., Jochem, A. and Spaepen, A., 2003, Posture, muscle activity and muscle fatigue in prolonged VDT work at different screen height settings. *Ergonomics*, **46**, pp. 714-30.

Sjøgaard, G., *et al.*, 1986, Intramuscular pressure, EMG and blood flow during low-level prolonged static contraction in man. *Acta Physiologica Scandinavica*, **128**, pp. 475-84.

Sjøgaard, G. and Jensen, B.R. 2006, Low-level static exertions. In: *Fundamentals and Assessment Tools for Occupational Ergonomics, Vol. I., edited by* Karwowski, W., (London: International Publishing Services).

Søgaard, K., *et al.*, 2003, Evidence of long term muscle fatigue following prolonged intermittent contractions based on mechano- and electromyograms. *Journal of Electromyography and Kinesiology*, **13**, pp. 441-50.

Sommerich, C.M., McGlothlin, J.D. and Marras, W.S., 1993, Occupational risk factors associated with soft tissue disorders of the shoulder: a review of recent investigations in the literature. *Ergonomics*, **36**, pp. 697-717.

Sundelin, G., 1993, Patterns of electromyographic shoulder muscle fatigue during MTM-paced repetitive arm work with and without pauses. *International Archives of Occupational and Environmental Health*, **64**, pp. 485-93.

Takala, E.P., 2002, Static muscular load, an increasing hazard in modern information technology. *Scandinavian Journal of Work, Environment and Health*, **28**, pp. 211-3.

Thorn, S., *et al.*, 2007, Trapezius muscle rest time during standardised computer work--a comparison of female computer users with and without self-reported neck/shoulder complaints. *Journal of Electromyography and Kinesiology*, **17**, pp. 420-7.

Veiersted, K.B., Westgaard, R.H. and Andersen, P., 1990, Pattern of muscle activity during stereotyped work and its relation to muscle pain. *International Archives of Occupational and Environmental Health*, **62**, pp. 31-41.

Visser, B. and van Dieen, J.H., 2006, Pathophysiology of upper extremity muscle disorders. *Journal of Electromyography and Kinesiology*, **16**, pp. 1-16.

Walker-Bone, K. and Cooper, C., 2005, Hard work never hurt anyone: or did it? A review of occupational associations with soft tissue musculoskeletal disorders of the neck and upper limb. *Annals of the Rheumatic Diseases*, **64**, pp. 1391-6.

Westgaard, R.H., Jansen, T. and Jensen, C., 1999, EMG of the neck and shoulder muscles: the relationship between muscle activity and muscle pain in occupational settings. In: *Electromyography in Ergonomics*, edited by Kumar, S. and Mital, A., (CRC Press), pp. 227-58.

CHAPTER TWO

Low-back problems in recreational self-contained underwater breathing apparatus divers: Prevalence and specific risk factors

K. Knaepen[1,2], E. Cumps[2], R. Meeusen[2] and E. Zinzen[1]

[1]Department of Movement Education and Sports Training, Vrije Universiteit Brussel, Brussels, Belgium
[2]Department of Human Physiology & Sports Medicine, Vrije Universiteit Brussel, Brussels, Belgium

1. INTRODUCTION

Participation in certain sports has been investigated as a potential contributor to low-back problems (LBP) (Sward *et al.,* 1991; Verni *et al.,* 1999; Bono, 2004). Bono (2004) revealed prevalence rates of LBP ranging from 1% to more than 30% among athletes (Bono, 2004). Alpine skiing (Peacock *et al.,* 2005), cross-country skiing (Bahr *et al.,* 2004), triathlon (Villavicencio *et al.,* 2006), (elite) rhythmic gymnastics (Hutchinson, 1999), running (Woolf and Glaser, 2004), wrestling (Granhead and Morelli, 1988), weight-lifting (Granhead and Morelli, 1988), rowing (Hickey *et al.,* 1997), tennis (Saraux *et al.,* 1999) and golf (Hosea and Gatt, 1996) are associated with particular high prevalence rates (Table 1). Yet prevalence rates of LBP among many other sports and among more recreational athletes (Bono, 2004) are not well known.

Table 1. Lifetime and one-year prevalence of low-back problems in different sports.

Sport	One-Year Prevalence	Lifetime Prevalence
alpine skiing (instructors) [Peacock *et al.* 2005]		75.0%
cross-country skiing [Bahr *et al.* 2004]	63.0%.	65.4%
gymnastics (elite) [Sward *et al.* 1991]		79.0%
rowing [Hickey *et al.* 1997]		27.5%
running [Woolf and Glaser 2004]		75.0%
tennis [Saraux *et al.* 1999]		40.0 %
triathlon [Villavicencio *et al.* 2006]		67.8%
walking [Woolf and Glaser 2004]		68.0%
weight lifting [Granhed and Morelli 1988]		23.0%
wrestling [Granhed and Morelli 1988]		59.0 %

To date, very few researchers have focused on low-back related problems in self-contained underwater breathing apparatus (SCUBA) divers. When addressing injuries in scuba diving, predominantly sport-specific injuries, such as barotraumas (i.e. middle ear squeeze, inner ear barotrauma, tympanic membrane rupture or pulmonary barotrauma), decompression sickness or arterial gas embolism rather than musculoskeletal injuries are discussed (Clenney and Lassen, 1996; Hunt, 1996; Pelletier, 2002; Van Tulder et al., 2002; Worf, 2002; Shroder et al., 2004). However, Jäger et al. (2002) mentioned the occurrence of acute LBP among scuba divers as a result of acute decompression sickness (DSC) type II (i.e. DSC with neurological and pulmonary symptoms) (Jäger et al., 2002). They warned that when acute LBP with progressive neurological deficits is found in a scuba diver, a differential diagnosis should be performed immediately and DSC type II should be included, next to sciatica or pseudoradicular syndrome (Jager et al., 2002). No other records of LBP in relation to scuba diving can be found in the literature. This lack of research is surprising, since Verni et al. (1999) pointed out in their study of lumbar pain and fin swimming (i.e. using a single fin): 'a functional overload of the vertebral column in water sports can occur when the dynamic equilibrium between mechanical stress on the spine and the physiological reactions of the spine to this stress, is interrupted' (Verni et al., 1999). They stated that in such a case, the response is often pathophysiological or pathological and LBP is the most common symptom. Moreover, Verni et al. (1999) emphasized the unique biomechanical features (i.e. the upper limbs and lumbo-sacral axis have a directional function and the lower limbs and the fin have a pushing function) that are present in the lumbo-sacral region during fin swimming (i.e. as opposed to classical swimming strokes). Not surprisingly, fin swimmers are prone to LBP which are often triggered by structural and mechanical spine problems (i.e. intrinsic factors) and/or by deficits in technique, training or equipment (i.e. extrinsic factors) (Verni et al., 1999).

Despite the lack of scientific evidence in the literature, one can assume that unique static and dynamic biomechanical features are also present in scuba divers. For instance, when considering a scuba diver in a static, horizontal position, the natural/anatomical curvatures of the spine are stressed: a cervical hyperlordosis and an augmented lumbar lordosis appear when the view is set horizontally. From a dynamic point of view, a similar situation as in fin swimming is distinguished in scuba diving: the "up and down" oscillation of the propulsive upper limbs and fins (i.e. two separate fins) by alternate front and back arching, will be absorbed by the lumbo-sacral region. The strain and shearing forces of this cyclic lumbo-sacral swing will act mainly on the posterior column of the spine, particularly on the arch and pars interarticularis and could cause the onset of LBP symptoms (Verni et al., 1999). Furthermore, during the preparation and reconditioning phase of scuba diving heavy loads (i.e. up to 30 kg) are handled which places the lumbar spine also at a certain risk. In addition, the weight belt (i.e. heavy weight on a small area) gives also reason for concern as it brings about a large local compression force on the lumbar spine during the performance phase.

The aim of this study was to investigate whether recreational scuba divers are at increased risk of LBP. For this purpose, a retrospective (i.e. over the preceding 12 months) self-assessment questionnaire was developed and assessed for reliability, to gather data on demographics, general and sport-specific risk factors of LBP, injuries, characteristics of scuba diving and prevalence and characteristics of LBP among active, recreational, Flemish scuba divers.

2. METHODS

2.1 Questionnaire 'Injuries in scuba diving'

2.1.1 Conceptualisation and development

A questionnaire was designed to (i) collect prevalence data (i.e. one-year prevalence and lifetime prevalence) on LBP, (ii) identify the characteristics of LBP and (iii) search for possible general and/or sport-specific risk factors for LBP, in a large sample of recreational scuba divers.

To develop a first draft of the retrospective self-assessment questionnaire and to add value to its content validity, an extensive, qualitative literature review was performed on definitions, classifications, characteristics and (general and sport-specific) risk factors of LBP. For the anatomical definition of LBP, the definition of Haanen (1984) was employed: "All sorts of pain in the region of the lumbar spine, the os sacrum and the M. gluteii (i.e. lumbar region); the boundaries of this region are: from the lower margin of the 12th rib to the iliacal articulations and the caudal vertebra (i.e. coccyx) and from the lateral part of the M. erector spinae to the spina illiaca posterior superior; the centre of this area is formed by the 4th lumbar vertebra" (Haanen, 1984). In addition, the widespread classification of LBP of Von Korff (1994) and the additions of de Vet et al. (2002) were opted for, to reach a uniform comprehension of the concept of what is colloquially referred to as 'low-back problems', in this questionnaire (Haanen, 1984; Von Korff, 1994; de Vet *et al.* 2002) (appendix 1). The definition of Haanen (1984) was communicated to the subjects by means of an anatomical figure showing the low-back region. The different classifications of LBP were explained to the subjects in brief words following the questions concerned.

Next to this procedure, five experts (i.e. two on LBP, two on scuba diving and one on clinimetrics) assessed the developed self-administered questionnaire to assure the construct validity. The final questionnaire contained 252 items, questioned in 45 self-report measures and grouped into six categories (Table 2).

2.1.2 Reliability

Test-retest reliability of the questionnaire was assessed for each item ($n = 252$), with a one-week interval (Leung *et al.*, 1999), in a small sample ($n = 24$) of Flemish recreational scuba divers. Depending on the nature of the variables (i.e. categorical or continuous), correlation coefficients (Cohen's (modified) Kappa, Kendall's tau-b correlation coefficient and Intraclass correlation coefficient), the

Wilcoxon Signed Ranks and the F-test were used as measures for reliability. Correlations were considered statistically significant if P < 0.050. Correlation values greater than 0.70 were considered good to excellent, values between 0.50 and 0.70 were moderate to fair, and values of less than 0.50 represented poor agreement beyond chance alone (Leung *et al.*, 1999; Resnik and Dobrzykowsk, 2003).

Table 2. Content of the self-assessment questionnaire based on a literature review and an expert evaluation.

A. Demographics: (7 questions)	- age [Kostova and Koleva 2001, Loney and Stratford 1999, Manchikanti 2000, Manek and MacGregor 2005, Shelerud 1998] - gender [Kostova and Koleva 2001, Manchikanti 2000, Manek and MacGregor 2005] - body weight [Han *et al.* 1997, Leboeuf-Yde 2004, Leboeuf-Yde 2000b, Leboeuf-Yde *et al.* 1999] - height [Han *et al.* 1997, Leboeuf-Yde 2004, Leboeuf-Yde 2000b, Leboeuf-Yde *et al.* 1999] - marital status - number of children - occupation
B. General risk factors: (12 questions)	- workload [Eriksen *et al.* 1999, Gerr and Mani 2000, Hartvigsen et al. 2001, Marras 2000, Matsui *et al.* 1997, Ratti and Pilling 1997, Ying *et al.* 1997] - smoking [Akmal *et al.* 2004, Eriksen *et al.* 1999, Goldberg *et al.* 2000, Leboeuf-Yde 1995, Leboeuf-Yde *et al.* 1998, Leboeuf-Yde 2002] - alcohol use [Leboeuf-Yde 2000a, Leboeuf-Yde 2000c] - pregnancy and parturition [Levangie 1999, Manchikanti 2000] - physical activity and exercise [Manchikanti 2000, Manek and MacGregor 2005, Mortimer *et al.* 2001] - general health [Hestbaek *et al.* 2003b, Leboeuf-Yde 2004, Manek and MacGregor 2005, Matsui *et al.* 1997] - psychosocial factors [Feyer *et al.* 2000, Hoogendoorn *et al.* 2000b, Jansen *et al.* 2004, Kerr *et al.* 2001, Mannion *et al.* 1996, Simmonds *et al.* 1996, Thorbjornsson *et al.* 2000]
C. Injuries (1 question):	- retrospective, overall injury registration
D. Prevalence of LBP: (4 questions)	- one-year prevalence - lifetime prevalence - history of LBP [Adams *et al.* 1999, Manchikanti 2000, Manek and MacGregor 2005, Papageorgiou *et al.* 1996, Thorbjarnsson *et al.* 1998, Vingard *et al.* 2000]
E. Characteristics of LBP: (7 questions)	- onset (i.e. sudden, gradual) - localisation (i.e. diffuse, local) - development (i.e. acute, subacute, chronic, recurrent, transient) - pain (i.e. intensity of pain) - disability (i.e. amount of disability)
F. Properties of scuba diving: (14 questions)	- diving experience (i.e., level, dives, years, hours) - diving habits (e.g., depth, weight system) - equipment (e.g., buoyancy control device, suit, weight) - diving injuries (e.g., barotrauma, decompression illness)

2.2 Data collection

A random selection of 10 clubs out of a total of 138 Flemish scuba diving clubs, all affiliated members of the Dutch League for Underwater Research and Sports (NELOS v.z.w. ®), was carried out. To anticipate possible drop-out of clubs, another ten clubs were randomly selected and added to a reserve list. The stratified randomization procedure accounted for the distribution of scuba diving clubs among all five Flemish regions to obtain a representative sample of Flemish scuba divers (Table 3). In a first stage, a standardized, cover email was sent to the selected clubs to explain the purpose of the study and to invite them to take part in the research. Subsequently, further information and agreements were made with the chairmen of the scuba diving clubs. Next, a researcher visited all ten clubs after a training session in a swimming pool and distributed the retrospective questionnaire among voluntary, Dutch speaking, recreational scuba divers of all levels and all ages. All the participants received a standardized explanation on the aims of the research project and consent was obtained after completion of the written self-assessment questionnaire. Confidentiality of responses was addressed on the introduction page of the questionnaire. The study was performed according to the Declaration of Helsinki for Medical Research involving human subjects and was part of a larger injury registration survey which has been approved by the Ethical Committee of the Vrije Univeristeit Brussel.

Table 3. Randomization procedure of the sample survey of recreational scuba divers.

	n	**%**
NELOS v.z.w. ® scuba diving clubs	138[†]	100%
Contacted scuba diving clubs	13	9.42%
Recruited scuba diving clubs:	10	7.25%
Brabant (Flemish part)	2	1.45%
West-Flanders	2	1.45%
East-Flanders	2	1.45%
Antwerp	3	2.17%
Limburg	1	0.72%
Total members of NELOS v.z.w. ®	> 10.000	100%
Total of recruited scuba divers	200	±2%

[†] This total of 138 NELOS scuba diving clubs counted for the year 2005

2.3 Data analysis

Statistical analysis was performed using the Statistical Package for Social Sciences (SPSS) 13.0 for Windows Statistical Software[â]. Descriptive statistics (i.e. frequency, mean ± SD) were used to calculate demographic characteristics, prevalence rates and characteristics of LBP in scuba divers. To estimate significant differences in general and sport-specific characteristics between scuba divers with and without LBP, a Chi² (χ^2) test for nominal data, a Mann-Whitney U-test for ordinal data and an independent t-test for normally distributed interval and/or ratio data, were computed. Distributions were checked for normality with a One-Sample-Kolmogorov-Smirnov Test. The Contingency and the Spearman's Rho correlation coefficients were used to reveal significant correlations between sport-specific characteristics of scuba diving and overall characteristics of LBP. Subsequently, based on the outcome of the comparison between the two groups, binary logistic regression analysis (i.e. backward stepwise logistic regression method) was conducted to explore the strongest sport-specific predictors of LBP in scuba divers. Statistical differences and correlations were considered significant when $P < 0.050$.

3. RESULTS

3.1 Reliability of the self-assessment questionnaire

Almost 58% (57.5%) of the items (n = 145) in the questionnaire had a high reliability   ($r \geq 0.70$) and a significant correlation ($P < 0.05$), 7.5% of the items (n = 19) had a moderate to fair reliability ($0.50 \leq r < 0.70$) and a significant correlation ($P < 0.050$) and 5.2 % (n = 13) had an insufficient correlation value ($r < 0.50$). For 29.8% (n = 75) of the data, correlation coefficients were undefined due to (i) a uniform answer of all 24 subjects or (ii) the item did not apply to any of the subjects (i.e. prevalence of the question/condition was zero). The Wilcoxon Signed Ranks and the F-test, revealed no significant differences ($p \geq 0.050$) between the test and retest data. Test-retest reliability of the self-assessment questionnaire was considered adequate considering the purpose of this study.

3.2 Study population
The study population consisted of recreational scuba divers from 10 randomly selected scuba diving clubs in Flanders and who voluntary participated in the study. A total of 200 scuba divers completed the retrospective questionnaire, of which 19 were excluded because they had not made their first open water dive yet or they did not complete the questionnaire properly. Of the 181 respondents, 138 (76.2%) were male and 43 (23.8%) were female. The mean age of all the scuba divers together was 39.1 (± 12.5) years; 40.3 (± 12.8) years for men and 35.0 (± 10.9) years for women. The mean BMI was 24.7 (± 3.67) kg/m²; 25.2 (± 3.52) kg/ m² for men and 22.8 (± 3.48) kg/m² for women.

3.3 Scuba diving

Table 4 shows the overall diving experience of the study population (n = 181). The majority of scuba divers (57.5%) were experienced to very experienced divers with at least 200 dives at a ratio of at least 50 dives/year. Almost 41% (40.9%) were low to middle experienced divers with less than 200 dives at a ratio of less than 50 dives/year.

Table 4. Overall diving experience of the study population (n =181).

Diving certificate	% (n =181)	Diving years (dy) (Md=Me)§	Dives (d) (Md=Me)	Diving hours /last 12 months (dh) (Md=Me)	Diving hours /year (dh) (Md=Me)
1 star diver	13.8 %	< 1 dy	0 < d ≤ 50	0 < dh ≤ 50	≤ 50 dh
2 star diver	27.1 %	< 1dy	50 < d ≤ 200	0 < dh ≤ 50	≤ 50 dh
3 star diver	31.5 %	1-5 dy	200 < d ≤ 600	0 < dh ≤ 50	50 < dh ≤ 200
4 star diver	12.2 %	6-10 dy	200 < d ≤ 600	0 < dh ≤ 50	≤ 50 dh
1 star instructor	7.7 %	6-10 dy	> 600 d	0 < dh ≤ 50	50 < dh ≤ 200
2 star instructor	6.1 %	6-10 dy	> 600 d	50 < dh ≤ 200	50 < dh ≤ 200
3 star instructor	/	/	/	/	/
none	1.6 %	< 1 dy	0 < d ≤ 50	0 < dh ≤ 50	/

§ Modus = Median

Most scuba divers (76.8%) preferred recreational dives with a depth of less than 31 metres. In all diving conditions (open water/wet suit, open water/dry suit, swimming pool) the use of a weight belt was more common than the use of free weights in the buoyancy control device or diving suit. When diving in a pool 86.7% used less than 4 kg of weights, when diving in open water with a wet suit most scuba divers (56.9%) used weights between 4 and 8 kg. When diving in open water with a dry suit the weights amounted to more than 8 kg in most scuba divers (22.1%). Over the preceding 12 months, 5.3% of the scuba divers suffered a typical scuba diving injury among which were headache (1.6%), tympanic membrane rupture (1.1%), poisoning (1.1%), middle ear squeeze (0.5%), inner ear barotrauma (0.5%), and hypothermia (0.5%).

3.4 Low-back problems

3.4.1 Prevalence of LBP (n = 181)

Within the sample of 181 recreational scuba divers a one-year prevalence of LBP of 50.3% and a lifetime prevalence of LBP of 55.8% was found. The one-year prevalence of injuries in general was 60.2%. Regarding the general injury registration, injuries/complaints/ problems of the lower back were the most commonly reported injuries (29.3%) among scuba divers over the preceding 12 months, followed by injuries of the shoulder (18.8%), knee (14.4%) and cervical spine (14.4%).

3.4.2 Characteristics of LBP (n = 91)

Thirty-three percent of the symptomatic scuba divers indicated that they consulted a physician for their LBP and received a specific diagnosis regarding the LBP among which herniated disc (10.1%) was the most common followed by degenerative disease (8.8%), inflammation (4.4%), muscle strain (2.2%), traumatic injury of the disc or facet joint (2.2%), spondylolysis (2.2%) and radiculopathy (1.1%). However, most of the symptomatic scuba divers (64.8%) did not receive a specific diagnosis from their physician or did not consult a physician for their LBP at all. The most diagnosed structural abnormalities of the LBP were lumbar hyperlordosis (5.5%), lack of strength of the abdominal muscles (5.5%), scoliosis (3.3%) and tilt of the pelvis (3.3%). Primary impairments due to LBP were predominantly pain (71.4%), rigidity (38.5%), disability (19.8%) and strain (13.2%) of the lower back.

Among the 91 scuba divers with LBP, 64.8% had a history of LBP. The onset of LBP was in 44.0% of the subjects gradual and in 22.0% of the subjects sudden. Next, LBP was mostly localized in the entire low back (50.5%) while LBP with radiation to the gluteal area or leg appeared in 24.2% of the recreational scuba divers. Following the definitions of Von Korff (1994), the LBP was in 25.3% acute or subacute, in 23.1% recurrent and in 15.4% chronic (Von Korff 1994). Almost 52% (51.6%) of the symptomatic scuba divers did not experience any relationship between their LBP and their diving activities, while 41.8% did: 26.4% during the preparation phase, 23.1% during the reconditioning phase (maximum 24 hours after), 20.9% during the reconditioning phase (longer than 24 hours after) and 19.8% during the dive itself.

3.4.3 General and sport-specific characteristics in relation to LBP

A comparison between the group scuba divers without LBP (n = 90) and the group scuba divers with LBP (n = 91) for general subject characteristics is shown in Table 5. No significant differences (P ≥ 0.050) could be found for gender, age, BMI, civil state, working hours, smoking, alcohol use and sports participation between the two groups. However, significantly more scuba divers with LBP reported a history of LBP (P < 0.001) and structural abnormalities of the lower back diagnosed by a physician (P < 0.001). Next to this, a significant difference was found for heavy workload (hours/day). Scuba divers with LBP were exposed to more hours of heavy work per day than scuba divers without LBP (P = 0.036). Also significantly more female scuba divers with LBP experienced a pregnancy (P = 0.022) and/or a parturition (P = 0.048) than female scuba divers without LBP.

Another significant difference (P = 0.021) was present for the general injury frequency (i.e. low back not taken into account) as scuba divers with LBP also reported more injuries of other parts of the body than scuba divers without LBP, more specific injuries of the hip/groin and the thoracic or cervical spine. Scuba divers with LBP tended to have more problems of the hip/groin (P= 0.024), the thoracic (P = 0.023) or cervical spine (P = 0.037) than scuba divers without LBP. Moreover, symptomatic scuba divers spent less hours a day sitting (P = 0.045), they bent more often forwards or backwards in one day (P = 0.015) and they are generally more fatigued (P = 0.031) than asymptomatic scuba divers. There were no significant differences (P ≥ 0.050) between the two groups for psychosocial factors such as stress, anger and vitality at home, at work or among friends.

Table 5. Comparison between groups for general characteristics.

	No LBP (n = 90)	LBP (n = 91)	p
Gender (n)			0.890
Female	21 (23.3%)	22 (24.2%)	
Male	69 (76.7%)	69 (75.8%)	
Body mass index (kg/m²)	24.3 ± 3.7	25.0 ± 3.6	0.201
Age (years)	38.0 ± 14.2	40.2 ± 10.6	0.237
Civil state (n)			0.410
Single / divorced	37 (41.1%)	32 (35.2%)	
Married / living together	53 (58.9%)	59 (64.8%)	
Amount of work - hours/week (n)			0.437
≤30	4 (4.4 %)	3 (3.3 %)	
>30	66 (73.3 %)	77 (84.6 %)	
Amount of work - hours/day (n)			
Light ≤4	12 (13.3 %)	18 (19.8 %)	0.130
>4	39 (43.3 %)	34 (37.4 %)	
Moderate ≤4	24 (26.7 %)	35 (38.5 %)	0.411
>4	11 (12.2 %)	12 (13.2 %)	
Heavy ≤4	8 (8.9 %)	17 (18.7 %)	0.036*
>4	8 (8.9 %)	23 (25.3 %)	
Smoking (n)			
Current	16 (17.8 %)	17 (18.7 %)	0.931
Past	21 (23.3 %)	28 (30.8 %)	0.159
Alcohol use (n)	81 (90.0 %)	83 (91.2 %)	0.780
Pregnancy (n)	8 (8.9 %)	16 (17.6 %)	0.022 *
Parturition (n)	8 (8.9%)	15 (16.5 %)	0.048 *
Sports participation (n) (besides scuba diving)	67 (74.4 %)	60 (65.9 %)	0.211
Structural abnormalities (n)	1 (1.1%)	19 (20.9%)	0.000 *
History of LBP (n)	10 (11.1%)	59 (64.8%)	0.000 *

* P < 0.05: significant difference between the group with LBP and the group without LBP

Figure 1 shows significant differences between scuba divers with and without LBP for two sport-specific characteristics. Scuba divers with LBP had a significant (P = 0.007) higher diving certificate than scuba divers without LBP. Moreover, they used a significantly (P = 0.003) higher amount of weight on their weight belts during indoor training sessions than scuba divers without LBP.

Also while diving outdoor with a dry suit symptomatic scuba divers used a significantly higher amount of weight (P = 0.044) than asymptomatic scuba divers.

Finally, when diving outdoor scuba divers with LBP made significantly more use of certain scuba diving devices than the scuba divers without LBP (Table 6), including a decompression parachute (P = 0.009), a dry suit (P = 0.046), a compass (P = 0.010) and a diving computer (P = 0.012).

No significant correlations between sport-specific characteristics of scuba diving and overall characteristics of LBP could be revealed.

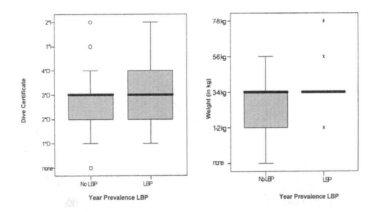

Figure 1. Boxplot for the dive certificate and the amount (in kg) of weight used during diving in a pool in scuba divers with and without LBP. The shaded box represents the range between which 50% of the data fall (i.e., interquartile range). The 'I' shaped whisker represents all of the data that fall within 3 standard deviations of the mean (i.e., the black horizontal bar). The circles (o) and stars (*) above the 'whiskers' represent outliers (i.e., individual scores very different from the overall scores in the shaded box).

Following simple binary stepwise logistic regression the strongest sport-specific predictors for LBP were diving certificate and the weight on the weight belts during outdoor diving with a dry suit (Table 7). The weight on the weight belt during indoor diving and during outdoor diving with a wet suit could not be considered as a significant predictor for LBP, as was diving level, diving experience, diving depth during outdoor dives, type of weight used on the weight belt, type of breathing gas used during outdoor dives, amount of indoor training sessions and diving injuries.

Multiple stepwise logistic regression for diving certificate and diving experience and for amount of weight used on the weight belt in different circumstances revealed no predictive models for LBP in scuba divers (Table 8).

Table 6. Comparison between grous for sport-specific characteristics.

	No LBP (n = 90)	LBP (n = 91)	p
Diving certificate (n)			0.007 *
≤ 2 star diver	44 (48.9%)	33 (36.3%)	
2 star diver < certificate ≤ 4 star diver	38 (42.2 %)	41 (45.1%)	
≥ 1 star instructor	8 (8.9%)	17 (18.7%)	
Diving years (n)			0.522
≤ 5 diving years	39 (43.3%)	36 (39.6%)	
5 < diving years ≤ 10	24 (26.7%)	23 (25.3%)	
> 10 diving years	27 (30.0%)	32 (35.2%)	
Dives (n)			0.120
0 < dives ≤ 100	37 (41.1%)	24 (26.4%)	
100 < dives ≤ 400	25 (27.8%)	36 (39.6%)	
> 400 dives	28 (31.1%)	31 (34.1%)	
Diving hours /last 12 months (n)			0.580
0 < diving hours ≤ 50	59 (65.6%)	58 (63.7%)	
> 50 diving hours	28 (31.1%)	27 (29.7%)	
Diving hours /year (n)			0.511
≤ 50 diving hours	55 (61.1%)	58 (63.7%)	
50 < diving hours ≤ 200	28 (31.1%)	29 (31.9%)	
> 200 diving hours	6 (6.7%)	2 (2.2%)	
Depth (n)			0.463
0 < depth ≤ -30 m	68 (75.6%)	71 (78.0%)	
-30 < depth ≤ -57m	21 (23.3%)	19 (20.9%)	
> -57 m depth	1 (1.1%)	0 (0.0%)	
Diving equipment (n)			
wet suit	90 (100%)	86 (94.5%)	0.079
dry suit	24 (26.7%)	37 (40.7%)	0.046 *
gloves	89 (98.9%)	90 (98.9%)	0.316
boots	87 (96.7%)	89 (97.8%)	0.313
knife	75 (83.3%)	85 (93.4%)	0.076
buoyancy control device	90 (100%)	90 (98.9%)	0.319
torch	87 (96.7%)	91 (100%)	
buddyline	78 (86.7%)	84 (92.3%)	0.402
compass	74 (82.2%)	87 (95.6%)	0.010 *
watch	60 (66.7%)	65 (71.4%)	0.720
depth gauge	75 (83.3%)	75 (82.4%)	0.560
pressure gauge	85 (94.4%)	82 (90.1%)	0.099
diving computer	71 (78.9%)	86 (94.5%)	0.012 *
decompression parachute	53 (58.9%)	70 (76.9%)	0.009 *

* P < 0.05: significant difference between the group with LBP and the group without LBP

Table 7. Simple logistic regression for sport-specific predictors of LBP.

Models	Model χ^2	-2 Log Likelihood	R^2 (Cox & Snell)	R^2 (Nagelkerke)	B (SE)	exp b
Model 1 (included)	4.773*	246.14	0.026	0.035		
constant					- 0.801 (0.405) *	0.449
diving certificate					0.475 (0.221) *	1.608
Model 2 (included)	4.498*	236.63	0.026	0.034		
constant					- 0.126 (0.181)	0.850
weight on weight belt during outdoor diving in wet suit					0.255 (0.122)*	1.290

```
 * P < 0.05
** P < 0.01
```

Table 8. Multiple logistic regression for possible sport-specific predictors of LBP.

Models	Model χ^2	-2 Log Likelihood	R^2 (Cox & Snell)	R^2 (Nagelkerke)	B (SE)	exp b
Model 3 (included)	3.324	247.59	0.018	0.024		
constant					-0.377 (0.261)	0.686
diving certificate by diving experience					0.101 (0.056)	1.106
Model 4 (included)	3.506	216.15	0.022	0.029		
constant					- 0.25 (0.182)	0.975
weight on weight belt during 3 different circumstances					0.092 (0.052)	1.097

```
 * P < 0.05
** P < 0.01
```

4. DISCUSSION

4.1 Prevalence of LBP

The present study revealed a one-year prevalence of LBP of 50.3% and a lifetime prevalence of LBP of 55.8% among recreational Flemish scuba divers. The one-year prevalence of LBP in this study (50.3%) is higher than recorded in the general population (Loney and Stratford, 1999; Marras, 2000; Santos-Eggimann et al., 2000; Walker, 2000; Van Tulder et al., 2002). The lifetime prevalence of LBP of 55.8% lies within the range (50% - 80%) of lifetime prevalence rates of LBP in the general population (Loney and Stratford, 1999; Marras, 2000; Santos-Eggimann et al., 2000; Van Tulder et al., 2002; Devereaux, 2004), although the mean age (i.e. 39.1 (± 12.5) years) of our subjects is slightly lower than generally found in literature (Walker, 2000). Moreover, comparisons with other epidemiological studies are difficult since there is no consensus on the definition or classification of LBP, a lack of methodological uniformity and results are often based upon self-administered data with no verification from an independent source (Van Mechelen *et al.*, 1992; Aagaard and Jorgensen, 1996; Borenstein, 1997; Loney and Stratford, 1999; Ozguler *et al.*, 2000; Van Tulder *et al.*, 2002; Devereaux, 2004; Manek and MacGregor, 2005). The lifetime prevalence of LBP in recreational scuba divers lies within the range of 23.0% - 79.0% as seen in other sports (Granhead and Morelli, 1988; Hosea and Gatt, 1996; Hickey *et al.*, 1997; Hutchison, 1999; Saraux *et al.*, 1999; Bahr *et al.*, 2004; Woolf and Glaser, 2004; Peacock *et al.*, 2005; Villavicencio *et al.*, 2006) (Table 1). However, when comparing these data one should again be careful because they are strongly subject to the definition and classification of LBP used, the methodology of the research design, the gender, type and level of sports, training intensity, training frequency or technique.

In the current study, a remarkable discrepancy could be found between the one-year prevalence of LBP (50.3%) and the one-year prevalence of injuries/complaints/problems of, what is commonly referred to as the lower back (29.0%) (i.e. the latter was assessed in the question on general injury registration). Although 29.0 % is the highest score among the general injury registration (i.e., compared to injuries of other parts of the body), one would expect this score to come close to the one-year prevalence of LBP as LBP can be considered as injuries/complaints/problems of the lower back. However, as 'injuries/complaints/problems of the lower back' is (deliberately) not properly defined in the questionnaire and LBP is, it is possible that subjects, in first instance, did not regard their LBP as an injury/complaint/ problem of the lower back. Thus, by not defining 'injuries/complaints/problems of the lower back', it is likely that minor LBPs were only registered in the specific questions on LBP and not in the general injury registration.

4.2 Characteristics of LBP

In this study the differential diagnosis of LBP in scuba divers included disc herniation, degenerative disease, inflammation, spondylolysis, muscle strain, traumatic injury of the disc or facet joint and radiculopathy. This inclusion is more or less in accordance with other findings in the literature: Baker and Patel (2005) and Dunn *et al.* (2006) indicated muscle strain, ligamentous sprain, disc herniation, degenerative disease, compression fracture and spinal stenosis as common conditions in adult athletes (Baker and Patel, 2005; Dunn *et al.*, 2006). Nachemson (1992) listed also degenerative disc disease and spondylolysis as the most common abnormalities associated with LBP in athletes (Nachemson, 1992). Spondylolysis is often regarded as the most serious injury of the lower back in sports involving repetitive hyperextension and axial loading (Rossi and Dragoni, 1990; Nyska *et al.*, 2000; Dunn *et al.*, 2006). These increased loads are transformed into shear forces acting mainly on the pars interarticularis of the vertebra (i.e. L5) (Rossi and Dragoni, 1990; Nyska *et al.*, 2000; Dunn *et al.*, 2006). The cases of spondylolysis (n = 2) reported in the current study could explained by scuba diving since strain and shearing forces on the lumbo-sacral region resulting from the "up and down" oscillation of the propulsive upper limbs act mainly on the posterior column of the spine, particularly on the arch and pars interarticularis (Verni *et al.*, 1999; Nyska *et al.*, 2000; Dunn *et al.*, 2006). Moreover, in swimming and fin swimming, which both have some similarities with scuba diving, clear associations between intensive leg workout and a chronic overload upon the posterior or spinal column were found (Goldstein *et al.*, 1991; Ferrell, 1999; Verni *et al.*, 1999; Nyska *et al.*, 2000).

Most LBP in scuba divers was non-specific, characterized by primary impairments as pain, rigidity, disability and strain of the lower back. This observation corresponds to findings from the general population where LBP is predominantly mechanical or non-specific (Kerr *et al.*, 2001; Ehrlich, 2003a, 2003b). Over the preceding 12 months a structural abnormality of the lower back (i.e. lumbar hyperlordosis, lack of strength of the abdominal muscles, scoliosis and deviation of the pelvis) was reported in 20.9 % of the symptomatic subjects and in 1.1% of the asymptomatic subjects ($P < 0.001$). Obviously, in the normal course of events asymptomatic subjects rarely consult medical services to search for structural abnormalities of their back and thus it would be premature to conclude that symptomatic scuba divers are more prone to structural abnormalities of the lower back. However, if these abnormalities are present in recreational scuba divers, it is possible that they contribute in some way to LBP as they can inhibit a correct spinal reaction to mechanical stress of the locomotor apparatus during effort (i.e. alternate front and back arching by the lower limbs and fins) (Verni *et al.*, 1999). Symptomatic scuba divers associated their LBP generally with the preparation (26.4%) and reconditioning (23.1%) phase of their diving activities during which heavy loads (i.e. up to 30 kg) are handled (i.e. lifting up and putting on the scuba diving cylinder and accessories). This is in accordance with findings from the general population, where repetitive handling of heavy loads is also associated with LBP (Pope *et al.*, 1995; Matsui *et al.*, 1997; Hoogendoorn *et al.* 2000a; Palmer *et al.*, 2003). Only 19.8% experienced LBP in relation to the dive itself. This does not support the hypothesis that during the dive itself unique static and dynamic biomechanical features of the scuba diver, put the lower back at a higher level of stress and thus at a higher risk of LBP. Notwithstanding, these results are based upon self-reported information and thus fundamental biomechanical research is needed to verify this hypothesis.

4.3 General characteristics in relation to LBP

When comparing the groups of scuba divers with and without LBP, no significant differences could be found for general subject characteristics (i.e. gender, age, BMI, civil state, working hours, smoking, alcohol use and sports participation). In the literature conflicting results are found on which general risk factors contribute to LBP (e.g., gender, BMI, age, smoking or alcohol use) (Matsui *et al.*, 1997; Thorbjornsson *et al.*, 1998; Adams *et al.*, 1999; Levangie, 1999; Feyer *et al.*, 2000; Thorbjornsson *et al.*, 2000; Vingard *et al.*, 2000; Kerr *et al.*, 2001; Kostova and Koleva, 2001; Hestbaek *et al.*, 2003b; Jansen *et al.*, 2004; Leboeuf-Yde, 2004).

The contradictory findings from the literature and the results from the current study, endorse the recent theory that risk factors are multidimensional and that complex interactions between risk factors of LBP exist since these can play a role as determinants, confounders or effect modifiers (Zinzen, 2002; Rubin, 2007). In this study, no such interactions between risk factors were examined. Some significant general risk factors for LBP (i.e. history of LBP, structural abnormalities, heavy workload, pregnancy and parturition, general fatigue and bending forwards or backwards) revealed in this study, are confirmed by concerted findings on risk factors of LBP in the literature: history of LBP (Papageorgiou *et al.*, 1996; Thorbjornsson *et al.*, 1998; Feyer *et al.*, 2000; Vingard *et al.*, 2000), structural abnormalities (Adams *et al.*, 1999; Nourbakhsh and Arab, 2002.), heavy workload (Ratti and Pilling, 1997; Ying *et al.*, 1997; Shelerud, 1998; Eriksen *et al.*, 1999; Gerr and Mani, 2000; Hartvigsen *et al.*, 2001), pregnancy and parturition (Levangie, 1999), general fatigue (Mannion *et al.*, 1996; Simmonds *et al.*, 1996) and bending forwards or backwards (Ying *et al.*, 1997; Hoogendoorn *et al.*, 2000a; Picavet and Schouten, 2000; Cole and Grimshaw, 2003; Jansen *et al.*, 2004). Furthermore, symptomatic scuba divers were also more vulnerable to other injuries (P = 0.021) (principally injuries of the hip/groin and the cervical or thoracic spine) which more or less supports the assumption of Hestbaek et al. (2003b) that diseases (headache/migraine, respiratory disorders, cardiovascular disease, or general health) seem to cluster in individuals and that LBP is part of this pattern (Hestbaek *et al.,* 2003a, 2003b).

4.4 Sport-specific characteristics in relation to LBP

The results of the current study revealed that symptomatic scuba divers had a higher dive certificate, indicating greater experience and leadership qualities. Presumably scuba divers with a higher dive certificate (i) have a higher responsibility during diving activities (i.e. guiding dives, teaching novices), (ii) they take along more equipment (i.e. extra regulator and octopus, diving computer, or decompression parachute) and (iii) they are allowed to dive in more aggravating conditions (i.e. greater depth, more flow, or poorer visibility). One or a combination of these factors could put scuba divers with a higher diving certificate at a higher level of (physical and/or psychological) stress and thus at a higher risk of LBP. Next to this, symptomatic scuba divers use a significantly higher amount of weight on their weight belts during indoor training sessions and during outdoor dives with a dry suit. More weight gives a larger local compression force on the lumbar spine which could also contribute to LBP. Nevertheless, this finding could not be confirmed for the use of weights during outdoor training sessions with a wet suit.

The absence of significant correlations between sport-specific characteristics of scuba diving and overall properties of LBP does not mean that scuba diving activities are not associated with LBP since not all sport-specific characteristics have been investigated (i.e. biomechanical features). Further research is necessary to elucidate a possible relationship between sport-specific characteristics and overall characteristics of LBP.

Simple and multiple logistic regression revealed two significant sport-specific predictors for LBP. A higher diving certificate increases the risk for LBP and more weight on the weight belt during outdoor diving with a dry suit also increases the risk of LBP. Both models are a significant fit of the data of LBP in scuba divers.

4.5 Study limitations

Certain weaknesses of the study may limit the generalizability of the findings. First of all, data were only gathered by means of a self-assessment questionnaire in a retrospective way and no verification from an independent source (i.e. researcher and/or physician) and/or objective measurements (i.e. physical and/or radiological examinations) was available. Secondly, although non-parametric statistics are assumption-free tests (i.e. they make no assumptions, or less restrictive assumptions, about the distribution of the data than parametric tests) and they have less statistical power than their parametric counterparts (Field and Hole, 2003; Field, 2005), in general non-parametric statistics were used to incorporate data statistically due to the nature of the questionnaire. Moreover, no statistical tests were applied to search for confounding factors or effect modifiers when comparing the two groups so data could be skewed. Next to this, underlying intrinsic as well as extrinsic factors could contribute to LBP in scuba divers. A challenging potential area of future research would lie in a static and dynamic biomechanical analysis of scuba diving. The self-assessment questionnaire used in this study had sufficient test-retest reliability and content and construct validity for its purpose. Nevertheless if, in future research, the questionnaire would be used to evaluate LBP in other sports, these findings about the properties of the questionnaire should not be generalized. It is recommended to test for validity and reliability once more in a sample of a different population to strengthen the scientific findings of this study. Moreover, the sport-specific part of the questionnaire has to be adapted each time to the sport involved. Next to this, an additional question on the presence of current LBP (point-prevalence) would be appropriate to get a more accurate idea of the prevalence of LBP in different sports and to be able to make comparisons to other findings in the literature (Saraux *et al.*, 1999; Wolf and Glaser, 2004; Peacock *et al.*, 2005).

5. CONCLUSION

According to the results from this epidemiological survey, LBP in scuba diving gives reason for moderate concern. Lifetime and one-year prevalence of LBP among recreational Flemish scuba divers were 55.8% and 50.3%, respectively. This study also sheds light on the characteristics of these LBP in recreational scuba divers. Only a few previously-reported general risk factors for LBP could also be found among these recreational scuba divers (i.e., history of LBP, structural abnormalities, heavy workload, pregnancy and parturition, general fatigue and bending forwards or backwards). Sport-specific risk factors for LBP found in this study are the diving certificate and the amount of weight used on the weight belt but further (biomechanical) research should clarify the underlying mechanisms. Nevertheless, the results of the current study could serve as a guide towards more accurate and specific research methods to study LBP in (recreational) scuba divers profoundly.

Acknowledgements

The authors would like to thank Dr. Van Bogaert (hyperbaric physician and company doctor) for bringing up the research hypothesis in the first place, Prof. Dr. William Duquet for his assistance in the data analysis and all recreational scuba divers who participated in this study.

References

Aagard, H. and Jorgensen, U., 1996, Injuries in elite volleyball. *Scandinavian Journal of Medicine and Science in Sports*, **6**, pp. 228-232.

Adams, M.A., Mannion, A.F. and Dolan, P., 1999, Clinical studies - personal risk factors for first-time low back pain - Personal risk factors for first-time low back pain (LBP) were assessed prospectively on five occasions during a 3-year follow-up period on 403 young health care workers. *Spine Hagerstown,* **24**, pp. 2497-2505.

Akmal, M., et al., 2004, Effect of nicotine on spinal disc cells: a cellular mechanism for disc degeneration. *Spine Hagerstown*, **29**, pp. 568-575.

Bahr, R., et al., 2004, Low back pain among endurance athletes with and without specific back loading--a cross-sectional survey of cross-country skiers, rowers, orienteerers, and non-athletic controls. *Spine Hagerstown*, **29**, pp. 449-54.

Baker, R.J. and Patel, D., 2005, Lower back pain in the athlete: common conditions and treatment. *Primary Care Clinics in Office Practice*, **32**, pp. 201-229.

Bono, C.M., 2004, Low back pain in athletes. *The Journal of the Bone and Joint Surgery*, **86A**, pp. 383-396.

Borenstein, D.G., 1997, Epidemiology, etiology, diagnostic evaluation, and treatment of low back pain. *Current Opinion in Rheumatology*, 9 (2), pp. 144-150.

Clenney, T.L. and Lassen, L.F., 1996, Recreational scuba diving injuries. *American Family Physician*, **53**, pp. 1761-1774.

Cole, M.H. and Grimshaw, P.N., 2003, Low back pain and lifting: a review of epidemiology and aetiology. *Work*, **21**, pp. 173-184.

De Vet, H.C., et al., 2002, Episodes of low back pain: a proposal for uniform definitions to be used in research. *Spine-Hagerstown*, **27**, pp. 2409-2416.

Devereaux, M.W., 2004, Low back pain. *Primary Care*, **31**, pp. 33-51.

Dunn, I.A., Proctor, M.R. and Day, A.L., 2006, Lumbar spine injuries in athletes.

Neurosurgical Focus, **21**, E4.

Ehrlich, G.E., 2003a, Back pain. *Journal of Rheumatology: Supplement*, **67**, pp. 26-31.

Ehrlich, G.E., 2003b, Low back pain. *Bulletin of the World Health Organisation*, **81**, pp. 671-676.

Eriksen, W., Natvig, B. and Bruusgaard, D., 1999, Smoking, heavy physical work and low back pain: a four-year prospective study. *Occupational Medicine*, **49**, pp. 155-160.

Ferrell, M.C., 1999. The spine in swimming. *Aquatic Sports Injuries and Rehabilitation*, 18(2), pp. 389-393.

Feyer, A.M., et al., 2000, The role of physical and psychological factors in occupational low back pain: a prospective cohort study. *Occupational and Environmental Medicine*, **57**, pp. 116-120.

Field, A., 2005, *Discovering Statistics Using SPSS*, (London: Sage Publications).

Field, A., and Hole, G., 2003, *How to Design and Report Experiments*, (London: Sage Publications).

Gerr, F., and Mani, L., 2000, Work-related low back pain. *Primary Care*, **27**, pp. 865-876.

Goldberg, M.S., Scott, S.C. and Mayo, N.E., 2000. A review of the association between cigarette smoking and the development of non-specific back pain and related outcomes. *Spine-Hagerstown*, 25 (8), 995-1014.

Goldstein, J., et al., 1991, Spine injuries in gymnasts and swimmers. *American Journal of Sports and Medicine,* **19**, pp. 463-468.

Granhead, H. and Morelli, B., 1988, Low back pain among retired wrestlers and heavyweight lifters. *American Journal of Sports Medicine*, **16**, pp. 530-533.

Haanen, H.C.M., 1984, *Een epidemiologisch onderzoek naar lage rugpijn*, (Holland: Boots Company B.V).

Han, T.S., et al., 1997, The prevalence of low back pain and associations with body fatness, fat distribution and height. *International Journal of Obesity and Related Metabolic Disorders*, **21**, pp. 600-607.

Hartvigsen, J., et al., 2001, The association between physical workload and low back pain clouded by the 'healthy worker' effect: Population-based cross-sectional and 5-year prospective questionnaire study. *Spine-Hagerstown*, **26**, pp. 1788-1791.

Hestbaek, L., Leboeuf-Yde, C. and Manniche, C., 2003a, Low back pain: what is the long-term course? A review of studies of general patient populations. *European Spine Journal*, **12**, pp. 149-165.

Hestbaek, L., Leboeuf-Yde, C. and Manniche, C., 2003b, Is low back pain part of a general health pattern or is it a separate and distinctive entity? A critical literature review of comorbidity with low back pain. *Journal of Manipulative and Physiological Therapeutics*, **26**, pp. 243-252.

Hickey, G.J., Fricker, P.A. and McDonald, W.A., 1997, Injuries to elite rowers over a 10-yr period. *Medicine and Science in Sports and Exercise*, **29**, pp. 1567-1572.

Hoogendoorn, W.E., et al., 2000a, Epidemiology - flexion and rotation of the trunk and lifting at work are risk factors for low back pain: Results of a prospective cohort study. *Spine Hagerstown,* **25**, pp. 3087-3092.

Hoogendoorn, W.E., et al., 2000b, Systematic review of psychosocial factors at work and private life as risk factors for back pain, *Spine-Hagerstown,* **25**, pp. 2114-2025.

Hosea, T.M., and Gatt, C.J., 1996, Back pain in golf. *Clinics in Sports Medicine,* 15, pp. 37-53.

Hunt, J.C., 1996, Diving the wreck: Risk and injury in sport scuba diving. *Psychoanalytic Quarterly,* **65**, pp. 591-622.

Hutchinson, M.R., 1999, Low back pain in elite rhythmic gymnasts. *Medicine and Science in Sports and Exercise,* **31**, pp. 1686-1688.

Jager, M., et al., 2002, Acute low back pain with progressive sensorimotor paralysis. Differential diagnosis and therapy of acute decompression disease. *Deutsche Medizinische Wochenschrift,* **127**, pp. 1188-1891.

Jansen, J.P., Morgenstern, H., Burdorf, A., 2004, Dose-response relations between occupational exposures to physical and psychosocial factors and the risk of low back pain. *Occupational and Environmental Medicine,* **61**, pp. 972-979.

Kerr, M.S., et al., 2001, Biomechanical and psychosocial risk factors for low back pain at work. *American Journal of Public Health,* **91**, pp. 1069-1075.

Kostova, V. and Koleva, M., 2001, Back disorders (low back pain, cervicobrachial and lumbosacral radicular syndromes) and some related risk factors. *Journal of Neurological Sciences,* **192**, pp. 17-26.

Leboeuf-Yde, C., 1995, Does smoking cause low back pain? A review of the epidemiologic literature for causality. *Journal of Manipulative and Physiological Therapeutics,* **18**, pp. 237-243.

Leboeuf-Yde, C., 2000a, Alcohol and low back pain: A systematic literature review. *Journal of Manipulative and Physiological Therapeutics,* **23**, pp. 343-346.

Leboeuf-Yde, C., 2000b, Body weight and low back pain. A systematic literature review of 56 journal articles reporting on 65 epidemiologic studies. *Spine-Hagerstown,* **25**, pp. 226-237.

Leboeuf-Yde, C., 2000c, Alcohol and low-back pain: a systematic literature review. *Journal of Manipulative and Physiological Therapeutics,* **23**, pp. 343-346.

Leboeuf-Yde, C., 2002, Smoking and low back pain. A systematic literature review of 41 journal articles reporting 47 epidemiologic studies. *Spine-Hagerstown,* **24**, pp. 1463-1470.

Leboeuf-Yde, C., 2004, Back pain-individual and genetic factors. *Journal of Electromyography and Kinesiology,* **14**, pp. 139-133.

Leboeuf-Yde, C., Kyvik, K.O. and Bruun, N.H., 1998, Low back pain and lifestyle: Part I: Smoking information from a population-based sample of 29.424 twins. *Spine-Hagerstown,* **23**, pp. 2207-2213.

Leboeuf-Yde, C., Kyvik, K.O. and Bruun, N.H.., 1999, Low back pain and lifestyle. Part II – Obesity. *Spine-Hagerstown,* **24**, pp. 779-782.

Leung, A.S.L., et al., 1999, Use of a subjective health measure on Chinese low back pain patients in Hong Kong, *Spine-Hagerstown,* **24**, pp. 961-966.

Levangie, P.K., 1999, Association of low back pain with self-reported risk factors among patients seeking physical therapy services. *Physical Therapy*, **79**, pp. 757-766.

Loney, P.L. and Stratford, P.W., 1999, The Prevalence of low back pain in adults: A methodological review of the literature. *Physical Therapy*, **79**, pp. 384-396.

Manchikanti, L., 2000, Epidemiology of low back pain. *Pain Physician*, **3**, pp. 167-192.

Manek, N.J., and MacGregor, A.J., 2005, Epidemiology of back disorders: prevalence, risk factors, and prognosis. *Current Opinion in Rheumatology*, **17**, pp. 134-140.

Mannion, A.F., Dolan, P. and Adams, M.A., 1996, Psychological questionnaires: Do "abnormal" scores precede or follow first-time low back pain? *Spine-Hagerstown*, **21**, pp. 2603-2611.

Marras, W.S., 2000, Occupational low back disorder causation and control. *Ergonomics*, **43**, pp. 880-902.

Matsui, H., et al., 1997, Risk Indicators of Low Back Pain Among Workers in Japan: Association of Familial and Physical Factors With Low Back Pain. *Spine Hagerstown*, **22**, pp. 1242-1247.

Mortimer, M., et al., 2001, Musculoskeletal Intervention Center, Sports activities, body weight and smoking in relation to low-back pain: a population-based case-referent study. *Scandinavian Journal of Medicine and Science in Sports*, **11**, pp. 178-184.

Nachemson, A.L., 1992, Newest knowledge of low back pain. A critical look. *Clinical Orthopaedics and Related Research*, **279**, pp. 8-20.

Nourbakhsh, M.R., and Arab, A.M., 2002, Relationship between mechanical factors and incidence of low back pain. *Journal of Orthopaedic and Sports Physical Therapy*, **32**, pp. 447-460.

Nyska, M., et al., 2000, Spondylolysis as a cause of low back pain in swimmers. *International Journal of Sports Medicine*, **21**, pp. 375-379.

Ozguler, A., et al., 2000, Individual and occupational determinants of low back pain according to various definitions of low back pain. *Journal of Epidemiology and Community Health*, **54**, pp. 215-220.

Palmer, K.T., et al, 2003, The relative importance of whole body vibration and occupational lifting as risk factors for low-back pain. *Occupational and Environmental Medicine*, **60**, pp. 715-721.

Papageorgiou, G.C., et al., 1996, Influence of previous pain experience on the episode incidence of low back pain: Results from the South Manchester Back Pain Study. *Pain-Journal of the International Association for the Study of Pain*, **66**, pp. 181-185.

Peacock, N., et al., 2005, Prevalence of low back pain in alpine ski instructors. *Journal of Orthopaedic and Sports Physical Therapy*, **35**, pp. 106-110.

Pelletier, J.P., 2002, Recognizing sport diving injuries. *Dimensions of critical care nursing*, **21**, pp. 26-27.

Picavet, J.S. and Schouten, H.S., 2000, Physical load in daily life and low back problems in the general population-The MORGEN study. *Preventive Medicine*, **31**, pp. 506-512.

Pope, M.H., Wilder, D.G. and Magnusson, M.L., 1999, A review of studies on seated whole body vibration and low back pain. *Proceedings of the Institution of Mechanical Engineers- Journal of engineering in Medicine*, **213**, pp. 435-446.

Ratti, N. and Pilling, K., 1997, Back pain in the workplace. *British Journal of Rheumatology*, **36**, pp. 260-264.

Resnik, L. and Dobrzykosk, E., 2003, Guide to outcomes measurement for patients with low back pain syndromes. *Journal of Orthopaedic and Sports Physical Therapy*, **33**, pp. 307-316.

Rossi, F. and Dragoni, S., 1990, Lumbar spondylolysis: occurrence in competitive athletes. *The Journal of Sports Medicine and Physical Fitness*, **30**, pp. 450-452.

Rubin, D., 2007, Epidemiology and risk factors for spine pain. *Neurological Clinics*, **25**, pp. 353-371.

Santos-Eggiman, B., et al., 2000, One-year prevalence of low back pain in two Swiss regions. *Spine-Hagerstown*, **25**, pp. 2473-2479.

Saraux, A., et al., 1999, Are tennis players at increased risk for low back pain and sciatica? *Revue du Rhumatisme: Joint, Bone, Spine Diseases*, **66**, pp. 143-145.

Shelerud, R., 1998, Epidemiology of occupational low back pain. *Occupational Medicine*, **13**, pp. 1-22.

Shroder, S., Lier, H. and Wiese, S., 2004, Diving accidents. Emergency treatment of serious diving accidents. *Anaesthesist*, **53**, pp. 1093-102.

Simmonds, M.J., Kumar, S. and Lechelet, E., 1996, Psychosocial factors in disabling low back pain: Causes or consequences? *Disability and Rehabilitation*, **18**, pp. 161-168.

Sward, L., et al., 1991, Disc degeneration and associated abnormalities of the spine in elite gymnasts. A magnetic resonance imaging study. *Spine-Hagerstown*, **16**, pp. 437-443.

Thorbjornsson, J.C., et al., 1998, Physical and psychosocial factors associated with low back pain: A 24-Year follow-up among women and men in a broad range of occupations. *Occupational and Environmental Medicine*, **55**, pp. 84-90.

Thorbjornsson, J.C., et al., 2000, Physical and psychosocial factors related to low back pain during a 24-year period. A nested case-control analysis. *Spine-Hagerstown*, **25**, pp. 369-371.

Van Mechelen, W., Hlobil, H. and Kemper, H.C.G., 1992, Incidence, severity, aetiology and prevention of sports injuries. *Sports Medicine*, **14**, pp. 82-99.

Van Tulder, M., Koes, B. and Bombardier, C., 2002, Low back pain. *Best Practice and Research Clinical Rheumatology*, **16**, pp. 761-775.

Verni, E., et al., 1999, Lumbar pain and fin swimming. *Journal of Sports Medicine and Physical Fitness*, **39**, pp. 61-65.

Villavicencio, A.T., et al., 2006, Back and neck pain in triathletes. *Neurosurgical Focus*, **21**, E7.

Vingärd, E., et al., 2000, To what extent do current and past physical and psychosocial occupational factors explain care-seeking for low back pain in a working population? *Spine Hagerstown*, **25**, pp. 493-500.

Von Korff, M., 1994, Studying the natural history of back pain. *Spine-Hagerstown*, **19**, pp. 2041S-2046S.

Walker, B.F., 2000, The Prevalence of low back pain: A systematic review of the literature from 1996 to 1998. *Journal of Spinal Disorders*, **13**, pp. 205-217.

Woolf, S.K. and Glaser, J.A., 2004, Low back pain in running-based sports. *Southern Medical Journal*, **97**, pp. 847-851.

Worf, N., 2002, Scuba-diving injuries. *Emergency Medicine Service*, **31**, pp. 147-149.

Ying, X.U., Bach, E. and Orhede, E., 1997, Work environment and low back pain: the influence of occupational activities. *Occupational and Environmental Medicine*, **54**, pp. 741-745.

Zinzen, E., 2002, Epidemiology: Musculoskeletal problems in Belgian nurses. In *Musculoskeletal Disorders in Health-Related Occupations*, edited by Reilly, T., (Amsterdam: IOS Press), pp. 41-64.

Appendix 1

Classification of LBP of Von Korff (1994) and De Vet *et al.* (2002):

- **Acute LBP**: back pain that is not recurrent or chronic (as defined below) and whose onset is recent and sudden
- **Subacute LBP**: back pain that is not recurrent or chronic (as defined below) and whose onset is progressive
- **Chronic LBP**: back pain present on at least half the days in a 12-month period in a single or in multiple episodes
- **Recurrent LBP**: back pain present on less than half the days in a 12-month period, occuring in multiple episodes over the year
- **Transient LBP**: an episode in which back pain is present on no more than 90 consecutive days and does not recur over a 12-month observation period
- **First onset of LBP**: refers to an episode of back pain that is the first occurence of back pain in the person's lifetime
- **Flare-up of LBP**: a phase of pain superimposed on a recurrent or chronic course; a distinctive period (i.e. usually a week or less) when back pain is markedly more severe than is usual for the patient
- **Episode of LBP**: a period of pain in the lower back lasting for more than 24 hours, preceded and followed by a period of at least one month without low back pain

A single bout of cold water immersion therapy has no beneficial effect on recovery from the symptoms of exercise-induced muscle damage

J.R. Jakeman, R. Macrae and R. Eston

School of Sport and Health Sciences, University of Exeter, Exeter, UK

1. INTRODUCTION

Athletes are often required to train and compete several times during a relatively short period, without allowing adequate time for a natural recovery (Cochrane, 2004).This is apparent in a variety of events, and is perhaps most evident in sports such as rugby union and association football (soccer), where individuals compete and train daily. On occasion, and depending on the period of season, players may compete on more that one occasion during a week, and can sometimes play three to four games in less than two weeks. Interventions to assist the recovery process are therefore crucially important, and frequently applied in sporting situations.

The effects of exercise-induced muscle damage (EIMD) have long been established, and the resultant functional damage can impair athletic performance (Byrne et al., 2004). Exercise induced muscle damage is characterised by reductions in isometric muscle strength, variations in joint range of motion, changes in limb girth, and muscle protein leakage into the blood (Brown et al., 1997). It is often most notably manifest through delayed-onset muscle soreness, and is frequently associated with unaccustomed eccentric exercise (Byrne et al., 2004). The mechanisms for EIMD and subsequent delayed-onset muscle soreness have been associated with the depletion of muscle glycogen, and mechanical disruptions which affect the contractile force-length properties of muscle, and the structure of the muscle sarcomeres (Appell et al., 1992; Eston et al., 1995; Malm, 2001; Gleeson et al., 2003; Morgan and Proske, 2004).

A single bout of cold water therapy has become a widely used muscle recovery method in competitive sport (Sellwood et al, 2007). Professional athletes are often subjected to whole-body immersion in cold water after activity in an effort to enhance recovery and reduce the effects of EIMD from exercise. The underlying assumptions of cryotherapy are that it can be used to treat musculoskeletal soreness and injury, as the process decreases local tissue temperature, constricts local blood vessels, and reduces inflammation and oedema in the area (Eston and Peters, 1999; Bailey et al., 2007; Sellwood et al., 2007; Vaile et al., 2008a, b). However, the effects of cryotherapy on the symptoms of

EIMD have yet to be convincingly clarified. Cold water immersion treatment strategies have been shown to demonstrate little or no beneficial effect on the markers of EIMD (Isabell *et al.*, 1992; Eston and Peters, 1999; Howatson *et al*, 2004; Sellwood *et al*, 2007). However, Bailey *et al.* (2007) found that cold water immersion had a beneficial effect on some of the indices of EIMD, including pain and muscle power output. The beneficial effects of hydrotherapy treatments, including cold water immersion on the recovery of isometric muscle force, and dynamic power has been recently shown by Vaile *et al.* (2008a). The authors have also observed that cold water therapy following a high-intensity exercise bout has a beneficial performance effect on a subsequent high-intensity bout performed 40-min later (Vaile *et al.*, 2008b).

The equivocal findings from the literature are to some extent confounded by the variety of recovery protocols used, and research has involved the effects of both repeated (Eston and Peters, 1999) and single bouts of cold water immersion (Bailey *et al.*, 2007; Sellwood *et al.*, 2007), and a wide variety of immersion temperatures and duration (Paddon-Jones and Quigley, 1997; Eston and Peters, 1998; Bailey *et al.*, 2007; Sellwood *et al.*, 2007). At present, there seems to be little consensus regarding the optimal protocol for cold water immersion to improve recovery from training or competitive sport (MacAuley, 2001; Cheung *et al.*, 2003). It has yet to be shown unequivocally that a single dose of cold water immersion therapy is an effective recovery modality following exercise.

The aim of this study was to assess the effectiveness, and consequently the appropriateness of a single bout of cold water immersion, a commonly used procedure in competitive sports such as rugby, association football, and hockey, as a treatment and prevention strategy for the symptoms of exercise induced muscle damage.

2. METHODS

2.1 Participants and Design

Twenty female athletes (age 19.9 ± 0.97 years, height 1.66 ± 0.05 m, mass 63.7 ± 10 kg) volunteered for the study. None had muscular or skeletal injuries prior to participation, and none had engaged in any specific plyometric or lower-limb training in at least six weeks prior to testing. Each participant provided written informed consent to participate in the study, which was approved by the home School's ethics committee. After baseline measurements, participants performed a plyometric exercise protocol to induce muscle damage, and were randomly allocated to a control or treatment group. Symptoms of exercise-induced muscle damage were compared to baseline values for a period of five days.

2.2 Exercise-Induced Muscle Damage

Muscle damage was induced using 10 sets of 10 counter-movement jumps (Twist and Eston, 2005). Participants were asked to perform maximal vertical jumps throughout the exercise, and were asked to adopt a 90^0 knee angle on landing, in order to facilitate muscle damage.

2.3 Assessment of Exercise-Induced Muscle Damage

The criterion measures for establishing exercise-induced muscle damage (EIMD) were creatine kinase activity, functional muscle strength, assessed through isokinetic dynamometry, and perceived limb soreness during an unweighted squat. These measurements were recorded at baseline, and at 1, 24, 48, 72, and 96 h, after the plyometric exercise protocol.

2.4 Creatine Kinase Activity

Samples of plasma creatine kinase were obtained using fingertip capillary blood sampling. The finger was cleaned using a sterile alcohol swab, and a capillary puncture was made using a Haemocue lancet (Haemocue, Sheffield, UK), with blood refrigerated at 4^0C until separated by centrifuge for analysis. Creatine kinase activity was analysed by spectrophotometry (Jenway, Dunmow, UK), in accordance with the manufacturer guidelines (Randox, Co. Antrim, UK).

2.5 Functional Muscle capacity

Concentric quadriceps muscle strength was assessed through isokinetic dynamometry (Biodex Medical Systems, New York, USA). Particiapants were instructed to complete a warm up and stretch of their own choosing, lasting approximately five minutes. This involved a gentle jog, and static stretching for the majority of participants. Participants then completed five maximal voluntary concentric muscle contractions at 45 degrees.s (0.9 rad.s^{-1}) were measured through a 90 degrees range of movement. The best of five concentric muscular contractions, expressed as a value of uncorrected peak torque, was used as the indicator of maximal quadriceps contractile ability.

2.6 Perceived Soreness

Perceptions of peak soreness were assessed during a 2-second, unweighted squat at 90 degrees knee flexion. Participants indicated on a 10-cm visual analogue scale (0= No pain, 10= Worst pain ever) their feelings of perceived soreness, with values being recorded with a pen mark at the appropriate site along the line.

2.7 Cold Water Immersion Treatment

Within ten minutes of completing the damaging exercise, a single bout of cold water immersion was administered to the treatment group. This bout consisted of a 10-min immersion at 10 degrees C ($\pm 1^{0}$C) Participants were seated with legs at 90 degrees to the torso, in a bath filled to the level of the superior iliac crest to ensure the whole lower limb was submerged. Temperature was maintained by the addition of cubed ice, in keeping with field testing methods, and was monitored using a thermometer.

2.8 Analysis

Data were analysed using a two-factor (group by time) analysis of variance using the SPSS 13.0 statistical software package. Assumptions of homogeneity of variance were tested using the Mauchly sphericity test. Where this was violated, the Greenhouse-Geisser value was used to adjust degrees of freedom to increase the critical value of the F-ratio. Significance level was set a-priori at P = 0.05.

3. RESULTS

3.1 Perceived Soreness

In both groups there was a significant main effect of time on perceived soreness ($F_{5,45}$ =14.6, P<0.01). No significant difference between groups ($F_{1,9}$ =0.009, P>0.05), and no significant group by time interaction on perceived soreness ($F_{5,45}$ =1.6, P>0.05), were observed (Figure 1).

3.2 Creatine Kinase Activity

Creatine kinase data were converted to natural logarithms to satisfy the assumptions of analysis of variance. There was a significant main effect of time on creatine kinase activity ($F_{5,40}$ = 3.105, P<0.05). No differences between groups ($F_{1,8}$ = 0.391, P>0.05) and no interaction effect on CK activity ($F_{5,40}$ = 0.129, P>0.05) were noted. The CK activity of both groups followed a similar temporal pattern, peaking at 24 h after the plyometric activity, returning to baseline values at 48 h (Figure 2).

3.3 Concentric Muscle Strength

There was a significant main effect of time on concentric muscle strength ($F_{5,45}$ = 7, P<0.05), no significant difference between groups ($F_{1,9}$ = 0.09, P>0.05) and no group by time interaction effect on concentric muscle strength ($F_{5,45}$ = 0.6, P>0.05). Concentric muscle strength variability followed a similar trend for both groups (Figure 3.).

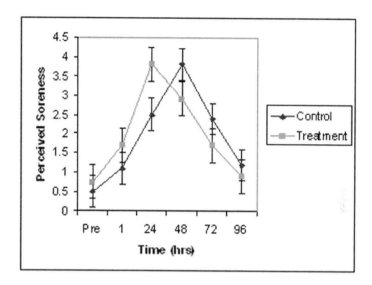

Figure 1. Variation of perceived soreness measured over time for control and treatment groups.

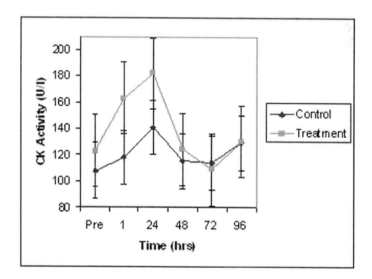

Figure 2. Variation of creatine kinase activity measured over time for control and treatment groups.

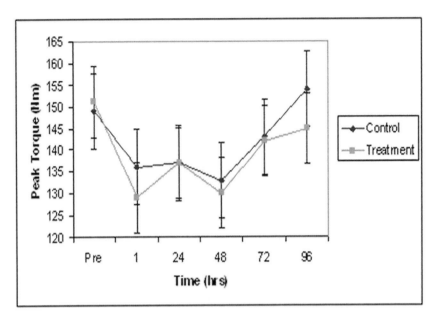

Figure 3. Variation of peak muscle torque (quadriceps) measured over time for control and treatment groups.

4. DISCUSSION

The aim of this study was to investigate the effectiveness of a cold water immersion therapy protocol, similar to one which may be used in field situations, as a recovery strategy following strenuous physical activity. Indicators of exercise-induced muscle damage (EIMD) suggest that the plyometric jumping protocol used in this study was sufficient to induce muscle damage. However, no group or interaction effects were highlighted on any of the variables investigated. The results of this study therefore suggest that a single bout of cold water immersion after plyometric activity is of no significant benefit to alleviate the symptoms of exercise-induced muscle damage.

Plasma creatine kinase (CK) was investigated in this study, as a widely accepted indicator of muscle damage (Eston and Peters, 1999; Zainuddin *et al.*, 2005; Sellwood *et al.*, 2005). The peak in CK activity at 24 h after the plyometric exercise protocol mirrors the typical temporal pattern associated with exercise-induced muscle damage in large muscle groups, and indicates that the damaging protocol was sufficiently severe to induce muscle damage (Fig. 2.). Although there was a significant change in CK activity across time, there was no overall difference between groups, and the changes in CK across time were similar for both groups. The single bout of cold water immersion treatment used in this study was therefore ineffective at reducing the CK activity response following EIMD.

Some studies have suggested that a single bout of cryotherapy treatment has a positive effect on subsequent exercise (Vaile *et al.*, 2008b), recovery of muscle power (Vaile *et al.*, 2008a), and perceptions of soreness (Bailey *et al.*, 2007), which imply an analgesic effect following immersion. However, the actual duration of analgesic effect is uncertain. It has been reported to remain for only a relatively short period of between three minutes to three hours (Meeusen and Lievens, 1986), which may account for the lack of reported effect in the treatment participants. Despite the underlying potential mechanisms of cryotherapy indicating that such a treatment may be beneficial to markers of EIMD, Isabell *et al.* (1992) observed that participants who received cryotherapy treatment had higher levels of perceived soreness than those in a control group. They suggested that the mechanisms of tissue repair and subsequent relief may be adversely affected by cold treatments, but further investigation is required to determine this. In agreement with other studies of similar design (e.g., Yackzan *et al.*, 1984; Paddon-Jones and Quigley, 1997) the current study indicated that perceived soreness increased significantly from baseline over the test period, confirming that the plyometric exercise protocol induced muscle damage (Twist and Eston, 2005). However, variations in perceived soreness were not moderated by group, suggesting that cold water immersion was not effective as a strategy for recovery from perceived soreness resulting from EIMD.

The decrease in maximal voluntary concentric contraction of the quadriceps group followed a similar temporal pattern and trend as perceived pain (Figure 3). This observation is in keeping with the majority of research in this area, and is representative of the magnitude of muscle damage induced during the period of testing (Eston *et al.*, 1995; Paddon-Jones and Quigley, 1997; Howatson *et al.*, 2005; Bailey *et al.*, 2007; Sellwood *et al.*, 2007). The maximal contractile capability of the quadriceps over time was not moderated by group, indicating that the cold water immersion treatment was not effective as a recovery strategy from EIMD. This outcome is in contrast with the observations of other researchers who have found beneficial effects using different forms of hydrotherapies (Gill *et al.*, 2006; Bailey *et al.*, 2007). However, it is important to note that Gill *et al.* (2006) used contrast water therapy as a treatment intervention, and Bailey *et al.* (2007) observed beneficial effects on the knee flexors only after a cold water immersion treatment, but no effect on the knee extensors, the prime movers for many movements, as well as the major stabilising muscle group for the knee.

5. CONCLUSION

The aim of this study was to assess the effectiveness of a single bout of cold water immersion as a treatment and prevention strategy for the symptoms of exercise-induced muscle damage. A single bout of cold water immersion therapy had no significant effects on a variety of markers of exercise-induced muscle damage. It is concluded that a single bout of cold water immersion therapy at ten degrees C for 10 min is not effective in providing an enhanced recovery rate for individuals who exhibit symptoms of exercise-induced muscle damage. Research in this area remains highly equivocal, with more investigation required to determine the effectiveness of other single-bout immersion protocols, and the effects of repeated treatments of cold water immersion on athletic individuals.

Acknowledgements

The contributions of participants from the University of Exeter are gratefully acknowledged.

References

Appell, H.J., Soares, J.M., and Duarte, J.A., 1992, Exercise induced muscle damage and fatigue. *Sports Medicine,* **13**, pp. 108-115.
Bailey, D.M., et al., 2007, Influence of cold water immersion on indices of muscle damage following prolonged intermittent shuttle running. *Journal of Sports Sciences,* **25**, pp. 1163-1170.
Brown, S.J., et al., 1997, Exercise-induced skeletal muscle damage and adaptation following repeated bouts of eccentric muscle contractions. *Journal of Sports Sciences,* **15**, pp. 215-222.
Byrne, C., Twist, C., and Eston, R., 2004, Neuromuscular function after exercise-induced muscle damage. *Sports Medicine,* **34**, pp. 49-69.
Cheung, K., Hume, P.A., and Maxwell, L., 2003, Delayed onset muscle soreness: treatment strategies and performance factors. *Sports Medicine,* **33**, pp. 145-164.
Cochrane, D.J., 2004, Alternating hot and cold water immersion for athlete recovery: A Review. *Physical Therapy in Sport,* **5**, pp. 26-32.
Eston, R.G., Mickleborough, J., and Baltzopoulos, V., 1995, Eccentric activation and muscle damage: biomechanical and physiological considerations during downhill running. *British Journal of Sports Medicine,* **29**, pp. 89-94.
Eston, R.G. and Peters, D., 1999, Effects of cold water immersion on the symptoms of exercise-induced muscle damage. *Journal of Sports Sciences,* **17**, pp. 231-238.
Gill, N.D., Beaven, C.M. and Cook, C., 2006, Effectiveness of post-match recovery strategies in rugby players. *British Journal of Sports Medicine,* **40**, pp. 260-263.
Gleeson, N., et al., 2003, Effects of prior concentric training on eccentric exercise induced muscle damage. *British Journal of Sports Medicine,* **37**, pp. 119-125.
Howatson, G., Gaze, D. and Van Someren, K.A., 2004, The efficacy of ice massage in the treatment of exercise induced muscle damage. *Scandinavian Journal of Medicine and Science in Sports,* **15**, pp. 416-422.
Isabell, W.K., et al., 1992, The effects of ice massage, ice massage with exercise, and exercise on the prevention and treatment of delayed onset muscle soreness. *Journal of Athletic Training,* **27**, pp. 208-217.
MacAuley, D.C., 2001, Ice therapy: How good is the evidence? *International Journal of Sports Medicine,* **22**, pp. 379-384.
Malm, C., 2001, Exercise-induced muscle damage and inflammation: fact or

fiction? *Acta Physiologica Scandanavica,* **171**, pp. 233-239.

Meeusen, R. and Lievens, P., 1986, The use of cryotherapy in sports injuries. *Sports Medicine*, **3,** pp. 398-414.

Morgan, P.L. and Proske, U., 2004, Popping sarcomere hypothesis explains stretch-induced muscle damage. *Clinical and Experimental Pharmacology and Physiology,* **31**, pp. 541-545.

Paddon-Jones, D.J. and Quigley, B.M., 1997, Effect of cryotherapy on muscle soreness and strength following eccentric exercise. *International Journal of Sports Medicine,* **18**, pp. 588-593.

Sellwood, K, L., et al., 2007, Ice-water immersion and delayed onset muscle soreness: a randomised control trial. *British Journal of Sports Medicine,* **41**, pp. 392-397.

Twist, C., and Eston, R., 2005, The effects of exercise-induced muscle damage on maximal intensity intermittent exercise performance. *European Journal of Applied Physiology*, **94**, pp. 652-658.

Vaile, J., et al., 2008a, Effect of hydrotherapy on the signs and symptoms of delayed onset muscle soreness. *European Journal of Applied Physiology,* **102**, pp. 447-455.

Vaile, J., et al., 2008b, Effect of cold water immersion on repeat cycling performance and thermoregulation in the heat. *Journal of Sports Sciences,* **26**, pp. 431-440.

Yackzan, L., Adams, C., and Francis, K.T., 1984, The effects of ice massage on delayed muscle soreness. *American Journal of Sports Medicine,* **12**, pp. 159-165.

Zainuddin, Z., et al., 2005, Effects of massage on delayed-onset muscle soreness, swelling and recovery of muscle function. *Journal of Athletic Training,* **40**, pp. 174-180.

CHAPTER FOUR

Towards a better understanding of ulnar wrist paraesthesia and entrapments in leisure and competitive sports

Steven Provyn, Peter Van Roy, Aldo Scafoglieri and Jan Pieter Clarys

Department of Experimental Anatomy (EXAN-LK),
Vrije Universiteit Brussel, Belgium

1. INTRODUCTION

Wrist problems are not uncommon in sports requiring prolonged grip function – e.g. tennis, cycling, motor sport, baseball, golf – and are often associated with entrapment syndromes and related phenomena, emanating from an intrinsic (e.g. ganglia) or an extrinsic (e.g. nerve irritation caused by repetitive loading, repetitive traumata) aetiolology (Murata *et al.*, 2003). To understand these syndromes and their related phenomena, it is necessary to clarify the existence of three different tunnels at the wrist. In clinical practice there seems to be a colloquial conviction that the wrist area contains two nervous and vascular passages only that can create entrapment symptoms, e.g. the carpal tunnel and the ulnar canal. Guyon's canal or the canalis ulnaris, according to the nomina anatomica (Berkovitz *et al.* 1998) was first described in 1861 by the French urologic surgeon Jean Casimir Felix Guyon and is only occasionally mentioned in basic anatomy atlases. If the canal is mentioned, the information that is presented in a lot of books and reports is very often vague or confusing, in spite of the many publications dedicated to this subject under clinical circumstances. Seldom does one observe so many contradictory descriptions of a particular anatomical region covering a restricted area of the wrist. Most of the anatomical books and ad hoc internet sites, describe extensively and precisely the carpal tunnel but, if in addition a second passage is described, one of the different "versions" of the canalis ulnaris is presented. A clear description of the carpus with its three different tunnels is never found. One thing is clear, Guyon's canal is considered to be the major cause of ulnar tunnel syndrome (Salgeback, 1977; Turner and Caird, 1977; Luethke and Dellon, 1992; Netscher and Cohen, 1997; Bozkurt *et al.*, 2005; Sturzenegger, 2005; Wang *et al.*, 2005). In an attempt to end the confusion, both for educational and for clinical purposes, the present study was undertaken in order to clarify the vagueness concerning the description of Guyon's canal by pointing out the existence of the three "tunnels" that can be distinguished in the palmar carpus.

2. METHODS

There were three different methodological approaches in the present study: (i) a dissection, (ii) an internet search and (iii) a bibliometric survey of educational textbooks or atlases and of scientific articles. Seventy-four wrists were dissected and photographed. The dissections were used to illustrate the findings. During an internet search in May 2006, different synonyms were found to define Guyon's canal. Each time a particular synonym was found, that synonym was entered into the scientific search engine. The same keywords were used on different internet sites and in the bibliometric survey of clinically and fundamentally oriented scientific papers (see Table 1). In the student textbooks and atlases of anatomy we verified the descriptive detail of the regional anatomy of the wrist.

3. RESULTS

3.1 Dissection

The dissection of 74 wrists confirmed evidence for 100% of Guyon's original description (Figure 1) but revealed a number of anatomical variants concerning the intrinsic and extrinsic hand musculature in the neighbourhood of the ulnar canal (cf. infra). The following quote represents a literal translation of the original description in French (Guyon, 1861),

"I call this space intra-aponeurotic as it seemed evident to me that its anterior wall is part of the fibrous bed of the area; its posterior wall is formed by the anterior ligament of the carpus. It cannot be said that there were any lateral layers; however, medially, the pisiform forms a sort of wall covered in aponeurotic tissue; above, below and distally, the anterior wall simply merges with the palmar aponeurosis, of which it is only a part. Its size, although not large, is about one centimetre to one and a half centimetres in each direction; above and below, it is approximately equal to the size of the fold at the wrist; distally, it stops roughly at the middle of the anterior ligament of the carpus, so that this little space is only part of the internal section and inspection through the skin takes this particular circumstance into account." (Guyon, 2006).

Several muscular variants were encountered during the dissection. The roof of the canal was reinforced by muscular variants at the level of entrance in 17 cases. The reinforcement was made up by a lacertuslike expansion of the M. flexor carpi ulnaris or by an additional tendon of the M. palmaris longus. In five other cases, the M. flexor carpi ulnaris revealed two tendons and in two other cases, the M. flexor carpi ulnaris presented a small additional muscular head, laterally from the pisiform bone. An additional muscle belly was also provided by the flexor digitorum superficialis muscle.

An additional head of one of the intrinsic hypothenar muscles (or a radial expansion) was seen in almost 20% of the cases. In four wrists, a small muscle could be seen between the pisiform and the hamulus of the hamate bone. In two wrists, the M. palmaris brevis sent an additional head towards the floor of the tunnel and in two other cases, the ulnar tunnel contained an additional head of the M. flexor carpi ulnaris. Other muscular variants dealt with the earlier described morphology that the hypothenar muscles may emanate directly from the carpal bones or via a common strong transverse fibrous tissue that represents a bow structure above the entrance of the canalis pisohamatum. Also variants of the ulnar nerve and artery could be observed. The ulnar artery may send a deep branch to the carpus via the canalis pisohamatum or via a more distal course.

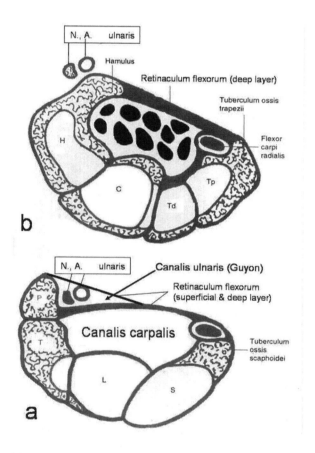

Figure 1. Transversal drawing of the proximal carpus (a), including the Canalis ulnaris and the distal carpus (b).

3.2 Bibliographic search

Of the 2559 hits on the internet, only the results written in English, French, German and Dutch clearly describing Guyon's canal were retained. Unfortunately, almost all information referred to personal details, observations about Guyon as a person that were irrelevant for this study. The results of the three different types of bibliographic survey e.g. books, web, papers, (N=117) are presented in Figure 2. In a total of 21.4% of these sources, Guyon's canal was erroneously presented as the pisohamatum tunnel. In another 25.6% of all sources, Guyon or its synonyms were not mentioned, nor described, nor drawn at all. Twenty four percent of sources mentioned a canalis ulnaris or a synonym but did not describe the canal or tunnel topography as such. Of the remaining 28.2% only 12% of the references presented a fully correct description (nomination plus topographic description including a clear illustration) and in 6.8% the Guyon tunnel or its synonym was indicated on a drawing or other image presentation only. The remaining 9.4% gave a correct description of the canalis ulnaris as part of the retinaculum flexorum but with no nomination. Within these A to F fragmented findings, with the exception of C 12% fully correct descriptions, the highest rate of erroneous descriptions is found in the scientific and clinical papers with 44% against 30% on the web and 20% in educational reference volumes.

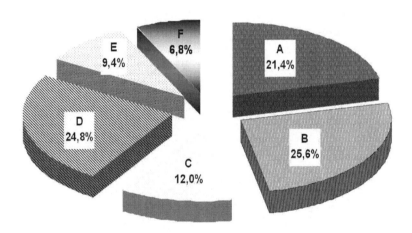

Figure 2. Presentation of Guyon's Canal according to different sources. A= Erroneous presentation; B= No mention of Guyon's canal or synonym; C= Fully correct description of Guyons's canal; D= Mention of Guyon's canal or synonym but no description; E= correct description but no mention of Guyon's canal or synonym; F= Mention of Guyon's canal or synonym but no drawing.

Most publications describing clinical, biomechanical or other aspects of the retinaculum flexorum area start with an anatomical description; however, confusion is created by the multiple designations/names used for the description of the ulnar canal. Table 1 shows the different terminology that authors have been using to define Guyon's canal but the ultimate element of confusion derives from the description of the pisohamatum tunnel. At a certain point the pisohamatum tunnel is assumed to be the Guyon's space or part of it. McFarlane *et al.* (1976) summarised as follows: – *"The term used by Enna et al. (1974) – the pisohamate tunnel – is a descriptive and an appealing alternative to the eponym "Loge de Guyon".* At the same time they recognised that dissection and surgery in the wrist area without recognising that the ramus profundus of the N. ulnaris ramus profundus passes to the hypothenar, leads to ignorance of a major mechanism of compression. Although Hayes *et al.* (1969) and Lotem *et al.* (1973) already referred to the importance of pressure neuropathy of the deep branch of the N. ulnaris to the hypothenar, one can deduct from the discussion and the drawings in McFarlane *et al.* (1976) that he combined the canalis ulnaris and the canalis pisohamatum into one continuing tunnel. "The proximal entrance of the so-called pisohamatum tunnel is in reality the correct canalis ulnaris (Guyon) and the distal exit (of the same tunnel) is the correct pisohamatum tunnel with slant walls, e.g. the os pisiforme, the hamulus with the ligamentum pisohamatum as its roof" (Lotem *et al.*, 1973). Thereafter a too liberal use of the term "pisohamate tunnel or canal" has led to confusion in particular in scientific articles and informative texts on the internet sites, resulting in erroneous representations, erroneous explanations and even in topographical manipulations to demonstrate Guyon's canal. The fact that 25.6% (Figure 2) of the references investigated do not describe Guyon's canal or its synonyms at all, may be another source of confusion. It needs no argumentation that both the correct topography of Guyon's canal (1861) and its denomination in the nomina anatomica (Berkovitz *et al.*, 1998) e.g. canalis ulnaris, should be used in every kinesiological or clinical study related to this area. Without pretending to be complete, Table 1 gives an overview of synonyms used for Guyon's canal, with the associated references.

A number of authors divided Guyon's canal into three zones (Shea and McClain, 1969; Eckman *et al.*, 1975). According to Cobb *et al.* (1996) Guyon's anatomical description was correct but not complete enough. In their study they showed that Guyon's canal has no attachment to the hook of the hamate bone (but Guyon did not present this statement). Lindsey and Watumull (1996) described this region more in detail and extended the canal both distally and proximally. They added that the roof of the canal is formed by the M. palmaris brevis, and that it is filled with adipose and fibrous tissue. The floor of the tunnel is made up of the Lig. carpi transversum, the Lig. pisohamatum, the Lig. pisometacarpalis and the M. opponens digiti minimi. To complete the medial wall, Lindsey and Watumull (1996) added the M. flexor carpi ulnaris and the M. abductor digiti minimi, while they described the lateral boundary as being formed by the tendons of the extrinsic flexors, the transverse carpal ligaments and the hook of the os hamatum. The least one can point out is that this is "unrealistically complicating" the anatomy of the wrist.

Studies related to muscular anomalies, neurological anomalies; vascular anomalies and pathological findings, often become difficult to understand because elementary knowledge is ignored, complicated or transformed (Uriburu *et al.*, 1976; Kilgore and Newmeyer, 1977; Denman, 1978). The confusion should be ended; anatomists, orthopaedic surgeons, radiologists and hand surgeons should agree on the distinction of three different tunnels in the wrist: (i) the carpal tunnel deep under the main sheet of the retinaculum flexorum and medial above (ii) the canalis ulnaris between two layers of the retinaculum flexorum that respectively originate from the top of the os pisiforme and from the top of the os triquetrum (e.g. bottom of the os pisiforme) (Figure 1). Both the canalis carpi (carpal tunnel) and the canalis ulnaris (Guyon) are oriented in the prolongation of the forearm (and the A., V., N. ulnaris). The length of the canalis ulnaris equals the length of the proximal row of the carpus. At its exit and the level of the distal row, a fibrous arch formation indicates the lateral-proximal edge of the ligamentum pisohamatum on top of the os pisiforme and the hamulus. (iii) The third tunnel is the canalis pisohamatum that allows the N. ulnaris ramus profundus into the hypothenar. All three tunnels are illustrated via dissection (Figure 3).

Table 1. Different designation for Guyon's canal used by authors.

Terminology	Authors
Guyon's Canal/ tunnel/ loge or space	De Vecchi and Moller, 1959; Shea and McClain, 1969; Eckman *et al.*, 1975; Uriburu *et al.*, 1976; Kilgore and Graham, 1977; Denman, 1978; Spinner, 1996; Pleet and Massey, 1978; Sunderland, 1981; Razemon, 1982; Weeks and Young, 1982; Bonnel and Vila, 1985; Wu *et al.*, 1985; Howard, 1986; Bergfield and Aulicino, 1988; Dellon and Mackinnon, 1988; Kuschner *et al.*, 1988; Olney and Hanson, 1988; Subin *et al.*, 1989; Dodds *et al.*, 1990; Thurman *et al.*, 1991; Zeiss *et al.*, 1992; Richards *et al.*, 1993; Sanudo *et al.*, 1993; Cobb *et al.*, 1994; König *et al.*, 1994; Pribyl and Moneim, 1994; Cobb *et al.*, 1996; Gonzalez *et al.*, 1996; Kang *et al.*, 1996; Lindsey and Watumull, 1996; Muller *et al.*, 1997; Tyrdal *et al.*, 1997; Koch *et al.*, 1998; Netscher and Cohen, 1998; Kothari, 1999; De Maeseneer *et al.*, 2005
Ulnar tunnel / canal Canalis Ulnaris	Merle d'Aubigné and Tubiana, 1956; Forshell and Hagstrom, 1975; Eckman *et al.*, 1975; Sunderland, 1981; Gross and Gelberman, 1985; Kuschner *et al.*, 1988; Subin *et al.*, 1989; Moneim, 1992; Zeiss *et al.*, 1992; Zeiss *et al.*, 1995; Spinner *et al.*, 1996; Olave *et al.*, 1997; Balogh *et al.*, 1999; Kothari, 1999; Murata *et al.*, 2003; Moutet, 2004; Sturzenegger, 2005.
carpal ulnar neurovascular space	Cobb *et al.*, 1994; Cobb *et al.*, 1996
Pisoretinaculair space	Denman, 1978
Pisohamate tunnel	Enna *et al.*, 1974; McFarlane *et al.*, 1976

Anatomically this canalis pisohamatum is the logical site for neuropathic compression. This point was already stated in the 1950s and the 1960s (Hayes *et al.*, 1969; Spinner *et al.*, 1996). A simplified drawing of the wrist will emphasise the three different canals or tunnels including the direction of the neurovascular passage, namely the canalis carpi (carpal tunnel), the canalis ulnaris (Guyon) and the canalis pisohamatum (Figure 4). A last criticism is that there is no mention of the canalis pisohamatum in the Terminologia Anatomica of the Federative Committee on Anatomical Terminology (FCAT) (Berkovitz *et al.*, 1998).

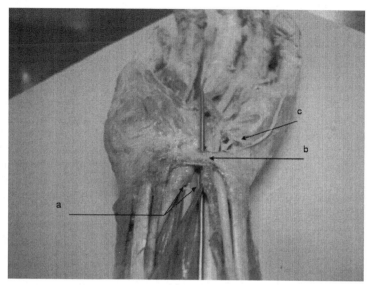

Figure 3. Dissection of the canalis carpi (a), canalis ulnaris (b), canalis pisohamatum (c).

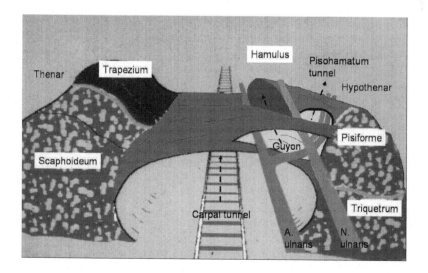

Figure 4. Simplified model of the three tunnels: carpal, ulnar and pisohamatum tunnel.

4. DISCUSSION AND CONCLUSION

In the literature, the canalis ulnaris (Guyon's canal) is often forgotten or described in an incomplete or erroneous manner. The canalis ulnaris should not be confused with the canalis pisohamatum which is another and third tunnel with other entrapment possibilities. The Lig. pisohamatum lies at the bottom of the pisohamate tunnel and should certainly not be mentioned as the roof of this tunnel, neither as the roof of Guyon's canal.

Clinical symptoms largely depend on the localization of ulnar nerve entrapments. If the entrapment is localized proximally in the Guyon's canal, both motor deficit (muscular weakness and muscular atrophy and so on) and sensory deficit (hyperalgesia, paraesthesia, numbness and so on) may occur if both the sensory and motor fascicles of the ulnar nerve are subjected to entrapment at this site. Motor deficits might be present in the hypothenar muscles, the interossei muscles, the two medial lumbricales muscles, the adductor pollicis and the deep (in other cases also the superficial head) of the flexor pollicis brevis muscle. In most cases sensory deficits are localized at the palmar side of the fifth and the medial half of the fourth finger. The cutaneous innervation of the ulnar nerve is also subjected to inter-individual differences; this can result in further extension of the clinical signs towards the entire palmar side of the fourth finger or at the third finger.

In the distal part of Guyon's canal, the ulnar nerve divides into a deep motor branch and a superficial branch. The superficial branch continues through Guyon's canal, the motor branch reaches the hypothemar muscles via the pisohamate tunnel and continues between the superficial and deep heads of the M. opponens digiti minimi. This motor branch is more subjected to entrapment than the superficial branch (McFarlane *et al.*, 1976). Entrapment often occurs at the level of the fibrous arch at the entrance of the pisohamate tunnel (Moneim, 1992). This may lead to motor deficit of the hypothenar muscles, the small intrinsic hand muscles, as well as in the adductor pollicis and flexor pollicis brevis muscles. When the entrapment occurs deeper inside the pisohamatum tunnel, distal to the branches reaching the hypothenar muscles, muscular deficit is restricted to the small intrinsic muscles of the hand and the thenar muscles innervated by the ulnar nerve. Finally, if the entrapment occurs proximally to the branches reaching the first dorsal interosseus muscle and the adductor pollicis muscle and the deep (eventually also the superficial) head of the flexor pollicis brevis muscle, muscular deficit is restricted to the lateral innervation area of this deep muscular branch (Wu *et al.*, 1985; Kothari, 1999).

After providing deep muscular innervation to the M. palmaris brevis, the superficial branch of the bifurcation of the ulnar nerve in Guyon's canal continues as a cutaneous nerve. An entrapment of the ulnar nerve at the distal level of the palmaris brevis muscle leads to sensory deficit of the medial palmar side of the hand only. The palmaris brevis muscle itself is innervated by one of the palmar cutaneous branches of the ulnar nerve. One has to realize that the dorsal cutaneous branch of the ulnar nerve towards the dorsal medial aspect of the hand and fingers usually originates proximally to the wrist and does not pass through the canal of Guyon.

The ulnar nerve may show several anatomical variants in its distal course. In some cases, only the motor branch passes through Guyon's canal (Lindsey and Watumull, 1996). In a number of anatomical specimens, a cutaneous branch towards the ulnar artery can be observed proximally to the wrist joint: i.e. Henle's nerve. In other cases a Martin-Gruber anastomosis between the ulnar and median nerve can be found. The complexity of the anatomy of the ulnar nerve challenges clinical reasoning (Wu *et al.*, 1985). Therefore, ending the existing confusion between the ulnar canal and the pisohamatum tunnel is helpful in increasing diagnostic accuracy and optimisation of the communication about the ulnar canal and pisohamatum tunnel syndromes.

References

Balogh, B., Valencak, J., Vesely, M., Flammer, M., Gruber, H. and Piza-Katzer, H., 1999, The nerve of Henle: an anatomic and immunohistochemical study. *Journal of Hand Surgery*, **24**, pp. 1103-1108.

Bergfield, T.G. and Aulicino, P.L., 1988, Variation of the deep motor branch of the ulnar nerve at the wrist. *Journal of Hand Surgery*, **13**, pp. 368-369.

Berkovitz, B., Heimer, L., Henkel, C. K., Ma, T. P., McNeal, J. E. and Moore, N.A., 1998, *Terminologia Anatomica: International Anatomical Terminology (Fcat)*, (New York: Thieme Stuttgart).

Bonnel, F. and Vila, R.M., 1985, Anatomical study of the ulnar nerve in the hand. *Journal of Hand Surgery*, **10**, pp. 165-168.

Bozkurt, M.C., Tagil, S.M., Ozcakar, L., Ersoy, M. and Tekdemir, I., 2005, Anatomical variations as potential risk factors for ulnar tunnel syndrome: a cadaveric study. *Clinical Anatomy*, **18**, pp. 274-280.

Cobb, T.K., Carmichael, S.W. and Cooney, W.P., 1994, The ulnar neurovascular bundle at the wrist. A technical note on endoscopic carpal tunnel release. *Journal of Hand Surgery*, **19**, pp. 24-26.

Cobb, T.K., Carmichael, S.W. and Cooney, W.P., 1996, Guyon's canal revisited: an anatomic study of the carpal ulnar neurovascular space. *Journal of Hand Surgery*, **21**, pp. 61-869.

Dellon, A.L. and Mackinnon, S.E., 1988, Anatomic investigations of nerves at the wrist: II. Incidence of fibrous arch overlying motor branch of ulnar nerve. *Annals of Plastic Surgery*, **21**, pp. 36-37.

De Maeseneer, M., Van Roy, P., Jacobson, J.A. and Jamadar, D.A., 2005, Normal MR imaging findings of the midhand and fingers with anatomic correlation. *European Journal of Radiology*, **56**, pp. 278-285.

Denman, E.E., 1978, The anatomy of the space of Guyon. *Hand*, **10**, pp. 69-76.

De Vecchi, J. and Moller, G., 1959, Carpal tunnel syndrome and syndrome of Guyon's space. *Boletín Sociedad de Cirugía del Uruguay*, **30**, pp.275-278.

Dodds, G.A., Hale, D. and Jackson, W.T., 1990, Incidence of anatomic variants in Guyon's canal. *Journal of Hand Surgery*, **15**, pp. 352-355.

Eckman, P.B., Perlstein, G. and Altrocchi, P.H., 1975, Ulnar neuropathy in bicycle riders. *Archives of Neurology*, **32**, pp. 130-2.

Enna, C.D., Berghtholdt, H.T. and Stockwell, F., 1974, A study of surface and deep temperatures along the course of the ulnar nerve in the pisohamate tunnel. *International Journal of Leprosy and Other Mycobacterial Diseases*, **42**, pp. 43-47.

Forshell, K.P. and Hagstrom, P., 1975, Distal ulnar nerve compression caused by ganglion formation in the loge de Guyon. Case report. *Scandinavian Journal of Plastic and Reconstructive Surgery*, **9**, pp. 77-79.

Gonzalez, M.H., Brown, A., Goodman, D. and Black, B., 1996, The deep branch of the ulnar nerve in Guyon's canal: branching and innervation of the hypothenar muscles. *Orthopedics*, **19**, pp. 55-58.

Gross, M.S. and Gelberman, R.H., 1985, The anatomy of the distal ulnar tunnel. *Clinical Orthopaedics and Related Research*, **196**, pp. 238-247.

Guyon, F., 2006, Note on the anatomical condition affecting the underside of the wrist not previously reported. 1861. *Journal of Hand Surgery*, **31**, pp. 147-148.

Hayes, J.R., Mulholland, R.C. and O'Connor, B.T., 1969, Compression of the deep palmar branch of the ulnar nerve. Case report and anatomical study. *Journal of Bone and Joint Surgery, British Volume*, **51**, pp. 469-472.

Howard, F.M., 1986, Controversies in nerve entrapment syndromes in the forearm and wrist. *Orthopedic Clinics of North America*, **17**, pp. 375-381.

Kang, H.J., Yoo, J.H. and Kang, E.S., 1996, Ulnar nerve compression syndrome due to an anomalous arch of the ulnar nerve piercing the flexor carpi ulnaris: a case report. *Journal of Hand Surgery*, **21**, pp. 277-278.

Kilgore, E.S., Jr. and Newmeyer, W.L., 3rd, 1977, In favor of standing to do hand surgery. *Journal of Hand Surgery*, **2**, pp. 326-327.

Koch, H., Haas, F. and Pierer, G., 1998, Ulnar nerve compression in Guyon's canal due to a haemangioma of the ulnar artery. *Journal of Hand Surgery*, **23**, pp. 242-244.

Konig, P.S., Hage, J.J., Bloem, J.J. and Prose, L.P., 1994, Variations of the ulnar nerve and ulnar artery in Guyon's canal: a cadaveric study. *Journal of Hand Surgery*, **19**, pp. 617-622.

Kothari, M.J., 1999, Ulnar neuropathy at the wrist. *Neurologic Clinics*, **17**, pp. 463-476, vi.

Kuschner, S.H., Gelberman, R.H. and Jennings, C., 1988, Ulnar nerve compression at the wrist. *Journal of Hand Surgery*, **13**, pp. 577-580.

Lindsey, J.T. and Watumull, D., 1996, Anatomic study of the ulnar nerve and related vascular anatomy at Guyon's canal: a practical classification system. *Journal of Hand Surgery*, **21**, pp. 626-633.

Lotem, M., Gloobe, H. and Nathan, H., 1973, Fibrotic arch around the deep branch of the ulnar nerve in the hand. Anatomical observations. *Plastic and Reconstructive Surgery*, **52**, pp. 553-556.

Luethke, R. and Dellon, A.L., 1992, Accessory abductor digiti minimi muscle originating proximal to the wrist causing symptomatic ulnar nerve compression. *Annals of Plastic Surgery*, **28**, pp. 307-308.

McFarlane, R.M., Mayer, J.R. and Hugill, J.V., 1976, Further observations on the anatomy of the ulnar nerve at the wrist. *Hand*, **8**, pp. 115-117.

Merle D' Aubigne, R. and Tubiana, R., 1956, Surgical restoration of hand. *La Revue du Praticien*, **11**, pp. 3561-3564.

Moneim, M.S., 1992, Ulnar nerve compression at the wrist. Ulnar tunnel syndrome. *Hand Clinics*, **8**, pp. 337-344.

Moutet, F., Forli, A. and Voulliaume, D., 2004, Pulley rupture and reconstruction in rock climbers. *Techniques in Hand and Upper Extremity Surgery*, **8**, pp. 149-155.

Muller, L.P., Kreitner, K.F., Seidl, C. and Degreif, J., 1997, Traumatic thrombosis of the distal ulnar artery (hypothenar hammer syndrome) in a golf player with an accessory muscle loop around Guyon's canal. Case report. *Handchir Mikrochir Plast Chir*, **29**, pp. 183-186.

Murata, K., Shih, J.T. and Tsai, T.M., 2003, Causes of Ulnar Tunnel Syndrome: A Retrospective Study of 31 subjects. *Journal of Hand Surgery*, **28**, pp. 647-651.

Netscher, D. and Cohen, V., 1997, Ulnar nerve compression at the wrist secondary to anomalous muscles: a patient with a variant of abductor digiti minimi. *Annals of Plastic Surgery*, **39**, pp. 647-651.

Netscher, D.T. and Cohen, V., 1998, Ulnar nerve entrapment at the wrist: cases from a hand surgery practice. *The Southern Medical Journal*, **91**, pp. 451-456.

Olave, E., Del Sol, M., Gabrielli, C., Prates, J.C. and Rodrigues, C.F., 1997, The ulnar tunnel: a rare disposition of its contents. *Journal of Anatomy*, **191**, pp. 615-616.

Olney, R.K. and Hanson, M., 1988, AAEE case report #15: ulnar neuropathy at or distal to the wrist. *Muscle Nerve*, **11**, pp. 828-832.

Pleet, A.B., and Massey, E.W., 1978, Palmaris brevis sign in neuropathy of the deep palmar branch of the ulnar nerve. *Annals of Neurology*, **3**, pp. 468-469.

Pribyl, C.R. and Moneim, M.S., 1994, Anomalous hand muscle found in the Guyon's canal at exploration for ulnar artery thrombosis. A case report. *Clinical Orthopaedics and Related Research*, **306**, pp. 120-123.

Razemon, J.P., 1982, The carpal tunnel syndrome (author's transl). *Journal de Chirurgie*, **119**, pp. 283-294.

Richards, R.S., Dowdy, P. and Roth, J.H., 1993, Ulnar artery palmar to palmaris brevis: cadaveric study and three case reports. *Journal of Hand Surgery*, **18**, pp. 888-892.

Salgeback, S., 1977, Ulnar tunnel syndrome caused by anomalous muscles. Case report. *Scandinavian Journal of Plastic and Reconstructive Surgery*, **11**, pp. 255-258.

Sanudo, J.R., Mirapeix, R.M. and Ferreira, B., 1993, A rare anomaly of abductor digiti minimi. *Journal of Anatomy*, **182**, pp. 439-442.

Shea, J.D. and McClain, E.J., 1969, Ulnar-nerve compression syndromes at and below the wrist. *Journal of Bone and Joint Surgery*, **51**, pp. 1095-1103.

Spinner, R.J., Lins, R.E. and Spinner, M., 1996, Compression of the medial half of the deep branch of the ulnar nerve by an anomalous origin of the flexor digiti minimi. A case report. *Journal of Bone and Joint Surgery*, **78**, pp. 427-430.

Sturzenegger, M., 2005, Entrapment neuropathies of the arm. *Schweizerische Rundschau für Medizin Praxis*, **94**, pp. 1161-1165.

Subin, G.D., Mallon, W.J. and Urbaniak, J.R., 1989, Diagnosis of ganglion in Guyon's canal by magnetic resonance imaging. *Journal of Hand Surgery*, **14**, pp. 640-643.

Sunderland, S., 1981, The anatomic foundation of peripheral nerve repair techniques. *Orthopedic Clinics of North America*, **12**, pp. 245-266.

Thurman, R.T., Jindal, P. and Wolff, T.W., 1991, Ulnar nerve compression in Guyon's canal caused by calcinosis in scleroderma. *Journal of Hand Surgery*, **16**, pp. 739-741.

Turner, M.S. and Caird, D.M., 1977, Anomalous muscles and ulnar nerve compression at the wrist. *Hand*, **9**, pp. 140-142.

Tyrdal, S., Solheim, L.F. and Alho, A., 1997, Anomalous nerve anatomy at the wrist? *Annales Chirurgiae et Gynaecologiae*, **86**, pp. 79-83.

Uriburu, I.J., Morchio, F.J. and Marin, J.C., 1976, Compression syndrome of the deep motor branch of the ulnar nerve. (Piso-Hamate Hiatus syndrome). *Journal of Bone and Joint Surgery*, **58**, pp. 145-147.

Wang, B., Zhang, Z., Li, K. and Xiang, T., 2005, Clinical review of thirty-nine cases of ulnar tunnel syndrome. *Zhongguo Xiu Fu Chong Jian Wai Ke Za Zhi*, **19**, pp. 737-739.

Weeks, P.M. and Young, V.L., 1982, Ulnar artery thrombosis and ulnar nerve compression associated with an anomalous hypothenar muscle. *Plastic and Reconstructive Surgery*, **69**, pp. 130-131.

Wu, J.S., Morris, J.D. and Hogan, G.R., 1985, Ulnar neuropathy at the wrist: case report and review of literature. *Archives of Physical Medicine and Rehabilitation*, **66**, pp. 785-788.

Zeiss, J., Jakab, E., Khimji, T. and Imbriglia, J., 1992, The ulnar tunnel at the wrist (Guyon's canal): normal MR anatomy and variants. *AJR American Journal of Roentgenology*, **158**, pp. 1081-1085.

Zeiss, J. and Jakab, E., 1995, MR demonstration of an anomalous muscle in a patient with coexistent carpal and ulnar tunnel syndrome. Case report and literature summary *Clinical Imaging*, **19**, pp. 102-105.

CHAPTER FIVE

Electromyographic comparisons of the rectus femoris, vastus lateralis and vastus medialis during squat and leg extension exercises at high and low intensities

Michael J. Duncan[1], Sean Dowdall[2] and David Sammin[2]

[1]Department of Sport and Exercise Sciences, University of Derby, UK
[2]Department of Physical Education and Sports Studies, Newman University College, Birmingham, UK

1. INTRODUCTION

The parallel squat and the leg extension are both popular resistance exercises which are reported to enhance the overall strength of the quadriceps muscle group (Earle and Baechle, 2000). According to Signorile et al. (1994) if the parallel squat is exclusively used for quadriceps development it can cause a significant imbalance between the vastus lateralis and vastus medialis muscles. This may be an important issue as muscle imbalance between the vastus lateralis and vastus medialis is a contributing factor in patellofemoral pain syndrome and osteoarthritis of the knee (McConnell, 1986; Galtier et al. 1995; Edwards et al., 2008). As a result it has been recommended by trainers and coaches that the leg extension is a necessary addition to a weight-training programme in order to reduce this imbalance. Empirical studies supporting the assertion that there is a difference in muscle activity between the parallel squat and leg extension are sparse. Furthermore, there appears to be no scientific evidence that leg extension exercise can correct this imbalance. It appears that the notion that supplemental leg extension exercises are needed alongside the parallel squat is based on an unjustified coupling of data produced by previous unrelated studies and long-standing beliefs about the relative activity levels of the various quadriceps muscles (Signorile et al., 1994).

Several studies have shown higher levels of muscular activity in the vastus medialis during the final phases of leg extension exercises (Wheatley and Jahnke, 1951; Lieb and Perry, 1971; Basmajian et al., 1972) and Lieb and Perry (1971) reported that the vastus medialis was important in patella alignment. The results of such research appears to have given rise to the hypothesis that the vastus medialis can only be optimally activated by performing the leg extension and performing the squat alone causes an imbalance between vastus medialis and vastus lateralis due to patella misalignment associated with this exercise. These hypotheses are questionable as none of these early studies used functional loads employed during training; indeed the study by Lieb and Perry (1971) used amputated limbs.

More recently, scientists have investigated electromyographic (EMG) patterns of muscle activity during both the squat and the leg extension. Dionisio *et al.* (2006) recently reported equal activation of both the vastus medialis and vastus lateralis during downward squatting. This research supports previous studies with similar findings (Signorile *et al.*, 1995; McCaw and Melrose, 1999; Boyden *et al.*, 2000; Caterisano *et al.*, 2002). These studies have however reported variations in squatting technique and reported no significant differences in muscle activation of the vastus medialis, vastus lateralis and rectus femoris during squatting with different foot positions (Boyden *et al.*, 2000), that gluteus maximus activity increases with increases in squat depth (partial, parallel, full depth) but vastus medialis, vastus lateralis and rectus femoris activity remains constant (Caterisano *et al.*, 2002) and that there is no change in quadriceps muscle activity during wide and narrow width squatting (McCaw and Melrose, 1999).

Despite this recent research, few researchers have directly compared muscle activation patterns between the squat and leg extension. Signorile *et al.* (1994) examined activation of the vastus medialis and vastus lateralis muscles during performance of one set of squat and leg extension exercises. There was no difference in vastus medialis and vastus lateralis activation for the squat and leg extension but EMG activity was greater for both muscles in the squat compared to the leg extension. On this basis, Signorile *et al.* (1994) questioned the value of supplemental leg extension exercise in maintaining muscular balance and preventing patellar misalignment. They stressed that the results of one study should not provide the basis for changing training or rehabilitation practices and that research is needed that examines specific changes in activity of the entire quadriceps group throughout the range of motion of each exercise at normal training intensities. Clearly further investigation of this issue is required to support the assertions made by Signorile *et al.* (1994) In addition, in their study, raw EMG signals were used for analysis and no normalisation procedure was employed. This is an important consideration as accurate documentation of the muscle activities between muscles or between conditions requires some form of normalisation (Marras and Davis, 2001; Farina, 2006). Secondly, Signorile *et al.* (1994) used the first and last repetitions from one set of each exercise at each participant's 10-repetition maximum (10 RM) value. This method may not have provided a sound basis for changing strength training practices and it would seem logical to examine changes in quadriceps muscle activity during the squat and leg extension across a range of training intensities. Therefore, the aim of this study was to investigate muscle activity of the rectus femoris, vastus lateralis and vastus medialis during the squat and leg extension at high and low training intensities.

2. METHODS

2.1 Participants

Following informed consent and institutional ethical approval, 10 males (mean age

± S.D. = 22.7 ± 3.4 years), with regular strength training experience, participated in the study. All participants had specific experience of performing both the squat and leg extension exercises and were free of any musculoskeletal pain or disorders. All participants were asked to refrain from vigorous exercise and maintain normal dietary patterns in the 24 h prior to testing. All testing took place within the institution's human performance laboratory.

2.2 Procedure

Participants took part in two experimental sessions. Within 48 h of determining 1RM for the squat and leg extension, each participant performed three repetitions of the squat and leg extension at 30% and 90% of 1RM. The 1RM for each exercise was determined according to methods advocated by Kraemer *et al.* (2006). Proper lifting technique was demonstrated for the participants before the 1RM assessment. The 1RM value was used to set the 30% and 90% 1RM intensities undertaken during the experimental session. During the second experimental condition each participant then completed three repetitions of each exercise at 30% and 90% of 1RM in a randomised order. The two intensities and two exercises were performed on the same day to minimize the time requirement for a given test session, increase participant adherence and permit a more accurate comparison of intraindividual EMG data in accordance with protocols outlined by Lagally *et al.* (2002; 2004). A minimum time gap of 20 min between exercise intensities was set to allow adequate time for muscle activity to return to baseline levels.

2.3 Lifting Procedures

Leg extension exercise was performed for the dominant leg using a Nova leg extension machine (Nova Inc, United Kingdom) and in accordance with protocols previously described for leg extension by Earle and Baechle (2000). The starting position required the participant to align the knee with the axis of the machine with the ankle behind and in contact with the foot pad. This was at approximately 110° of knee flexion. The back was pressed firmly against the back pad. One complete repetition began in the starting position and consisted of raising the roller pad by extending the knee whilst keeping the torso and back firmly pressed against the back pad and without allowing the hips or buttocks to lift off the seat and then returning back to the starting position (Earle and Baechle, 2000). A trained researcher was present during all test sessions to ensure proper range of motion. Any lift that deviated from proper technique was not counted and the subject was asked to return to the starting position and repeat the lift. During all intensities, repetition frequency was paced by a metronome set at 60 cycles.min^{-1}. This cadence resulted in one complete repetition every 4 s with concentric and eccentric phases comprising 2 s each.

The squat exercise was performed using a Nova Smith machine (Nova Inc, United Kingdom) and in accordance with protocols previously described for the parallel squat by Earle and Baechle (2000). The starting position required the participant to step under the bar and position the feet parallel to each other. The bar was placed in a balanced position on the upper back and shoulders. One complete repetition began in the starting position and consisted of a downward phase and an upward phase. In the downward phase the hips and knees slowly flexed, whilst keeping the torso to floor angle relatively constant, to a point where the thighs were parallel to the floor. In the upward phase the hips and knees extended, maintaining torso to floor angle, until the starting position was reached (Earle and Baechle, 2000). During all intensities, repetition frequency was paced by a metronome set at 60 cycles.min^{-1}. This cadence resulted in one complete repetition every 4 s with concentric and eccentric phases comprising 2 s each.

2.4 Measurement of Muscle Activity

Muscular activation of the rectus femoris, vastus medialis and vastus lateralis muscles were measured using surface electromyography. Passive bipolar surface electrodes of 30-mm diameter (Blue Sensor, Ltd, Denmark, Ag/AgCl) were placed on the contracted belly of each of the muscles, in line with the muscle fibre direction, with an interelectrode distance of 1.5 cm. Before electrode placement, the area was cleaned with isopropyl alcohol, shaved and abraded in order to reduce skin impedance. Reference electrodes were also placed on a bony prominence on the right wrist. All electrodes remained in place until data collection was completed in the three intensities. Efforts were made to minimize crosstalk across the muscles of interest by selecting appropriate electrode size and interelectrode distance (Winter, 1996; Merletti, 1999). Manual muscle testing during pilot work confirmed that crosstalk in the muscles of interest was minimised as much as possible. Electromyographic activity was differentiated by pre-amplifiers and recorded via an on-line, cable ME6000 system (MEGA Electronics, LTD, Finland), with an input impedance of less than $10^{15}/0.2$ ohm/pF, a common mode rejection ratio at 60 Hz of greater than 110 dB, a noise level of 1.2 mV, a gain of 10 + 2% and a bandwidth range from 0 Hz – 500 Hz. Muscle activity was sampled at 1000 Hz via a 16bit DAQ-516 A/D card and stored on a Sony Vaio laptop computer using MegaWin software, version 1.2 (MEGA Electronics, LTD, Finland). The raw EMG data were filtered using a high and low pass Butterworth filter (5-500 Hz) and visually checked for artefacts, which were excluded from subsequent analysis. The EMG data were sampled at 100 ms with a 50% frame overlap. This moving average approach in which the time windows overlap ensures that the EMG curve follows the trend of the underlying rectified EMG, but without the variable peaks that are evident in the rectified EMG. This method has been recommended for use in dynamic contractions (Burden, 2008). On completion of the experimental protocol the Root Mean Square (RMS) of the filtered EMG signal was full-wave rectified and normalised as a % of the 1RM value for each exercise. All EMG data collected were subsequently expressed as a percentage of this value and used in the analysis.

2.5 Data Analysis

Any differences in muscle activation between the squat and leg extension and across the two intensities were investigated using a 2 (exercise) X 2 (intensity) ways, repeated measures analysis of variance (ANOVA). Where any significant differences were found, Bonferroni's multiple comparisons were used to determine where these differences lay. Descriptive statistics were also calculated and the level for statistical significance was set at P < 0.05. The Statistical package for Social Sciences, version 13 (SPSS inc, Chicago, IL, USA) was used for all analysis.

3. RESULTS

Results from repeated measures ANOVA for each muscle group indicated significant exercise X intensity interactions for rectus femoris (F $_{1,8}$ = 54.4, P< 0.01, partial η^2 = 0.872), vastus lateralis (F$_{1,8}$ = 18.2, P< 0.01, partial η^2 = 0.694) and vastus medialis (F$_{1,8}$ = 9.7, P< 0.05, partial η^2 = 0.518) These indicated, for all muscle groups, that muscle activity at 90% of 1RM was similar for the squat and leg extension but at 30% 1RM, leg extension muscle activity was significantly lower than muscle activity during the squat.

In addition, there were significant main effects for exercise for rectus femoris (F$_{1,8}$ = 48.5, P< 0.01, partial η^2 = 0.872), vastus lateralis (F $_{1,8}$ = 8.1, P<0.05, partial η^2 = 0.501) and vastus medialis (F$_{1,8}$ = 25.1, P< 0.01, partial η^2 = 0.758) and significant main effects for intensity for rectus femoris (F$_{1,8}$ = 54.4, P< 0.01, partial η^2 = 0.859), vastus lateralis (F$_{1,8}$ = 41.5, P< 0.01, partial η^2 = 0.838) and vastus medialis (F$_{1,8}$ = 84.3, P< 0.01, partial η^2 = 0.913). For all muscle groups muscle activity was greater during the squat exercise and at 90% 1RM compared to leg extension and 30%1RM respectively. Mean ± S.D. of normalised (%) root mean square EMG across intensities and exercises for the three muscles involved is presented in Table 1.

4. DISCUSSION

The main aim of this study was to investigate quadriceps muscle activity during the squat and leg extension at high and low training intensities. Prior studies that have compared muscle activation patterns between the squat and leg extension have been limited in number and the only study that appears to have examined this issue has used the first and last repetitions of each exercise at values corresponding to 10RM as their dependant variables (Signorile *et al.*, 1994). Furthermore, in the study by Signorile *et al.* (1994) these values were not normalised. The use of non-normalised EMG values in this study does not provide a relative reference point that is consistent across muscles, exertion and participants (Marras and Davis, 2001) and subsequently call into question the security of Signorile and co-workers' conclusions. Conversely, the present study compared normalised muscle activity of the squat and leg extension at high and low intensities. This is an important consideration when prescribing exercise for strength training, rehabilitation and prehabilitation as differing levels of muscle activation across exercises and intensities could provide guidance to practitioners as to which exercises to use, which intensities may best be suited for a particular programme of objective (e.g. rehabilitation) and whether supplemental exercises might be needed for additional strength development.

Table 1. Mean ± S.D. of normalised (%) Root Mean Square EMG of the Rectus Femoris, Vastus Lateralis and Vastus Medialis during the squat and leg extension exercises at high and low intensities.

	Squat					
	Rectus Femoris		Vastus Lateralis		Vastus Medialis	
	M	S.D.	M	S.D.	M	S.D.
30% 1RM	79.2	10.9	79.3	6.7	71.9	7.6
90% 1RM	113.9	19.5	102.6	20.6	107.6	18.2
	Leg Extension					
	Rectus Femoris		Vastus Lateralis		Vastus Medialis	
	M	S.D.	M	S.D.	M	S.D.
30% 1RM	33.4	3.9	44.6	12.4	41.8	7.6
90% 1RM	109.2	22.8	101.8	19.8	100.3	17.2

The main novel findings of this study were the significant exercise by intensity interactions in muscle activation for all three muscles investigated. These interactions indicated that for the rectus femoris, vastus lateralis and vastus medialis, muscle activity was significantly greater during the squat than the leg extension at 30% 1RM. However, there were no significant differences in rectus femoris, vastus lateralis and vastus medialis muscle activity during the squat and leg extension at 90% 1RM. These results provide some support for assertions made by Signorile *et al.* (1994) that supplemental leg extension exercises are required when athletes use the squat in an exercise programme in order to prevent muscular imbalances in the quadriceps muscle group as similar levels of muscle activation were found in the vastus medialis and vastus lateralis at 90% 1RM for both exercises. The results of this study also refute claims recently made by Dionisio *et al.* (2006) that muscle activity is greater in the vastus medialis and vastus lateralis compared to rectus femoris when performing the squat.

The results of this study demonstrate that as resistance exercise intensity increases so does muscle activation. This supports a range of prior studies that have documented increased muscle activation with increased resistance exercise intensity across a range of intensities and exercises (Lagally *et al.*, 2002; Lagally *et al.*, 2004; Duncan *et al.*, 2006). In addition, the significant exercise main effect indicated that overall, muscle activation was greater during the squat than the leg extension. This finding is also in line with prior research (Signorile *et al.*, 1994) and is not surprising given that the absolute load for the squat, exceeded that of the leg extension at both 30% and 90% 1RM.

In relation to rehabilitation and/or prehabilitation the results of this study may be important as they indicate that, at lower intensities, the squat is associated with greater muscle activation than the leg extension. Therefore, if increased quadriceps muscle activation is the goal of a training/rehabilitation programme this is maximised at low intensities by performing the squat. However, both the squat and the leg extension elicit similar muscle activation profiles at high intensity. This at least partially supports the notion that closed chain kinetic exercises such as the squat may be more influential in aiding recovery of anterior and posterior joint injuries where low intensity exercise is often prescribed (Palmitier *et al.*, 1991; Signorile *et al.*, 1994). Additionally, these findings may also have relevance for the general public in providing an insight into muscle activation across exercises and intensities and so provide an insight into prehabilitative strategies that could reduce the risk of lower-body injury (Caterisano *et al.*, 2002).

Future studies are also warranted to confirm the findings of this study. Research that examines a broader range of exercise intensities may be influential in helping develop effective training and rehabilitation programmes as would examining any differences in the timing of muscle activation across exercises and intensities. In addition, the participants in this study were all familiar with both the squat and the leg extension exercises. Further study of muscle activation patterns in participants who are unfamiliar with these exercises may also help shape strategies for rehabilitation.

References

Basmajian, J., Harden, T. and Regeons, E., 1972, Integrated actions of the four heads of the quadriceps femoris: An electromyographical study. *Anatomical Records*, **172**, pp.15-20.

Boyden, G., Kingman, J. and Dyson, R., 2000, A comparison of quadriceps electromyographic activity with the position of the foot during the parallel squat. *Journal of Strength and Conditioning Research*, **14**, pp. 379-382.

Burden, A., 2008, Surface electromyography. In: *Biomechanical Evaluation of Movement in Sport and Exercise*, edited by Payton, C. and Bartlett, R., (London: Routledge), pp. 77-102.

Caterisano, A., Moss, R., Pellinger, T., Woodruff, K., Lewis, V., Booth, W. and Khadra, T., 2002, The effect of back squat depth on the EMG activity of 4 superficial hip and thigh muscles. *Journal of Strength and Conditioning Research*, **16**, pp. 428-432.

Dionisio, V. C., Almeida, G. L., Duarte, M. and Hirata, R. P., 2006, Kinematic, kinetic and EMG patterns during downward squatting. *Journal of Electromyography and Kinesiology.* Epub ahead of print.

Duncan, M.J., Al-Nakeeb, Y. and Scurr, J., 2006, Perceived exertion is related to muscle activity during leg extension exercise. *Research in Sports Medicine*, **14**, pp. 179-189.

Earle R. W. and Baechle, T. R., 2000, Resistance training and spotting techniques. In: *Essentials of Strength Training and Conditioning*, edited by Baechle, T.R. and Earle, R.W., (Champaign, IL: Human Kinetics), pp. 343-392.

Edwards, L., Dixon, J., Kent J., Hodgson, D. and Whittaker, V., 2008, Effect of shoe heel height on vastus medialis and vastus lateralis electromyographic activity during sit to stand. *Journal of Orthopaedic Surgery and Research*, 3, 2 (Online). Available from: http://www.josr-online.com/content/3/1/2.

Farina, D., 2006, Interpretation of the surface electromyogram in dynamic contractions. *Exercise and Sports Science Reviews*, **34**, pp. 121-127.

Galtier, B., Buillot, M. and Vanneuville, G., 1995, Anatomical basis for the role of vastus medialis muscle in femoro-patellar degenerative athropathy. *Anatomical and Radiologic Anatomy*, **17**, pp. 7-11.

Kraemer, W. J., Ratamess, N. C., Fry, A. C., and French, D. N., 2006, Strength testing: Development and evaluation of methodology. In *Physiological Assessment of Human Fitness*, edited by Maud, P.J. and Foster, C., (Champaign, IL: Human Kinetics), pp. 119-150.

Lagally, K. M., Robertson, R. J., Gallacher, K. I., Goss, F. L., Jakicic, J. M., Lephart, S. M., McCaw, S. T. and Goodpaster, B., 2002, Perceived exertion, electromyography, and blood lactate during acute bouts of resistance exercise. *Medicine and Science in Sports and Exercise*, **34**, pp. 552-559.

Lagally, K. M., McCaw, S. T., Young, G. T., Medema, H. C. and Thomas, D. Q., 2004, Ratings of perceived exertion and muscle activity during the bench press exercise in recreational and novice lifters. *Journal of Strength and Conditioning Research*, **18**, pp. 359-364.

Lieb, F. J. and Perry, J., 1971, Quadriceps function: An electromyographical study under isometric conditions. *Journal of Bone and Joint Surgery*, **53-A**, pp. 749-758.

Marras, W. S. and Davis, K. G., 2002, A non-MVC EMG normalization technique for the trunk musculature. *Journal of Electromyography and Kinesiology*, **11**, pp. 1-9.

McCaw, S. T. and Melrose, D. R., 1999, Stance width and bar load effects on leg muscle activity during the parallel squat. *Medicine and Science in Sports and Exercise*, **31**, pp. 428-436.

McConnell, J., 1986, The management of chondromalacia pattalae: A long term solution. *Australian Journal of Physiotherapy*, **32**, pp. 215-223.

Merletti, R., 1999, Standards for reporting EMG data. *Journal of Electromyography and Kinesiology (All Volumes)*.

Palmitier, R. A., Kai-Nan, A., Scott, S. G. and Chao, E. Y. S., 1991, Kinetic chain exercise in knee rehabilitation. *Sports Medicine*, **11**, pp. 402-413.

Signorile, J. F., Kwiatkowski, K., Caruso, J. F. and Robertson, B., 1995, Effect of foot position on the electromyographical activity of the superficial quadriceps muscles during the parallel squat and knee extension. *Journal of Strength and Conditioning Research*, **9**, pp. 182-187.

Signorile, J. F., Weber, B., Roll, B., Caruso, J. F., Lowensteyn, I. and Perry, A. C., 1994, An electromyographical comparison of the squat and knee extension exercises. *Journal of Strength and Conditioning Research*, **8**, pp. 178-183.

Wheatley, M. and Jahnke, W., 1951, Electromyographical study of the superficial thigh and hip muscles in normal individuals. *Archives of Physical Medicine*, **32**, pp. 508-515.

Winter, D., 1996, EMG interpretation. In: *Electromyography in Ergonomics*, edited by Kumar, S. and Mital, A., (London: CRC Press), pp. 109-125.

CHAPTER SIX

The influence of scapula position on the function of shoulder girdle muscles

M.H. Alizadeh[1], S. Ahmadizad[2] and H. Rajabi Nush-Abadi[1]

[1] Faculty of Physical Education and Sport Science, University of Tehran, Iran
[2] Department of Exercise Physiology, University of Shahid Beheshti, Tehran, Iran

1 INTRODUCTION

In previous studies researchers have reported that the position of the scapula plays a critical role in achieving the appropriate upper-extremity function and proper posture (Kibler, 1998; Voight and Thomson, 2000). Kibler (1998) has reported that an abnormal position of the scapula may decrease the normal function of shoulder. The kinetic chain of shoulder girdle function and scapular position has been attributed to muscular connections between the spine, scapula, clavicle, and humerus. Since the scapula provides a stable base for glenohumeral movements, it seems reasonable to expect that normal pattern of glenohumeral rhythm may be affected if the position of the scapula is altered or muscles become weak or fatigued (Voight and Thomson, 2000). Shoulder pathology and abnormal scapular motion may be linked to imbalanced muscle activity rather than global weakness of scapulothoracic muscles (Warner et al., 1992). In this regard Voight and Thomson (2000) believed that when shoulder function is inefficient, it can not only result in decreased neuromuscular performance but also may predispose the individual to shoulder injury. It seems that insufficient stability of the scapula may result in significant external loading and strain on the shoulder joint.

An inappropriate position of the scapula may also influence the link between scapular and adjacent segments or even non-adjacent segments. Kebaetse et al. (1999) reported that individuals with slouched postures had decreased scapular upward rotation, posterior tilting and increased superior translation. Greenfield et al. (1995) compared scapular position and spinal posture variables between healthy and unhealthy subjects with shoulder overuse injuries and found no significant difference between patients and the healthy subjects except that patients had greater forward head posture and less shoulder scapular abduction.

Retraction and protraction are two important translation motions of the scapula that occur around the curvature of the thoracic wall (Kibler, 1998). Depending on the position of the arm, protraction may proceed in a slightly upward or downward motion while retraction is mainly in a curvilinear fashion around the wall. Kibler (1998) demonstrated that the translation of the scapula may occur over distances of 15 to 18 cm and it therefore needs a large amount of muscle activity to control the movement or position of the scapula.

Despite considerable interest in upper-extremity function, most of the literature has focused on the role of scapular position on shoulder impingement syndrome (Voight and Thomson, 2000; McClure *et al.*, 2004) or the relation between thoracic alignment and scapular position (Kuhn *et al.*, 1995; Kebaetse *et al.*, 1999; Terry and Chopp, 2001). However, the relationship between the position of the scapula and the performance or ability of shoulder girdle muscles has not been addressed yet. Many individuals with protracted posture may be at risk of altering the role of the scapula and decreasing the function of shoulder girdle muscles. Therefore, the present study was designed 1) to investigate the relationship between the position of the scapula and the performance of shoulder girdle muscles, and 2) to compare the endurance of shoulder girdle muscles in two groups of subjects with protected and retracted scapula.

2. METHODS

2.1 Participants

Fifty healthy male pupils who participated in the Tehran University's summer club were chosen and divided into two groups based on the scapula distance measurements. Those who had a distance of more than 14 cm were rated as subjects with protected scapula (height, 1.458 ± 0.039 m; mass, 41.44 ± 4.75 kg) and those with a distance less than 12 cm were rated as the subjects with retracted scapula (height, 1.443 ± 0.046 m; mass, 39.24 ± 6.37 kg). The distances for protracted group and retracted groups were 14.99 ± 0.92 and 11.01 ± 0.62, respectively. The study was approved by the University's Human Ethics and Research Degree Committees.

2.2 Study design and protocol

Kibler's Lateral Scapula Slide Test (LSST) (Corrie *et al.*, 2001) was used to measure the distance between the vertebrae and the inferior angle of the scapula. This test evaluates the position of the scapula in relation to a fixed point on the spine. The test was performed in the relaxed standing position with the arm relaxed at the sides in which the inferomedial angle of the scapula was palpated and marked on both sides (Figure 1). The reference point on the spine was the nearest spinous process. The measurements from the difference point on the spine to the medial angle of the scapula were obtained on both sides. The accuracy of marking the inferomedial angle of the scapula for assessing scapula asymmetry and the ability of shoulder girdler muscles has been shown by a high correlation (r = 0.91) between this test and the radiographic evaluation of the same point, when marked by a lead shot, for different positions (Kebaetse *et al.*, 1999). The measurements were taken three times with an intervening one-minute rest.

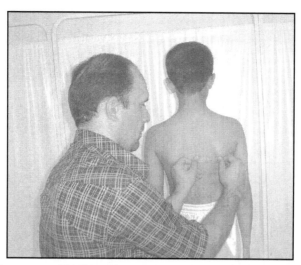

Figure 1. Measurement of the distance between the vertebrae and the inferior angle of scapula.

The performance of shoulder girdle muscles was examined using a modified pull-up test recommended by AAPHERD (McSweigin *et al.*, 1989). The modified pull-up is an important part of any physical performance programme and has been shown to be an effective method for measuring upper-body muscle endurance (Cotton, 1990; Rutherford and Corbin, 1994; Baker and Newton, 2004). The subjects assumed a supine position with shoulders directly under a bar set one to two inches (2.5-5.0 cm) above the subject's reach (Figure 2a). An elastic band was placed seven inches (17.8 cm) below and parallel to the bar. The subjects grasped the bar with an overhand grip (palms away from the body) and the pull-up began in a "down" position with arms and legs straight, buttocks off the floor, and only the heels touching the floor (Figure 2b). The subject then pulled up until his chin was above the elastic band, and returned to full arm extension. The subjects were asked to keep the body straight and they were allowed to repeat pulling the body up without time limit (Figure 2b) and the numbers of pull-ups was counted.

2.3 Statistical analysis

All statistical analyses were performed using SPSS version 11.0 (SPSS, Chicago, USA). Descriptive statistics were computed and distributions of the data were assessed for normality. Pearson correlation was used to examine the relationship between the distance of the scapular and the performance of shoulder girdle muscles. Independent *t*-tests were used to assess differences between modified pull-up performance for the protracted and the retracted subjects. Values are presented as mean ± SD. The acceptable probability level was set at $P < 0.05$.

Contemporary Sport, Leisure and Ergonomics

a

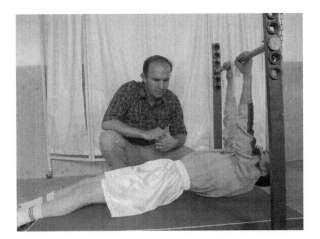

b

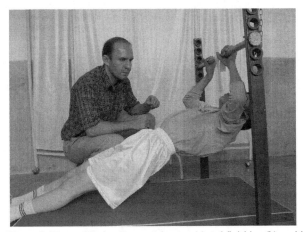

Figure 2. The modified pull-up test the start (a) and finishing (b) positions.

3. RESULTS

Mean (±SD) values for modified pull-up performance (number of pull-ups) for protracted and the retracted groups were 11.12±3.98 and 17.28±6.65, respectively. When the modified pull-up scores were compared for protracted and retracted individuals using independent t-tests, the results showed a significant difference between the performances of the two groups ($P < 0.01$). Statistical analysis of the data revealed a significant negative correlation between modified pull-up performance and the distance of the scapula (r = -0.43, P = 0.002).

4. DISCUSSION

The present study indicated a significant correlation between the position of the scapula and modified pull-up performance. This investigation focused primarily on the performance of shoulder girdle muscles which might be affected by the position of the scapula. This approach is based on a belief that normal scapula position may assist suitable regional muscle performance by generating adequate force and activation of the rhomboids and trapezius muscles. The lower and middle divisions of the trapezius and rhomboid muscles are key contributors to normal and abnormal scapular motion and control. Smith *et al.* (2004) reported that the rhomboid has an important role in normal shoulder function. They also emphasized that these muscles play a crucial role in the evaluation of shoulder and upper-limb disorders. These explanations may also be supported by the fact that among different shoulder girdle muscles, the insertions of the trapezius and rhomboids into the scapular vertebral border and inferior angle result in larger movement of the arms for production of scapular downward rotation and posterior tipping. Once the scapula is in protracted position the function of trapezius and rhomboid muscles may be affected because they can not be involved as the prime mover of the scapulothoracic muscles (Kibler, 1998). The trapezius and rhomboids are also unique among the shoulder girdle muscles because they have the ability to contribute to all components of the normal three-dimensional (3-D) movement of the scapula on the thorax during elevation of the arm particularly in the pull-up (Kebaetse *et al.*, 1998).

The results of this study also showed significant differences between modified pull-up performance of the protracted and the retracted individuals. One explanation for this may be the action of stabilizer muscles such as trapezius and rhomboid muscles on the scapula in modified pull-up performance. The action of trapezius and rhomboid muscles is the adduction of the scapula towards the spinal column as well as scapula downward rotation from the upward rotated position. These muscles also fix the scapula in adduction when the muscles of the shoulder joint adduct or extended the arm. When the subjects hung from the horizontal bar, like in the position of modified pull-ups, and the body was suspended by the hands, the scapula tends to be pulled away from the top to the chest. By starting chinning movement, trapezius and rhomboids draw the medial border of the scapula down and back toward the spinal column. Trapezius and rhomboid muscles working together, produce adduction with some slight elevation of the scapula in the movements and muscles action involved in chinning. Therefore, in individuals with protracted scapula the middle and lower fibres of these muscles are in a lengthened position and they may be unable to fix the scapula in adduction to provide a powerful action in chinning while the muscles of the shoulder joint adduct or extend in the arm. One reason for different muscle performance in the protracted and the retracted individuals may be the insufficient action of trapezius and rhomboid muscles in motion of the scapula toward the midline of the body or moving the scapula back toward the spine compared with retracted individuals during retraction of the shoulder girdle. Apart from the importance of performance of the shoulder girdle muscles, insufficient muscle function due to the alteration of scapula may increase the risk of shoulder injury and promote a round-shoulder posture. Since an imbalance of muscle activity and global weakness of shoulder girdle muscles have been proposed as contributing to shoulder pathology by many investigators (Ludewig and Cook 2000; Ludewig *et al.*, 2004) this may be a possible reason for therapeutic exercise protocols in prevention and rehabilitation of shoulder dysfunction and posture improvement.

Although care was taken on measuring the position of the scapula, it would be more desirable to measure glenohumeral orientation and other muscle functions while the performance is being measured. Scapula measurements were carried out under static conditions in a relaxed standing posture, which may not accurately represent what occurs during dynamic activities in the modified pull-up test. We also chose to assess the extremes of protracted posture, which may not necessarily reflect habitual postures assumed by an average person in many activities of daily living. Extreme postures are often assumed during more challenging activities or those activities where motion at other joints (knee, hip or lumbar) may be constrained. According to our findings the weakness of surrounding musculature of scapulothoracic joint and glenohumeral movement may be related to the position of the scapula. Voight and Thomson (2000) indicated that when the scapula fails to perform its stabilization role, shoulder function is inefficient. Perhaps the scapula position can be considered as index or indicator not only in anticipating decreased muscular performance but also in predisposing the individuals to faulty posture as well as shoulder injury (Folsom-Meek *et al.*, 1992).

In this study conclusions are drawn on how performance of shoulder muscles relates to scapula position and scapula orientation; however, these variables were not measured simultaneously. The results of the present study suggest that individuals who have a protracted position because of poor posture may be at a disadvantage (Smith *et al.*, 2004). Excessive protraction of the scapula may be a functional deformity of posture, but may also be a transient position induced by the demands of a particular task, such as keeping the hands forward for a long period or sitting in a slouched position. It should be taken into consideration that altering scapula position may affect scapular kinematics and therefore decrease the performance of shoulder girdle muscles.

Acknowledgements

This research was supported in part by Iranian Sports Science Research Centre (SSRC). The authors wish to thank Mahdieh Akouchekian for her valuable contribution in collecting data and support throughout the present study. We also thank volunteers for their enthusiastic participation in this study.

References

Baker, D.G. and Newton, R.U., 2004, An analysis of the ratio and relationship between upper body pressing and pulling strength. *Journal of Strength Conditioning Research,* **18**, pp. 594-598.

Corrie, J.O., Andrea, B.T., Christine, E.H. and Craig, R.D., 2001, Measurement of scapular asymmetry and assessment of shoulder dysfunction using the lateral scapular slide test: A reliability and validity study. *Physical Therapy,* **81**, pp. 799-809.

Cotton, D.J., 1990, An analysis of the NCYFS II modified pull-up test. *Research Quarterly Exercise Sport,* **61**, pp. 272-274.

Folsom-Meek, S.I., Herauf, J. and Adams, N.A., 1992, Relationships among selected attributes and three measures of upper body strength and endurance in elementary school children. *Perceptual Motor Skills,* **75**, pp. 1115-1123.

Greenfield, B., Catlin, P.A., Coats, P.W., Green, E., McDonald, J.J. and North, C., 1995, Posture in patients with overuse injuries and healthy individuals. *The Journal of Orthopaedic and Sports Physical Therapy,* **21**, pp. 287-295.

Kebaetse, M., McClure, P. and Pratt, N.A., 1999, Thoracic position effect on shoulder range of motion strength, and three-dimensional scapular kinematics. *Archives of Physical Medicine Rehabilitation,* **80**, pp. 945-950.

Kibler, B., 1998, The role of the scapula in athletic shoulder function. *The American Journal of Sports Medicine,* **26**, pp. 325-337

Kuhn, J.E., Plancher, K.D. and Hawkins, R.J., 1995, Scapular winging. *The Journal of the American Academy of Orthopaedic Surgeons,* **3**, pp. 319-325.

Ludewig, P.M. and Cook, T.M., 2000, Alterations in shoulder kinematics and associated muscle activity in people with symptoms of shoulder impingement. *Physical Therapy,* **80**, pp. 276–291.

Ludewig, P.M., Hoff, M.S., Osowski, E.E., Meschke, S.A. and Rundgrist, P.J., 2004, Relative balance of serratus anterior and upper trapezius muscle activity during push-up exercises. *American Journal of Sports Medicine,* **32**, pp. 484-492.

McClure, P.W., Bialker, J., Neff, N., Williams, G. and Karduna, A., 2004, Shoulder function and 3-dimensional kinematics in people with shoulder impingement syndrome before and after a 6-week exercise program. *Physical Therapy,* **84**, pp. 834-848.

McSweigin, P., Pemberton, C., Petray, C. and Going, S. 1989, Physical Best: The AAPHERD Guide to Physical Fitness Education and Assessment, (Reston VA: AAPHERD).

Rutherford, W.J. and Corbin, C.B., 1994, Validation of criterion-referenced standards for tests of arm and shoulder girdle strength and endurance. *Research Quarterly for Exercise and Sport,* **65**, pp. 110-119.

Smith, J., Padgett, D.J., Kaufman, K.R., Harrington, S.P., An, K. and Irby, S.E., 2004, Rhomboid muscle electromyography activity during 3 different manual muscle tests. *Archives of Physical Medicine Rehabilitation,* **85**, pp. 987-992.

Terry, G.C. and Chopp, T.M., 2001, Functional anatomy of the shoulder. *Journal of Athletic Training,* **35**, pp. 248-255.

Voight, L.M. and Thomson, B.C., 2000, The role of the scapula in the rehabilitation of shoulder injuries. *Journal of Athletic Training*, **5**, pp. 364-372.

Warner, J.J., Micheli, L.J., Arslenian, L.E., Kennedy, J. and Kennedy, R., 1992, Scapulothoracic motion in normal shoulders and shoulders with glenohumeral instability and impingement syndrome. A study using Moire topographic analysis. *Clinical Orthopaedics and Related Research*, **285**, pp. 191-199.

CHAPTER SEVEN

The impact of ankle taping upon range of movement and lower-limb balance before and after dynamic exercise

Katy Clay[1] and Keith George[2]

[1]Physiotherapy Department, Royal Liverpool University Hospital, Liverpool, UK
[2]Research Institute for Sport and Exercise Sciences, Liverpool John Moores University, Liverpool, UK

1. INTRODUCTION

Inversion ankle injuries are among the most common injuries in sport (Bernier and Perrin, 1998). Ankle taping is used by sports participants in an attempt to reduce the incidence of ankle injury (Hamer *et al.,* 1992). Whilst taping has been advocated for many years, the evidence-base for prophylactic ankle taping reducing injury occurrence is limited (Firer, 1990). The proposed efficacy of ankle taping in protecting the lateral ligament complex has been historically related to restriction of range of motion (ROM). It is evident, however, that most if not all of the ROM restriction is rapidly lost with dynamic exercise (Greene and Hillman, 1990; Callaghan, 1997; Cordova *et al.*, 2000; Meana *et al.*, 2008) although available evidence is not unanimous (Ricard *et al.*, 2000). It has been proposed that mechanical restriction of movement may be less important than the proprioceptive and sensory input derived from the ability of taping to stimulate mechanoreceptors in the skin, muscle, tendons and joint for injury prevention (Firer, 1990). Previous research has suggested that taping promotes earlier muscle activation at the ankle joint when presented with rapid inversion movement (Wilkerson, 1991) although this paper also suggested taping could impede performance of athletic skills. Consequently there has been substantial but contradictory data as to whether taping restricts ROM, promotes proprioception and muscle activity or alters gross performance (Verbrugge, 1996; Pederson *et al.*, 1997; Refshauge *et al.*, 2000; Riemann *et al.*, 2002; Arnold and Docherty, 2004).

Although there is evidence of the impact of exercise on taping and restriction of ankle ROM (Meana *et al.*, 2008), relatively few authors have evaluated the extent to which ankle taping may impact upon functional control and balance performance at the ankle joint (Wilkerson, 2002) beyond studies of rapid inversion. The extant literature is exemplified by Robbins *et al.* (1995) who investigated the effect of ankle taping on proprioception before and after exercise, concluding that taping improves foot position awareness. However, they employed a nonfunctional test which requires corroboration with alternate methods.

The purpose of this study was, therefore, to confirm that dynamic exercise altered restriction of ankle joint ROM after taping. Further we studied the impact of taping alone and in combination with brief dynamic exercise on lower-limb balance during a dynamic weight-bearing task. We hypothesised that ankle taping improves lower-limb balance at the same time as restricting ankle ROM and that an acute bout of dynamic exercise whilst resulting in an increase in ankle ROM further improves balance performance.

2. METHODS

2.1 Subjects

Fifteen physically active Physiotherapists with mean ± SD age of 29±5 years with full bilateral ankle ROM and no history of existing or pre-existing lower-limb, neurological or visual pathology or allergy to zinc oxide tape participated in the study. None of the subjects had previously undertaken balance training or used "basketweave" ankle taping. All provided written informed consent to participate. The study was approved by both the Liverpool (Adult) Local Research NHS Ethics Committee and Manchester Metropolitan University Ethics Committee.

2.2 Design

The study employed a counterbalanced, repeated measures cross-over design. Subjects were initially randomly allocated to either a control (CON) or taping (TAPE) trial for initial testing by a computer generated randomisation list and then returned at a later date (between 7 and 14 days at the same time of day) to be tested in the alternative trial. Blinding to trial allocation was not possible due to the awareness of both subject and researcher to the use of tape.

Ankle ROM for plantarflexion, dorsiflexion, inversion and eversion in the dominant ankle was assessed in both trials. These tests were performed in a randomised order at each assessment. Following ROM assessment subjects performed lower-limb balance trials on the Biodex Stability system (Biodex Medical Systems, Shirley NY). In the TAPE trial the subjects then received a "basketweave" tape of their ankle joint before repeating ROM and lower-limb balance tests. For the CON trial, subjects remained seated for the same duration as it took to apply the ankle taping and were then reassessed for ROM and lower-limb balance ability. In both trials the subjects then performed 20 min of dynamic lower-limb exercise, incorporating activities used in warm-up prior to sporting activity. In both trials, the exercise was followed immediately by a final assessment of ankle ROM and lower-limb balance ability.

2.3 Protocols

Ankle dorsiflexion, plantarflexion, inversion and eversion were measured actively using a goniometer (Electro-Medical Supplies Ltd, Oxford) with the subject in a half-lying position. Three trials were performed and the best scores recorded. For dorsiflexion and plantarflexion the proximal arm of the goniometer was aligned with the head of the fibula whilst the distal arm used the lateral aspect of the fifth metatarsal as a reference point. For inversion and eversion the fulcrum of the goniometer was centred over the anterior aspect of the ankle midway between the malleoli. The proximal arm of the goniometer was aligned with the anterior midline of the lower leg using the tibial tuberosity for reference and the distal arm was aligned with the midline of the second metatarsal (Norkin and White, 1995). Whilst a simplistic and non-invasive estimate, this method is common in clinical practice and results of reliability studies on goniometry, whilst diverse (Gajdosik and Bohannon, 1987), report intra-tester reliability to be high (Boone *et al.*, 1978; Goodwin *et al.*, 1992) and hence only one assessor was used in this study.

The tape used was zinc oxide (Neuromed Ltd, Wigan). Basket-weave ankle taping using a combination of anchors, stirrups and Gibney strips (Macdonald, 1994) was chosen in this study due to its popularity in prophylactic taping in sportspeople. The taping was applied consistently by a single experienced physiotherapist.

The 20-min exercise bout was lower-limb dominant and progressive in nature. The aim of the exercise bout was to stress the ankle joint dynamically in all planes of movement and match the sort of movements used in warm-up exercises in those sports people who may use prophylactic taping. The specific exercise content is reported in Table 1.

Table 1. Exercises contained in the 20-min bout of dynamic lower-limb exercise.

- Standing on one leg with eyes open (2 min)
- Standing on one leg with eyes closed (2 min)
- Standing on one leg picking object up off floor (2 min)
- Walking on heels (2 min)
- Walking on toes (2 min)
- Kicks with Theraband (2 min)
- Standing on one leg on wobble board (2 min)
- Standing on one leg on wobble board throwing/catching ball (2 min)
- Hopping: spot/ side to side/ forwards and back /diagonal (2 min)
- Target wobble board (2 min)

The lower-limb balance tests were performed on a computerised balance platform that records deflection of the platform from the horizontal which reflects perturbation of centre of pressure/gravity distribution. The platform challenges an individual to maintain balance whilst standing on a moving surface that tilts up to twenty degrees from the horizontal in all planes. The amount of stiffness in the platform is controlled electronically and two levels of performance were chosen: level 2 (difficult) and level 8 (easy) in a random order. Data were calculated for overall, anterior-posterior and medial-lateral stability (°). Subjects were tested barefoot, balancing on their dominant leg and with eyes closed to stress proprioceptive cues from the lower-limb. After familiarisation three trials were performed and recorded at each level at each assessment point with a minimum of 30 s recovery between trials. Each trial lasted for twenty seconds and failed trials (overbalancing) were discarded and repeated. Foot position coordinates were recorded for consistency between tests. This protocol negated learning effects (Lephart *et al.*, 1995; Pincivero *et al.*, 1995) and reliability is acceptable for clinical testing when tests are conducted without visual input and using a less stable base (Hinman, 2000; Cachupe *et al.*, 2001).

2.4 Data Analysis

The independent variables for this study were: 1-Trial (TAPE vs. CON) and 2-Timeline (baseline, post-taping or equivalent rest and post-exercise). The dependent variables were stability scores from the lower-limb balance tests and ankle ROM. Means and standard deviations were calculated and differences were compared by two-way repeated measures analysis of variance (ANOVA) using the Statistical Package for the Social Sciences for Windows (SPSS 11.5) and post-hoc Tukey tests where required.

3. RESULTS

3.1 ROM

Changes in dorsiflexion, plantarflexion, inversion and eversion are presented in Table 2. Patterns of change were consistent for ROM in all directions. For dorsiflexion a significant main effect for group ($P=0.029$), timeline ($P=0.001$) as well as a significant interaction ($P=0.001$) was found. Specifically, dorsiflexion did not change between repeat tests in the CON trial ($P>0.05$), whereas it was reduced by ~5° (94%) after taping which was different from baseline and the CON trial ($P<0.05$). After exercise in the TAPE trial, dorsiflexion increased by ~3° but was still different to baseline and the control trial ($P<0.05$).

Data analysis for plantarflexion revealed a significant main effect for group ($P=0.003$), timeline ($P<0.001$) as well as a significant interaction ($P<0.001$). Plantarflexion did not alter between repeated assessments in the CON trial ($P>0.05$) but was significantly reduced by ~18° (37%) after taping that was different from baseline and the CON trial ($P<0.05$). Plantarflexion increased by ~14° after exercise in the TAPE trial and was not different to baseline or the CON trial ($P>0.05$). Analysis of inversion data revealed a significant main effect for group ($P<0.001$), timeline ($P<0.001$) as well as a significant interaction ($P<0.001$).

In the CON trial inversion did not alter between repeat assessments (P>0.05) but was significantly reduced by ~13 ° (50%) after taping that was different from baseline and the CON trial (P<0.05). Inversion increased by ~7° after exercise in the TAPE trial but was still different to baseline and the CON trial (P<0.05). Analysis of eversion data revealed a non-significant main effect for group (P=0.255), a significant main effect for timeline (P=0.03) but a non-significant interaction (P=0.066). Qualitatively the pattern of changes was similar to other ROM data. Eversion ROM did not alter between repeat tests in the CON trial with no increase but conversely a drop in eversion ROM occurred after taping (~8°, 53%) that was almost completely removed after exercise (~7 °).

Table 2. The impact of taping and dynamic exercise upon ankle range of movement at baseline, post tape and post exercise.

	Dorsiflexion (°)		Plantarflexion (°)		Inversion (°)		Eversion (°)	
	Control	Tape	Control	Tape	Control	Tape	Control	Tape
Baseline	5.33±5.16	5.53±5.0	47.3±11.92	47.4±12.17	28.9±8.54	27.0±11.77	13.7±8.54	15.1±6.27
Post Tape	5.33±5.16	0.33±5.81	47.7±11.92	29.6±8.59	28.8±8.59	13.5±6.94	13.7±6.11	7.0±4.93
Post Exercise	5.33±5.16	3.33±4.88	47.7±11.92	43.9±8.24	28.8±8.59	21.0±6.32	13.7±6.11	14.0±5.07

3.2 Lower-limb balance tests

Overall, anterior-posterior and medial-lateral stability scores were not different across repeat tests in either the CON or TAPE trial at the easier of the levels (level 8). Level 2 (difficult) stability data are presented in Table 3. Analysis of the overall stability index revealed a non-significant main effect for group (P=0.433), a significant main effect of time (P=0.018) as well as non-significant interaction effect (P=0.293). Trends for the data demonstrated a small decline in the TAPE trial after taping (~1°) with a further small decline after exercise (~1°). In the CON trial no change was seen in overall stability index data until after exercise when again a decline of ~1° was noted. Anterior-posterior stability index revealed a non-significant main effect for group (P=0.666), a significant main effect of time (P= 0.028) as well as a non-significant interaction effect (P=0.119). Trends for these data, again, demonstrated a small decline in the TAPE trial after taping (~1°) with a further decline after exercise (~1°). In the CON trial a small increase in anterior-posterior stability index data was observed following rest followed by a small decrease (~1°) after exercise. Analysis of the medial-lateral stability index at level

2 revealed a non-significant main effect for group (P=0.332), a non-significant main effect of time (P=0.070) as well as a non-significant interaction effect of (P=0.715). Again trends demonstrated a small decline in the TAPE trial after taping (~0.4°) with a further small decline after exercise (~0.3°). In the CON trial no change was evident until after exercise when a decline of ~1° was noted. Performance at level 8 for all directions and overall stability were similar across the timeline and between trials because of the relative ease of the stability test.

Table 3. The impact of taping and dynamic exercise upon overall stability index, medial-lateral stability index and anterior-posterior stability index.

	Overall Stability Index (°)		Medial-Lateral Stability Index (°)		Anterior-Posterior Stability Index (°)	
	Control	Tape	Control	Tape	Control	Tape
Baseline	7.93±4.84	8.02±2.73	4.89±3.39	4.41±1.52	6.29±3.68	6.81±2.50
Post Tape	8.27±3.95	7.19±2.04	4.90±2.90	3.99±1.04	6.51±3.33	6.07±2.13
Post Exercise	7.02±2.31	6.01±1.76	4.05±1.36	3.67±1.28	5.72±2.38	4.80±1.67

4. DISCUSSION

This study confirmed that basket weave ankle taping significantly restricted ankle joint ROM and that as little as 20 minutes of dynamic exercise, mimicking warm-up type activities, effectively removed most of this ROM restriction. This finding supports a range of previous work. For example, Rarick *et al.* (1962) demonstrated a 40 % reduction in the effectiveness of tape in limiting ROM after 15 min of standard vigorous exercise including jumping, pivoting and running. Our data also agree with those of Frankeny *et al.* (1993) who obtained a similar reduction of 50 % after 15 min of standard vigorous exercises including jumping, pivoting and running. Recent work by Meana *et al.* (2008) also supports a loss of mechanical restriction from ankle taping after 30 min of dynamic exercise. Whilst there is evidence that not all pre-exercise restriction is lost (Ricard *et al.*, 2000) it is not clear what constitutes a clinically relevant restriction of ROM. The conclusion from the current and previous research is that athletes who tape prophylactically to restrict ankle ROM have probably lost most of this "support" after a warm-up and before they compete or play. The findings from our current study also provided some indication that the restrictions to plantarflexion and eversion are lost earlier than dorsiflexion and inversion. This may be explained by the nature of the exercise stimulus provided but may be worthy of further investigation.

As a consequence of this removal of ankle ROM restriction, it becomes pertinent to assess other parameters that may be relevant to the injury mechanism in lateral ankle ligament sprains. Indeed Lohrer *et al.* (1999) suggested that ROM measurements are of limited functional relevance because they are often assessed in an unloaded foot; thus the collection of data from a functional tilt platform was suggested as being more valuable because the pathophysiological inversion injury movement is similar (Lohrer *et al.*, 1999).

Whilst there are functional and proprioceptive parameters that could be assessed with ankle taping (e.g. early activation of muscles with rapid ankle inversion) we chose a dynamic performance test. Data generated during lower-limb balance tests do not support a statistically significant impact of ankle taping upon performance before or after exercise. The increases in overall, medial-lateral or anterior-posterior stability indices were small and likely within the error of measurement of the stability indices. The clinical consequences of these changes are likely to be small. These data support postural sway indices derived from a static balance test on a forceplate as used by Tropp *et al.* (1984). Conversely, Robbins *et al.* (1995) observed increased joint position sense at the ankle after taping. The obvious conclusion from these studies is that dynamic or weight-bearing tests are less influenced by taping than unloaded tests such as ankle joint repositioning. Cutaneous receptors may well have a variable role on performance and are dependent upon the nature of the task undertaken. Due to the extra information provided by tissue, joint and limb loading in weight-bearing trials, the role of cutaneous information is down-regulated.

In neither trial did the bout of dynamic exercise itself improve lower-limb balance performance to any clinically relevant level. As exercise induces many metabolic changes in muscles, joints and systemically (temperature increase, change in pH) (Pocock and Richards, 2004), the concept that exercise per se may improve balance is plausible. It may be, however, that the exercise stimulus provided in the current study whilst having an effect on taping restriction of ROM was of insufficient intensity and duration to have a significant metabolic effect. Future work may examine the impact of other acute exercise bouts on lower-limb balance.

4.1 Clinical implications, limitations and further research

Prophylactic taping techniques utilised by both athletes and physiotherapists are often governed by personal preference and experience as well as the lack of comparative studies between different taping methods (Callaghan, 1997). Further studies are needed to compare the restriction to ankle ROM offered by different ankle taping methods in a functional/ dynamic position and in different population groups. Certainly data available would support the contention that taping to reduce ankle ROM during performance in healthy ankles is not valuable clinical practice after as little as 20 minutes of dynamic warm-up exercise.

Previous studies have attempted to identify methods of measuring proprioception in a clinically useful way. Although proprioceptive exercise has been shown to be an important part of injury rehabilitation, there is currently no objective method to measure either deficits in proprioception that occur after injury or recovery of deficits after rehabilitation (Testerman and Van Der Griend, 1999). Whilst the lower-limb balance testing employed in the present study is a common test of balance performance, it of course is limited somewhat in its application to dynamic sport performance where many ankle injuries occur. We appreciate the complexity of the anatomy and function of the ankle joint complex, its measurement as well as the varied assessment techniques for broad concepts such as proprioception and advise the need for on-going research.

5. CONCLUSIONS

The hypothesis that the restriction in ankle ROM caused by taping is lessened by twenty minutes of dynamic balance exercise was supported by the results obtained. Conversely, the hypotheses that taping and a bout of dynamic exercise increase lower-limb balance within the setting of this study cannot be substantiated by the current data.

References

Arnold, B.L. and Docherty, C.L., 2004, Bracing and rehabilitation – what's new? *Clinics in Sports Medicine*, **23**, pp. 83-95.

Bernier, J.N. and Perrin, D.H., 1998, Effect of coordination training on proprioception of the functionally unstable ankle. *Journal of Orthopaedic Sports Physical Therapy*, **27**, pp. 264-274.

Boone, D.C., Azen, S.P., Lin, C.M., Spence, C., Baron, C. and Lee, L., 1978, Reliability of goniometric measurements. *Physical Therapy*, **58**, pp. 1355-1360.

Cachupe, W.J.C., Shifflett, B., Kahanov, L. and Wughalter, E.H., 2001, Reliability of Biodex Balance System measures. *Measurement in Physical Education and Exercise Science*, **5**, pp. 97-108.

Callaghan, M.J., 1997, Role of ankle taping and bracing in the athlete. *British Journal of Sports Medicine*, **31**, pp. 102-108.

Cordova, M.L., Ingersoll, C.D. and LeBlanc, M.J., 2000, Influence of ankle support on joint range of motion before and after exercise: a meta-analysis. *Journal of Orthopaedic Sports Physical Therapy*, **30**, pp. 170-177.

Firer, P., 1990, Effectiveness of taping for the prevention of ankle ligament sprains. *British Journal of Sports Medicine*, **24**, pp. 47-50.

Frankeny, J.R., Jewett, D.L., Hanks, G.A. and Sebastianelli, W.J., 1993, A comparison of ankle taping methods. *Clinical Journal of Sport Medicine*, **3**, pp. 20-25.

Gajdosik, R.L. and Bohannon, R.W., 1987, Clinical measurement of range of motion. *Physical Therapy*, **67**, pp.1867-1872.

Goodwin, J., Clark, C., Deakes, J., Burdon, D. and Lawrence, C., 1992, Clinical methods of goniometry: a comparative study. *Disability and Rehabilitation*, **14**, pp. 10-15.

Greene, T.A. and Hillman, S.K., 1990, Comparison of support provided by a semirigid orthosis and adhesive ankle taping before, during and after exercise. *American Journal of Sports Medicine*, **18**, pp. 498-506.

Hamer, P.W., Munt, A.M., Harris, C.D. and James, N.C., 1992, The influence of ankle strapping on wobbleboard performance, before and after exercise. *Australian Physiotherapy*, **38**, pp. 85-92.

Hinman, M.R., 2000, Factors affecting reliability of the Biodex Balance System: A summary of four studies. *Journal of Sport Rehabilitation*, **9**, pp. 240-252.

Lephart, S.M., Pincivero, D. and Henry, T., 1995, Learning effects and reliability of the Biodex stability system. *Biodex Medical Systems Information Booklet, Sports Medicine and Neuro Lab University of Pittsburgh, USA*.

Lohrer, H., Alt, W.A. and Gollhofer, A., 1999, Neuromuscular properties and functional aspects of taped ankles. *American Journal of Sports Medicine*, **27**, pp. 69-75.

Macdonald, R., 1994, *Taping Techniques: Principles and Practice*, (Oxford: Butterworth-Heinemann).

Meana, M., Alegre, L.M., Elvira, J.L. and Aguado, X, 2008, Kinematics of ankle taping after a training session. *International Journal of Sports Medicine*, **29**, pp. 70-76.

Norkin, C.C. and White, D.J., 1995, *Measurement of Joint Motion: A Guide to Goniometry*, (Philadelphia: F.A. Davis Company).

Pederson, T.S., Ricard, M.D., Merrill, G., Schulthies, S.S. and Allsen, P.E., 1997, The effects of spatting and ankle taping on inversion before and after exercise. *Journal of Athletic Training*, **32**, pp. 29-33.

Pincivero, D.M., Lephart, S.M. and Henry, T.J., 1995, Learning effects and reliability of the Biodex Stability System. *Journal of Athletic Training*, **S35**, pp. 48.

Pocock, G. and Richards, C.D., 2004, *Human Physiology: The Basis of Medicine*, (Oxford: Oxford University Press).

Rarick, G.L., Bigley, G., Karst, R. and Malina, R.M., 1962, The measurable support of the ankle joint by conventional methods of taping. *Journal of Bone and Joint Surgery*, **44A**, pp. 1183-1190.

Refshauge, K.M., Kilbreath, S.L. and Raymond, J., 2000, The effect of recurrent ankle inversion sprain and taping on proprioception at the ankle. *Medicine and Science in Sports and Exercise*, **32**, pp. 10-15.

Ricard, M.D., Sherwood, S.M., Schulthies, S.S. and Knight, K.L., 2000, Effects of tape and exercise on dynamic ankle inversion. *Journal of Athletic Training*, **35**, pp. 31-37.

Riemann, B.L., Schmitz, R.J., Gale, M. and McCaw, S.T., 2002, Effect of ankle taping and bracing on vertical ground reaction forces during drop landings before and after treadmill jogging. *Journal of Orthopaedic Sports Physical Therapy*, **32**, pp. 628-635.

Robbins, S., Waked, E. and Rappel, R., 1995, Ankle taping improves proprioception before and after exercise in young men. *British Journal of*

Sports Medicine, **29**, pp. 242-247.

Testerman, C. and Van Der Griend, R., 1999, Evaluation of ankle instability using the biodex stability system. *Foot and Ankle International*, **20**, pp. 317-321.

Tropp, H., Ekstrand, J. and Gillquist, J., 1984, Factors affecting stabilometry recordings of single limb stance. *American Journal of Sports Medicine*, **12**, pp. 185-188.

Verbrugge, J.D., 1996, The effects of semirigid Air-Stirrup bracing versus adhesive ankle taping on motor performance. *Journal of Orthopaedic Sports Physical Therapy*, **23**, pp. 320-325.

Wilkerson, G.B., 1991, Comparative biomechanical effects of the standard method of ankle taping and a taping method designed to enhance subtalar stability. *American Journal of Sports Medicine*, **19**, pp. 588-595.

Wilkerson, G.B., 2002, Biomechanical and neuromuscular effects of ankle taping and bracing. *Journal of Athletic Training*, **37**, pp. 436-445.

Part II

Occupational ergonomics

A practical cooling strategy for reducing the physiological strain associated with firefighting activity in the heat

D. Barr, W. Gregson, L. Sutton and T. Reilly

Research Institute for Sport and Exercise Sciences, Liverpool John Moores University, Liverpool, UK

1. INTRODUCTION

Firefighters are often required to perform prolonged periods of strenuous work under conditions of high environmental heat strain (Romet and Frim, 1987; Ilmarinen *et al.*, 1997; Smith *et al.*, 1997; Smith and Petruzzello, 1998; Smith *et al.*, 2001; Rossi, 2003). Field studies have shown that firefighting activities such as victim search and rescue, stair and ladder climbing, and carrying equipment entail large outlay of energy expenditure (Lemon and Hermiston, 1977; Gledhill and Jamnik, 1992; Bilzon *et al.*, 2001; von Heimburg *et al.*, 2006). The combined effects of such activity and the addition of protective clothing utilized by firefighters lead to high degrees of thermoregulatory and cardiovascular strain (Rossi, 2003). Strategies for rapidly reducing the physiological strain associated with occupational activity may therefore have important implications for the health and safety of firefighters since elevations in body temperature hinder both physical and mental performance (Hancock *et al.*, 2007).

In order to meet the demands of firefighting activity there is a need for practical cooling strategies that are quick and easy to use, and possess sufficient capacity to reduce thermal stress in a relative short period of time. Hand and forearm immersion (House, 1996; House *et al.*, 1997; Selkirk *et al.*, 2004; Giesbrecht *et al.*, 2007) and ice vests (Bennett *et al.*, 1995; Smolander *et al.*, 2004) have previously been shown to be effective in reducing the physiological strain in firefighters during moderate intensity work. Little research to date, has examined the influence of cooling strategies on the physiological responses during repeated bouts of strenuous activity that more closely reflect the highly demanding nature of firefighting. Firefighters may be required to re-enter structural fires following short recovery periods (~15 min). Such time periods have, however, been shown to be insufficient in reducing the thermoregulatory strain (Ilmarinen *et al.*, 1997; Ilmarinen *et al.*, 2004). Therefore, during repeated bouts of strenuous work, more aggressive cooling strategies may be required to maintain the degree of physiological strain within safe limits. Thus, the aims of this study are to examine whether a practical cooling strategy consisting of hand and forearm immersion combined with wearing an ice vest is effective in aiding physiological recovery following a bout of strenuous work in the heat, and to explore whether any benefits remain during a subsequent bout of strenuous work.

2. METHODS

2.1 Participants

Nine professional firefighters (mean ± *s:* age 41 ± 7 years, body mass 85.7 ± 7 kg, height 1.75 ± 4.7 m, maximal oxygen uptake ($\dot{V}O_{2max}$) 45 ± 5 ml×kg×min^{-1} and % body fat 20 ± 4%) volunteered to participate in this study. Each firefighter was familiarized with the experimental procedures prior to participation and gave written consent to participate. This study was approved by the Research Ethics Committee of Liverpool John Moores University.

2.2 Experimental design

Before the experimental trials each firefighter completed a pre-test assessment for the determination of $\dot{V}O_{2max}$ and body composition. Maximal oxygen uptake was assessed during continuous incremental treadmill running to volitional exhaustion (HP Cosmos, Pulsar, Germany). Oxygen uptake was measured breath-by-breath using an online gas analyzer (Oxycon Pro, Jaeger). Body composition was assessed using dual-energy X-ray absorptiometry (DXA). Each DXA scan (Hologic QDR Series Discovery A, Bedford, MA) was performed using the Whole-body fan beam mode.

On the morning of the exercise trials the firefighters arrived at the laboratory at 09:30 h having refrained from exercise, alcohol, caffeine and tobacco during the previous 24 hours. The trials were conducted at least one week apart in a randomised order. All trials were conducted at the same time of day to avoid the circadian variation in internal body temperature (Reilly and Brooks, 1986).

On arrival at the laboratory, participants consumed 100 ml of cold water (~11°C) whilst the temperature provided by an ingestible temperature sensor was monitored. If the temperature varied by ≤0.1°C, it was deemed that the ingestible temperature sensor was sufficiently sited down the gastrointestinal tract for the experimental protocol to continue. A measurement of nude body mass was then obtained (Seca, 704 electronic scale; Germany). Following the attachment of skin thermisters, heart rate chest strap and core temperature data logger, the firefighters donned their protective clothing and self-contained breathing apparatus (SCBA). The firefighters were then transferred to an environmental chamber (dry bulb temperature 49.6 ± 1.3°C; relative humidity 13 ± 2%) where they completed a 20-min bout of walking at an treadmill speed of 5 km h^{-1} and gradient of 7.5% (Love *et al.*, 1996; Graveling *et al.,* 1999). This protocol has been shown to produce similar thermoregulatory responses and respiratory demands similar to that encountered during extreme hard work firefighting activities (Rayson *et al., 2004;* Barr *et al.*, 2008). At the cessation of the 20-min exercise period the firefighters were moved from the environmental chamber and were administered one of two conditions for a 15-min period; Cool or Control. Prior to each 15-min period in both conditions the firefighters removed their helmets, anti-flash hoods, SCBA, tunics, and gloves. Cool required that the firefighters wore an ice vest while sitting in specially adapted chairs with water filled reservoirs fitted to the arm rests in which the firefighters immersed their hands and forearms up to the elbow in water (Temperature ~19°C) taken from a mains supply. The Control condition required that the firefighters remained seated for a 15-min period. During the 15-min period in both conditions the firefighters rested in a room with an ambient temperature of 21 ± 0.5 °C, which was controlled and is representative of a typical summer's day in the UK. During the recovery period in both conditions the firefighters drank a controlled amount of water (5 ml×kg^{-1} body weight) which was stored overnight at room temperature (~19°C). Following the 15-min rest period the firefighters were returned to the environmental chamber and completed a 20-min bout of treadmill walking of the same intensity and duration as the first bout. Following this second bout the firefighters' nude body mass was measured and changes in body mass loss, corrected for fluid intake were taken to represent fluid loss.

Hand and forearm immersion was achieved with the use of a recovery chair (Kore Kooler Rehab Chair, Morning Pride, USA) fitted with a reservoir to each arm which was filled with water. The firefighters were seated in the chair and immersed each arm in the water up to the elbows. The ice vest (Cool Zone, JT Solutions, Norfolk. UK) contained two ice packs (dimensions 35.5 x 35 cm) in micromesh pockets at the front and rear. Ice vests were kept frozen at -30°C until use.

Core body temperature was recorded throughout each trial using a disposable temperature sensor pill (CorTemp, Human Technologies Inc., St Petersburg, FL) which was ingested by the firefighters on the evening prior to each trial. The sensor pill was connected remotely to a data logger worn in a 'bum bag' around the waist of the participant. Prior to use, each pill was tested against a thermometer (Digitron 2080R, Silam Instruments Ltd, Torquay, UK) in two separate digitally controlled water baths (Grant W14, Grant Instruments, Cambridge, UK) in which temperatures were maintained at 30°C and 45°C. Each pill was placed in the water bath and allowed to stabilize for approximately five minutes. The pills falling outside a degree of accuracy of ±0.1°C were not used in the present study. Skin temperature was monitored throughout exercise by attaching insulated contact-type surface temperature skin thermisters (EUS-U-VL1-0, Grant Instruments, Cambridge) to four sites as specified by BS EN ISO 8996 (2004) (neck, right scapula, left hand, and right anterior calf). Each skin thermister was attached using sweat-proof tape (Transpore, 3M, England), and was connected to a data logger (Squirrel 1000 Series, Grant Instruments, Cambridge, UK). Both skin and core temperature were recorded at 5-s intervals. Mean skin temperature was calculated using the weights formulated by ISSO 8996 (2004). Heart rate was continuously monitored throughout the exercise test using a heart rate monitor (Polar S610, Kempele, Finland).

Sweat loss was calculated from the change in nude body mass before and after each trial taking into account fluid intake during each recovery period. The '6 to 20' scale for ratings of perceived exertion (RPE) was used to determine the participants' subjective effort (Table 1) (Borg, 1982). A 9-point thermal sensation scale was used to determine participants' perception of thermal sensation (Young *et al.*, 1987). These measurements were taken every five minutes during each bout. A '6 to 20' Total Quality Recovery scale was used to determine the participants' perception of the quality of 'recovery' during the exertion (Kentta and Hassmen, 1998).

Table 1. Subjective measurement scales.

Perceived Exertion Scale	*Recovery Quality Scale*	*Thermal Sensation Scale*
6	6	0.0 unbearably cold
7 very, very light	7 very, very poor recovery	0.5
8	8	1.0 very cold
9 very light	9 very poor recovery	1.5
10	10	2.0 cold
11 light	11 poor recovery	2.5
12	12	3.0 cool
13 somewhat hard	13 reasonable recovery	3.5
14	14	4.0 neutral (comfortable)
15 hard	15 good recovery	4.5
16	16	5.0 warm
17 very hard	17 very good recovery	5.5
18	18	6.0 hot
19 extremely hard	19 extremely good recovery	6.5
20 maximal exertion	20 maximal recovery	7.0 very hot
		7.5
		8.0 unbearably hot

2.4 Statistical analysis

Descriptive statistics (means $\pm s$) are reported for dependant variables. A two–way (time*condition) analysis of variance (ANOVA) with repeated measures was used to determine any treatment differences for core body temperature, heart rate, skin temperature, RPE, perceptions of thermal sensation and total quality recovery. Student paired T-tests were performed to detect differences in between conditions in sweat loss. All statistical analyses were performed using Statistical Package for the Social Sciences (SPSS) software. Level of statistical significance was set at $P<0.05$.

3. RESULTS

3.1 Core temperature

Core temperature increased significantly over time during both bout 1 ($F_{1.49,}$ $_{11.93}$=252.72; *P*<0.01) and bout 2 ($F_{1.37,10.01}$=183.29; *P*<0.01) in both conditions (Figure 1). At the onset of the 15-min recovery period, core temperature was 38.2 ± 0.32°C and 38.15 ± 0.19°C in the Cool and Control condition, respectively ($F_{1.8}$=1.15; P=0.32). During the recovery period there was a pronounced decrease in core temperature in Cool compared with Control (0.03 ± 0.01°C·min^{-1} vs -0.01 ± 0.01°C·min^{-1}) ($F_{1.21,8.52}$=8.22; P=0.01). At the end of the recovery period core temperature was 37.72 ± 0.34°C and 38.21 ± 0.17°C in the Cool and Control, respectively. Core temperature remained significantly lower during bout 2 in Cool compared to Control ($F_{1.8}$=13.79; P=0.006). Core temperature values at the end of bout 2 were 38.7±0.32°C and 39.28±0.28°C in the Cool and Control conditions, respectively.

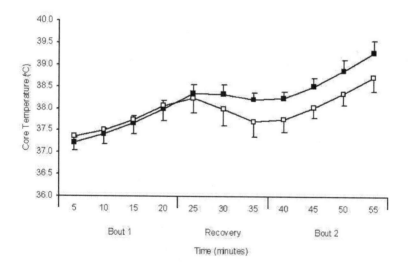

Figure 1. Core temperature responses of firefighters during work and recovery period in the Cool (o) and Control (■) conditions (n = 9, mean ± *s*). A main effect for condition, time and an interaction (P < 0.01) was observed between condition and exercise time.

3.2 Mean skin temperature

Mean skin temperature significantly increased throughout bout 1 ($F_{1.57,12.60}$=276.33; P<0.01) and bout 2 ($F_{2.47,19.77}$=412.74; P<0.01) (Figure 2). Mean skin temperature was similar in both conditions during bout 1 ($F_{1,8}$=3.34; P=0.11). Mean skin temperature at the end of bout 1 was 37.42 ± 0.51°C and 37.82 ± 0.67°C in the Cool and Control conditions. During the recovery period mean skin temperature decreased significantly ($F_{1.91,15.26}$=101.44; P<0.01) and was significantly lower ($F_{1,8}$=29.06; P<0.01) in the Cool condition compared to Control. Mean skin temperature remained significantly lower during bout 2 in the Cool condition compared to Control ($F_{1,8}$=38.93; P<0.01). At the end of the recovery period skin temperature was 31.22 ± 1.04°C and 33.31 ± 1.0°C in the Cool and Control condition, respectively.

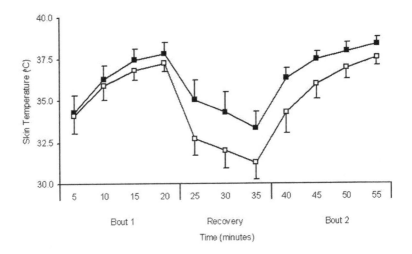

Figure 2. Mean skin temperature responses of firefighters during work and recovery period in the Cool (o) and Control (■) conditions (n = 9, mean ± *s*). A main effect for condition, time and an interaction (P < 0.01) was observed between condition and exercise time.

3.3 Heart rate

Heart rate increased significantly during bout 1 ($F_{1,8}$=1.88; P<0.01) and bout 2 ($F_{1.71,13.71}$=100.87; P<0.01) (Figure 3). Heart rate was similar between conditions during bout 1 ($F_{2.02,16.19}$=125.55; P=0.20). Heart rate at the end of bout 1 was 160 ± 6 beats·min⁻¹ (89 ± 8% % of age predicted maximum) and 165 ± 15 beats·min⁻¹ (93 ± 7% of age predicted max) for Cool and Control, respectively. Heart rate decreased significantly during the 15-min recovery period in both conditions

($F_{3.44,27.59}$=75.45; *P*<0.01). This decrease was significantly greater in Cool compared to the Control trial ($F_{1,8}$=47.37 *P*<0.01). Heart rate was 18 beats·min⁻¹ lower at the end of the recovery period in the Cool condition. Heart rate remained significantly lower during bout 2 in Cool ($F_{1,8}$=41.46; P=0.02). Final heart rates were 172 ± 16 beats•min⁻¹ (97 ± 8% of age predicted maximum) and 179 ± 14 beats•min⁻¹ (100 ± 8% of age predicted maximum) in the Cool and Control conditions, respectively.

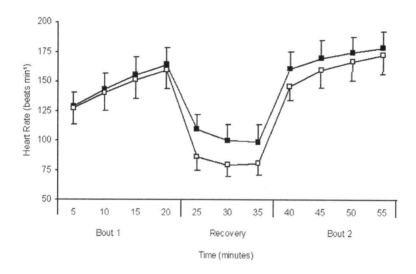

Figure 3. Heart rate responses of firefighters during work and recovery period in the Cool (o) and Control (■) conditions (n = 9, mean ± *s*). A main effect for condition, time and an interaction (P < 0.01) was observed between condition and exercise time.

3.4 Subjective measurements

The firefighters' perceived exertion increased significantly during bout 1 ($F_{1.17,10.61}$=20.03; *P*<0.01) and bout 2 ($F_{2.82,25.44}$=226.01; *P*<0.01) under both conditions. Ratings of perceived exertion were similar during bout 1 in the Cool and Control conditions ($F_{1,9}$=0.01; P=0.91). The firefighters perceived bout 1 to be 'hard' in both conditions There was a trend for a lower perceived exertion during bout 2 in the Cool condition ($F_{1,9}$=4.43; P=0.06). The firefighters perceived bout 2 to be 'very hard' in both conditions.

Perception of thermal sensation increased significantly from 'neutral' to 'hot' during bout 1 ($F_{2.05,18.44}$=12.29; *P*<0.01) and was similar under both conditions ($F_{1,9}$=0.23; P=0.63). At the end of the 15-min recovery period in the Cool condition the firefighters perceived themselves to be 'neutral' compared to 'warm' in the Control trial. As time during each recovery period progressed, the firefighters' perception of the quality of recovery increased in both conditions ($F_{1,11}$=46.73; $P<0.01$) and was similar between conditions ($F_{1,11}$=0.42; P=0.53). The firefighters' perception of recovery was 'very good recovery' and 'good recovery' in the Cool and Control conditions, respectively. During bout 2, the firefighters' perception of thermal sensation increased significantly ($F_{4,32}$=187.98; $P<0.01$) and was significantly lower following Cool compared to Control ($F_{1,8}$=187.98; P = 0.024). The firefighters' perception of thermal sensation at the end of bout 2 was 'very hot' in both conditions.

3.5 Sweat Loss

There was no significant difference in total sweat loss between the Cool (1.37 ± 0.36 l) and Control (1.45 ± 0.45 l) conditions ($t_{(7)}$=0.62; P=0.55).

4. DISCUSSION

The aims of present study were twofold: (i) to investigate whether a practical cooling strategy accelerates the physiological recovery in firefighters following a bout of strenuous work while wearing protective clothing in a hot environment, and (ii) to explore if the physiological benefits of the cooling strategy were evident during a subsequent bout of strenuous work in the heat.

The results of the present study are in agreement with previous studies which have demonstrated that the combined effects of strenuous work in firefighters' protective clothing in hot environmental conditions induce considerable thermoregulatory and cardiovascular strain. Maximum heart rate responses in the present study are similar to that reported during strenuous firefighting simulations performed under live-fire conditions (Smith *et al.*, 1997; Smith and Petruzzello, 1998; Smith *et al.*, 2001; Rossi, 2003). The fifteen minute period of passive recovery (Control condition) was insufficient in reducing the thermoregulatory strain as indicated by the lack of reduction in core temperature. This outcome was despite the fact that the firefighter consumed fluids during this period. This finding suggests that such strategies employed by firefighters are ineffective in accelerating the physiological recovery following bouts of strenuous firefighting activity. Core temperature continued to increase during the first 6 min of recovery in the Control trial. This rise is likely due to the continued circulation and the increased return of warm blood to the core following the cessation of exercise (Rowell, 1974) and also partly to the time lag in response of the ingestible temperature sensor (Lee *et al.*, (2000). This finding is in agreement with previous investigations which have employed a passive recovery period of a similar duration and under similar environmental conditions between two bouts of strenuous work in a hot environment (Ilmarinen *et al.*, 1997; Ilmarinen *et al.*, 2004; Selkirk *et al.*, 2004).

The cooling strategy employed in the present study was effective at accelerating the thermoregulatory and cardiovascular recovery in firefighters during the recovery period. At the end of the 15-min recovery period core temperature and heart rates were 0.5°C and 18 beats·min[-1] lower in the Cool compared to the control condition. The core temperature responses in the Cool condition in the present study are greater than those reported in previous studies which have used hand and forearm immersion alone following a single bout of activity in the heat. Both House and Groom (1998) and Selkirk *et al.* (2004) reported a 0.3°C reduction in rectal temperature following 20 min of hand and forearm immersion in water of similar temperature to that used in the present study. These findings suggest that cooling via application of an ice vest in conjunction with hand and forearm immersion is more effective in mediating decrements in physiological strain during repeated bouts of strenuous firefighting activity than hand and forearm immersion alone.

Using hand and forearm immersion, Carter *et al.* (2007) reported a greater decline in core temperature (0.8°C) following simulated firefighting activities compared with the cooling-induced changes observed in the present study. This increased level of cooling, however, is likely to reflect differences in the duration of recovery and the level of removal of protective clothing and ambient temperatures used in the recovery period. Carter *et al.* (2007) employed a 20-min recovery period compared to 15-min used in the present study. The firefighters in the study by Carter *et al.* (2007) were also required to roll their over-trousers down, whereas in the present study the firefighters did not alter the configuration of their over-trouser. Furthermore, and probably most importantly during the recovery period in the present study, firefighters rested at around 21°C, compared to 15°C in the study by Carter *et al.* (2007) which was performed during the month of December, suggesting that during the winter months of the year in the UK, passive cooling may be an adequate recovery strategy following firefighting activities. This factor may also explain why Carter *et al.* (2007) observed a relatively large decline in core temperature (0.6°C) during the 20-min recovery period in the control condition.

Bennett *et al.* (1995) previously reported that applying an ice vest for 40 min prior to activity led to a 0.7 °C reduction in rectal temperature following 30 min of moderate intensity treadmill walking in the heat (40°C) whilst wearing protective clothing. Smolander *et al.* (2004) reported 'slight' reductions in rectal temperature (no data provided) during 40 min of moderate intensity treadmill walking in the heat (40°C) in full firefighters' protective clothing when ice vests were donned 5 min prior to activity. The results from these two studies suggest that the level of benefit obtained from wearing ice vests appears to be related to the duration of wearing an ice vest prior to activity. However, since the timing of emergencies is unknown, relatively short periods of preparation may only be permitted prior to initial entry into the heat. The cooling strategy adopted in the present study may therefore offer a more practical solution compared with pre-activity cooling in situations where repeated bouts of strenuous firefighting activity occur.

Like previous researchers (House, 1996; Selkirk *et al.*, 2004) we found that the physiological benefits of the cooling strategy were evident during a subsequent bout of work. Compared to Control, the firefighters in the Cool condition experienced a lower core temperature, mean skin temperature, and HR during the second bout of exercise. These changes can only have a positive effect on the health and safety of a firefighter, and although no performance measures were used in this study, the changes observed in the Cool condition are likely to have a positive effect on job performance. Selkirk *et al.* (2004) reported that hand and forearm immersion increased tolerance time by 60% compared to passive cooling by reducing the overall rate of increase in rectal temperature. Based on the rate of rise of core temperature in the present study, the firefighters could have worked for a further 12 min before reaching the same final core temperature as in the control condition. When expressed as a percentage this figure equates to an increased tolerance time of 60%.

The firefighters' perception of thermal sensation was significantly lower at the end of the 15-min recovery period ($P<0.05$). Thermal sensation was also attenuated during the second bout of exercise in the Cool condition. This observation is consistent with Selkirk *et al.* (2004) who reported that hand and forearm immersion reduced the perception of thermal sensation during a subsequent bout of firefighting in the heat. The differences in the responses of thermal sensation during the second bout corresponded closely with skin temperature responses, suggesting that the firefighters may have been emphasizing their skin temperature when reporting thermal sensation. These observations support findings from previous research which indicate that alterations in thermal sensation parallel changes in skin temperature in warm environments (Wang *et al.*, 2006). As time during each recovery period progressed, the firefighters' perception of the quality of recovery increased in both conditions and was similar between conditions. Despite there being no difference in the firefighters' perception of the quality of recovery, thermal sensation was significantly lower at the end of the 15-min recovery period in the Cool condition. This lack of difference in the firefighters' perception of the quality of recovery could be due to the restricted choices available on the recovery scale and therefore the sensitivity of this scale to detect any difference.

It is concluded that a 15-min recovery period using hand and forearm immersion combined with wearing an ice vest markedly reduces thermoregulatory and cardiovascular strain in firefighters following a period of strenuous work in the heat. These thermoregulatory and cardiovascular adjustments are also maintained during a subsequent bout of strenuous work. The findings suggest the present novel cooling strategy is effective in promoting marked reductions in physiological strain during strenuous occupational challenges regularly experienced by firefighters.

References

Barr, D., Gregson, W. and Reilly T., 2008, Reduced physiological strain during firefighting activities using a practical cooling strategy. In *Contemporary Ergonomics*, edited by Bust, P. (London: Taylor and Francis), pp. 485-490.

Bennett, B.L., Hagan, R.D., Huey, K.A., Minson, C. and Cain, D., 1995, Comparison of two cooling vests on heat strain while wearing a firefighting ensemble. *European Journal of Applied Physiology*, **70**, pp. 322-328.

Bilzon, J.L., Scarpello, E.G., Smith, C., Ravenhill, N.A. and Rayson, M., 2001, Characterization of metabolic demands of simulated shipboard Royal Navy fire-firefighting tasks. *Ergonomics*, **44**, pp. 766-780.

BS EN ISO 9886. 2004, *Ergonomics, Evaluation of thermal strain by physiological measurements*. International organization for Standardization. Geneva. ISBN: 058043494X

Borg, G.A.V., 1982, Psychological bases of perceived exertion. *Medicine and Science in Sport and Exercise*, **14**, pp. 48-58.

Carter, J.M., Rayson, M.P., Wilkinson, D.M., Richmond, V. and Blacker, S., 2007, Strategies to combat heat strain during and after firefighting. *Journal of Thermal Biology*, **32**, pp. 109-116.

Giesbrecht, G.G., Jamieson, C. and Cahill, F., 2007, Cooling hyperthermic firefighters by immersing forearms and hands in 10°C and 20°C water. *Aviation, Space and Environmental Medicine*, **78**, pp. 561-567.

Gledhill, N. and Jamnik, V.K., 1992, Characterization of the physical demands of firefighting. *Canadian Journal of Sports Science*, **17**, pp. 207-213.

Graveling R, Johnson J, Butler D, Crawford J, Love R, MacLaren W, and Ritchie, P., 1999, *Study of the degree of protection afforded by fire-fighters clothing*. FRDG publishing, Report 1/99.

Hancock, P.A., Ross, J.M. and Szalma, J.L., 2007, A meta-analysis of performance response under thermal stressors. *Human Factors*, **49**, pp. 851-77.

House, J.R., 1996, Reducing heat strain with ice-vests or hand immersion. In *Proceedings of the 7th International Conference on Environmental Ergonomics*, edited by Shapiro, Y., Epstein, Y. and Moran, D (Jerusalem, Israel) October 1996, pp. 347-350.

House, J.R. and Groom., J.A.S., 1998, The alleviation of heat strain using water perfused forearm cuffs. In *Eighth International Environmental Ergonomics Conference*, edited by Hodgdon, J.A., Heaney, J.H. and Buono, M.J. (San Diego, USA), pp. 255-258.

House, J.R., Holmes, C. and Allsopp, A.J., 1997, Prevention of heat strain by immersing the hands and forearms in water. *Journal of Royal Naval Medical Services*, **83**, pp. 26-30.

Ilmarinen, R., Lindholm, H., Koivistionen, K. and Hellsten, P., 2004, Physiological evaluation of chemical protective suit systems (CPSS) in hot conditions. *International Journal of Occupational Safety and Ergonomics*, **10**, pp. 215-226.

Ilmarinen, R., Louhevaara, V., Griffins, B. and Kunemund, C., 1997, Thermal responses to consecutive strenuous fire-fighting and rescue tasks in the heat. In *Environmental Physiology, Recent Progress and New Frontiers*, edited by Shapiro, Y., Moran, D.S. and Epstein, Y. (London), pp. 295-298.

Kentta, G. and Hassmen, P., 1998, Overtraining and recovery. *Sports Medicine*, **26**, pp. 1-16.

Lee, S., Williams, W. and Schneider, S., 2000, Core temperature measurement during supine exercise: esophageal, rectal, and intestinal temperatures. *Aviation, Space and Environmental Medicine*, **71**, pp. 939-945.

Lemon, P.W.R. and Hermiston, R.T., 1977, The human energy cost of firefighting.

Journal of Occupational Medicine, **19**, pp. 558-562.

Love, R., Johnstone, J., Crawford, J., Tesh, K., Graveling, R., Richie, P., Hutchinson, P. and Wetherill, G., 1996, Study of the physiological effects of wearing breathing apparatus. FRDG Report 13/96.

Rayson, M., Wilkinson, D., Carter, J., Richmond, V., Blacker, S., Bullock, N., Robertson, I., Donovan, K., Graveling, R. and Jones, D., 2004, Physiological assessment of firefighting in the built environment. Report on behalf of the Office of the Deputy Prime Minister, December 2004. ISBN 1 851127615.

Reilly, T. and Brooks, G.A., 1986, Exercise and circadian variation in body temperatures. *International Journal of Sports Medicine*, **7**, pp. 358-362.

Romet, T.T. and Frim, J., 1987, Physiological responses to firefighting activities. *European Journal of Applied Physiology*, **56**, pp. 633-638.

Rossi, R., 2003, Firefighting and its influence on the body. *Ergonomics*, **46**, pp. 1017-1033.

Rowell, L.B., 1974, Human cardiovascular adjustments to exercise and thermal stress. *Physiological Reviews*, **54**, pp. 75-159.

Selkirk, G.A., McLellan, T. and Wong, J., 2004, Active versus passive cooling during work in warm environments while wearing firefighter's protective clothing, *Journal of Occupational and Environmental Hygiene*, **1**, pp. 521-531.

Smith, D.L. and Petruzzello, S.J. 1998, Selected physiological and psychological responses to live-fire drills in different configurations of firefighting gear, Ergonomics, 41, pp. 1141-54.

Smith, D.L., Petruzzello, S.J., Kramer, J.M. and Misner, J.E., 1997, The effects of different thermal environments on the physiological and psychological responses of firefighters to a training drill. *Ergonomics*, **40**, pp. 500-10.

Smith, D.L., Manning, T.S. and Petruzzello S.J., 2001, Effect of strenuous live-fire drills on cardiovascular and psychological responses of recruit firefighters. *Ergonomics*, **44**, pp. 244-54.

Smolander, J., Kuklane, K., Gahved, D., Nilsson, H. and Holmer, I., 2004, Effectiveness of a light-weight ice-vest for body cooling while wearing fire fighters protective clothing in the heat. *International Journal of Occupational Safety and Ergonomics*, **10**, pp. 111-117.

von Heimburg, E.D., Rasmussen, A.K.R. and Medbo, J.I., 2006, Physiological responses of firefighters and performance predictors during a simulated rescue of hospital patients. *Ergonomics*, **49**, pp. 111-126.

Wang, D., Zhang, H., Arens, E. and Huizenga, C., 2006, Observations of upper-extremity skin temperature and corresponding overall body thermal sensation and comfort. *Building and Environment*, **42**, pp. 3933-3943.

CHAPTER NINE

Effects of simulated firefighting on the responses of salivary cortisol, alpha-amylase and psychological variables

F. Perroni[1,2], A. Tessitore[1,3], G. Cibelli[4], C. Lupo[1], E. D'Artibale[1], C. Cortis[1,5], L. Cignitti[2], M. De Rosas[4] and L. Caprinica[1]

[1]Department of Human Movement and Sport Science, IUSM, Rome, Italy
[2]Department of Physical Training of Italian Fire Fighting Corp, Rome, Italy
[3]Department of Human Physiology and Sports Medicine, Vrije Universiteit of Brussel, Belgium
[4]Department of Biomedical Sciences, University of Foggia, Italy
[5]Department of Health Sciences, University of Molise, Campobasso, Italy

1. INTRODUCTION

Considerable information is available that substantiates firefighting as one of the most physically demanding and hazardous of all civilian occupations (Kales et al., 2007), implying variable working conditions and unpredictable and heavy physical demands (Bos et al., 2004). During the 1999-2002 period, the Italian Fire Fighting Corp (VV.FF., 2004) reported 685387 requests for intervention (1% rescues, 6% structural collapses, 7% car accidents, 7% water accidents, 26% fire accidents, 46% various, and 7% unnecessary) performed under high time pressure (mean delay from alarm: 11 minutes; mean duration of intervention: 45 minutes), and emotional stress related to both the victims' (safe: n=20661; injured: n=10081; deaths: n=4453) and the firefighters' personal safety (injuries: n=261; deaths: n=24). To reduce the risk of injuries, firefighters wear protective clothing, which is typically heavy (i.e., 11-23 kg), thick, multi-layered and bulky. Furthermore, it exacerbates the challenge of thermoregulation, decreasing the rate of heat exchange due to limited water vapour permeability across the clothing layers (Smith et al., 1997, 2001, 2005; Smith and Petruzzello, 1998; Baker et al., 2000; McLellan and Selkirk, 2004) They also wear a self-contained breathing apparatus (SCBA), which causes a 25% increase in energy expenditure (Louhevaara et al., 1995) and 22% and 75% reductions of tolerance time at low and high working intensity, respectively (White and Hodous, 1987).

Although the physical demands and the actual workload of firefighting are difficult to quantify, several authors have reported that firefighting activity results in near maximal heart rate (Smith et al., 1997, 2001, 2005, Smith and Petruzzello, 1998), decreases in stroke volume (Smith et al., 2001), increases in core temperature (Smith et al., 1997; 2001, 2005; Smith and Petruzzello, 1998), increases in blood lactate levels (Smith et al., 1997) and in psychological distress

(Smith *et al.,* 1997, 2001, 2005; Smith and Petruzzello, 1998; Ray *et al.,* 2006). Despite the number of studies that have been conducted on physiological aspects of firefighting, limited data are available concerning the hormonal responses to these activities (Smith *et al.,* 2005; Ray *et al.,* 2006).

Studying resting blood catecholamine levels of firefighters (Ray *et al.,* 2006) and the effect of simulated firefighting tasks on blood ACTH and cortisol (Smith *et al.,* 2005), the authors claimed that the interpretation their results might profit from further research in the area of psychobiological alterations of firefighting interventions to provide effective devise for firefighters. However, blood withdrawal might be unsuitable during firefighting interventions and firefighters might be reluctant to participate in experimental settings when such invasive methods are applied. Therefore, salivary samples might be favoured to reflect stress-related changes to exercise, considering that salivary cortisol (sC) is an indicator of the activation of the hypothalamus pituitary adrenal axis (for a review see Viru and Viru, 2004) and salivary alpha-amylase (sA-A) of the autonomic nervous system (Chatterton *et al.,* 1996; Walsh *et al.,* 1999; Li and Gleeson, 2004; Kivlighan and Granger, 2006). Results indicate that concomitant measurement of sC and sA-A activity seems to offer a unique possibility to complement a more comprehensive evaluation of stress responses. In particular, studies using psychological stressors indicated an increased sA-A activity also in response to social behaviour (Nater *et al.,* 2005, 2006; Gordis *et al.,* 2006), emotions (Van Stegeren *et al.,* 2006), and written examinations (Chatterton *et al.,* 1996; Skosnik *et al.,* 2000), while no studies have focused on changes as a result of a stressful and high physically demanding working activity such as firefighting.

Therefore, this study aimed to examine the stress-related changes of a simulated firefighting intervention, considering both the hormonal and psychological components of stress in relation to the load of the previous exercise.

2. METHODS

2.1 Approach to the Problem

The University Institutional Review Board approved the study designed to investigate the stresses of firefighting. The participants performed three experimental sessions, separated by seven days. The first session was designed to determine each firefighter's height, body mass, and maximal oxygen consumption ($\dot{V}O_{2max}$). The second session was designed during a resting day to provide information on the participants' sA-A, sC, and anxiety profiles, while the third experimental session was designed to evaluate the effect of a simulated intervention on physiological, psychological, and hormonal aspects of firefighting. Since firefighting garments have an impact on the exercise load due to the added thermoregulatory stress and the extra weight, to allow comparisons between laboratory and field data (Baker *et al.,* 2000) all the tests were performed with participants wearing their own complete National Fire Protection Agency (NFPA) standard protective firefighting turnout gear. This ensemble included gloves, boots, Nomex flash hood, helmet, respirator and self-contained breathing apparatus (SCBA). Standard issue cotton station long pants or shorts and a cotton T-shirt were worn beneath the turnout gear, along with underwear, and socks. The total weight of the ensemble was approximately 23 kg.

2.2 Participants

Twenty male Italian firefighters (age: 32±1 years) engaged in the residential Italian Fire Fighting Corp, provided their written informed consent to participate in this study. Prior to the evaluation, each individual answered the AAHPERD exercise/ medical history questionnaire to ascertain his activity level, educational background, dietary habits, tobacco smoking and alcohol consumption, medication use and history of physical activity. The Body Mass Index (BMI) was used to assess weight relative to height and is calculated by dividing body mass in kilograms by height in squared metres.

2.3 Measurement of Maximal Aerobic Power

In a laboratory setting (ambient temperature: 21±2 °C; relative humidity: 55±5%) the firefighters performed a graded incremental exercise test to volitional exhaustion on a treadmill. During the test heart rate (HR) was monitored (Sport Tester, Polar Electro, Kempele, Finland) and oxygen consumption ($\dot{V}O_2$), carbon dioxide production ($\dot{V}CO_2$), and ventilation were recorded on a breath-by-breath basis (K4b^2, COSMED, Italy). The K4 Cosmed flow meter was calibrated with a 3-litre syringe (Hans Rudolph Inc, Dallas), and the gas analyzer was calibrated with known gas mixtures (O_2: 16.0% and 20.9%; CO_2: 5.0% and 0.03%). Prior to the test participants performed a 5-min warm-up walking at 6 km·h^{-1} with a 0% of slope. This $\dot{V}O_{2max}$ test consisted of 3-min stages starting at treadmill speed of 8 km·h^{-1} followed by a 2 km·h^{-1} increase in treadmill speed every 3 min until the participant reached volitional exhaustion. Then, a 5-min active recovery was allowed at a walking pace of at 5 km·h^{-1} with at 0% slope. The $\dot{V}O_{2max}$ was identified at the occurrence of a plateau or an increase <1 ml·kg^{-1}·min^{-1} of $\dot{V}O_2$ despite further increases in the exercise intensity, a respiratory gas exchange ratio higher than 1.15, blood lactate concentration [La] higher than 9 mM, and a HR ± 5 beat·min^{-1} of predicted HR$_{max}$ (220-age). Where the test ended before the attainment of the $\dot{V}O_{2max}$, peak oxygen consumption ($\dot{V}O_{2peak}$) was calculated by averaging the final 30-s values of the exercise test. Determination of [La] at rest, at the end of exercise, and after 3, 6 and 9 minutes of recovery was carried out immediately after the collection of untreated capillary blood from a fingertip and immediately analysed using an Accusport Lactate Analyser (Roche, Basel, Switzerland), with a 0.992 single-trial intraclass reliability (Bishop, 2001).

2.4 Psychological measures

The following questionnaires were used to investigate the role of psychological factors in the present stress paradigm: the 20-item State (STAI Y-1) and Trait (STAI Y-2) Anxiety Inventories (Spielberger, 1983) and the 58-item Profile of Mood States (POMS) questionnaire (McNair *et al.,* 1991). The two separate STAI self-report scales are validated for measuring the temporary condition of state anxiety and the individual predisposition to anxiety, respectively. Scores increase in response to psychological stress, and decrease as a result of relaxation. The POMS reflects the individual's mood on six primary dimensions (i.e., Depression-dejection, Tension-anxiety, Anger-hostility, Vigour-activity, Fatigue-inertia, and Confusion-bewilderment). The respondent is required to describe the intensity of his feelings using a Likert 5-point scale (i.e., not at all = 0; somewhat =1; moderately so =2; very much so =3; very very much so =4). The firefighters were administered with the STAI Y-2 questionnaire during a rest day, while the STAI-Y-1 and POMS questionnaires were completed immediately before and after their simulated firefighting intervention. The questionnaires were completed individually, although an investigator was present to provide assistance if required.

2.5 Saliva samples

To avoid the effects of the circadian rhythm and variations in food intake, at the same time of the day the firefighters collected the baseline (morning: 07:00 h:min; afternoon: 17:00-18:30 h:min) saliva samples using a cotton swab and sa-liva collecting tube (Salivette, Sarstedt, Germany). Saliva samples were collected also during the experimental session, immediately before (pre-stress), and a 30 and 90-min after the simulated firefighting intervention (after stress). To reduce possible stress-related confounding factors, all the subjects were advised to avoid any relevant stressors (sports, physical activities and/or other relevant stressful situations) starting 24 h before the first saliva collection. During the first saliva collection, contamination with food debris was avoided by rinsing the mouth with water and by delaying the collection for 15 min after rinsing to prevent sample dilution. For all saliva samples, the absence of blood contamination was checked with a salivary blood contamination kit (Salimetrics LLC, State Col-lege, PA, USA). Following saliva collection, saliva collecting tubes were centrifuged at 3000 $rev\cdot min^{-1}$ for 15 min at 4 °C, saliva samples were then stored at $-80°C$ until they were assayed. All samples were tested in the same series to avoid any variations between tests. Concentrations of sC and sA-A were measured using two different commercially available kits. For salivary free cortisol an enzyme immunoassay kit (Salimetrics, LLC, US) was used. For sA-A a kinetic reaction assay kit (Salimetrics, LLC, US) was used. Cortisol intra-assay coefficient of variation of 3.5±0.5% and an inter-assay reproducibility of 5.08±1.33% were accepted. Amylase intra-assay coefficient of variation and inter-assay reproducibility of

5.47±1.49% and 4.7±0.15% were accepted. A standard plate reader (Power Wave XS, Bio-Tech Instruments, US) was used for salivary determination by 450 nm and 405 nm filters for sC and sA-A, respectively.

2.6 Simulated firefighting rescue intervention

The following four exercise modes were selected as representative of the specific activity of firefighting performed in a circuit with no rest in between: 1) climb a firemen's ladder and descend a 3-floor building carrying a 20 kg (child) dummy (Child Rescue); 2) run for 250 metres; 3) complete a maze in a dark chamber (Find an Exit); and 4) run for 250 metres. The firefighters reported to the fire service training centre at 08:00 hours to have their baseline measurements completed (ambient temperature: 13±1 °C; relative humidity: 63±1%). During the simulated firefighting rescue test, participants' heart rates were continuously recorded (Polar, Kempele, Finland) as average values of 5 s. Intensities of effort were subsequently calculated and expressed as percentages of individual's HR_{max}. In the case of the firefighter ending the $\dot{V}O_{2max}$ test before the attainment of the individual's $\dot{V}O_{2max}$, his theoretical HR_{max} was calculated using the formula (220-age). Heart rates were grouped in four intensity categories: 1) maximal effort (>95% HR_{max}); 2) high-intensity (86–95% HR_{max}); 3) low-intensity (76–85% HR_{max}); 4) and active recovery (65–75% HR_{max}). Arterialised capillary blood was taken from a fingertip before, and at 3, 6 and 9 minutes after the end of the test and analysed immediately for [La] measurement (Accutrend Lactate, Roche, Base, Switzerland). A 6 to 20 Borg scale for rating of perceived exertion (RPE) was used to determine the firefighters' perceived exertion (Borg, 1998) and a 0 to 11 scale (i.e., nothing at all and 11 maximum pain) for perceived lower limb muscle pain (RMP) were administered approximately 15 min after the test to ensure that the firefighters' perception referred to the whole test rather than the most recent exercise intensity (Impellizzeri et al., 2004). Following the last saliva sampling participants were strongly encouraged to drink water ad libitum.

2.7 Statistical Analysis

A 0.05 level of confidence was selected throughout the study. To evaluate the exercise load imposed by the simulated firefighting rescue intervention, differences in frequencies of occurrence of HR counts during the five phases of the test were verified by means of a Chi-square test. An ANOVA for repeated measures was used to assess differences between STAI Y-1 and STAI Y-2, and pre-test and post-test sA-A, sC, and POMS values. Cohen's effect sizes with respect to pre-exercise values were calculated to provide meaningful analysis for comparisons from small groups. Values ≤0.2, from 0.3 to 0.6, <1.2, and >1.2 were considered trivial, small, moderate, and large, respectively. A correlation matrix was calculated between peak values of HR, [La], sC, sA-A, anxiety and POMS.

3. RESULTS

Anthropometric characteristics of the firefighters were: height 1.77 ± 0.06 m; body mass 77.2 ± 8.7 kg; BMI 24.7 ± 2.1. Even though all the participants showed high RER values (1.15 ± 0.1), it was not possible to assume that they reached their $\dot{V}O_{2max}$, given the absence of a $\dot{V}O_2$ plateau, HR_{max} values corresponding to $90\pm5\%$ (176 ± 9 beat·min^{-1}) of their theoretical HR_{max}, and peak [La] values of 8.8 ± 2.0 mM. Thus, the term $\dot{V}O_{2peak}$ was used and was 43.1 ± 4.9 ml.kg^{-1}.min^{-1}.

The total duration of the simulated firefighting intervention was 704 ± 135 s. Table 1 reports the time and the mean HR recorded during the four exercise modes. The firefighters employed the longest time to find an exit in the dark chamber (438 ± 117 s), and the shortest time to climb a firemen's ladder and to descend a 3rd floor building carrying a 20-kg child dummy (82 ± 25 s), Mean HR increased during the simulated intervention and the highest values were recorded during the last 250 m run (170 ± 14 beat·min^{-1}); during this 250-m run the highest frequency of occurrence (80%) of peak HR values (184 ± 9 beat·min^{-1}) was also registered. Intensities of exercise were $2.7\pm0.9\%$, $6.3\pm2.0\%$, $29.5\pm5.4\%$, and $48.1\pm5.5\%$ of total time for active recovery, low-intensity, high-intensity, and maximal effort, respectively. A difference ($c^2= 8.42$; P=0.038) emerged between percentages of occurrence of HR lower ($37.5\pm29.0\%$) and higher ($62.5\pm29.0\%$) than 85% of HR_{max}. After 30 minutes of intervention completion HR was still significantly ($F_{(2, 38)}=631.24$, P<0.0001) higher (108 ± 15 beat·min^{-1}) than baseline values (66 ± 8 beat·min^{-1}). At the end of the simulated intervention, peak [La] values (9.2 ± 2.9 mM) were frequently observed at 3 (frequency of occurrence: 57%) and 6 minutes of resting (frequency of occurrence: 33%).

	Time (s)	*HR (beat·min^{-1})*	*Frequency of occurrence of HR_{peak} values (%)*
Child Rescue	81.8 ± 25.3	145 ± 21	0
250 m Run	92.8 ± 24.6	166 ± 9	0
Find an Exit	437.5 ± 116.6	165 ± 10	20
250 m Run	91.5 ± 23.7	170 ± 14	80

Table 1. Means ± standard deviations of the time required to perform the four exercise modes and the relative frequency of occurrence of HRpeak values.

The firefighters' RPE was 16 ± 2 (i.e., "hard"), ranging from 11 (i.e., "fairly light") to 19 (i.e., "very, very hard") and RMP was 3 ± 2, ranging from 0 (i.e., "no pain"), to 8 (i.e., "very strong"). Baseline STAI Y-1 (30.5 ± 5.3) and STAI Y-2 (30.8 ± 4.9) scores were highly correlated ($r=0.91$, $P<0.0001$). No significant change in state anxiety scores was observed as a result of the simulated firefighting intervention. With the exception of Vigour, high values of the POMS subscale are considered negative. Therefore, when the scores are profiled a desirable confirmation resembles an iceberg. However, given the lack of normative data across the entire spectrum of stressful occupations, the present data were used only for comparative purposes of pre- and post-simulated intervention. No difference emerged between pre- and post-intervention in POMS subscales (Tension-anxiety $=6\pm1$, Confusion-bewilderment $=5\pm1$, Fatigue-inertia $=3\pm1$, Anger-hostility $=3\pm1$, Depression-dejection $=2\pm1$, and Vigour-activity $=17\pm1$). When values were converted to T-scores, they ranged between 37 and 53.

According to studies of a similar nature, hormonal values are reported in means $\pm$ SEM. During the rest day, sA-A was 64.2 ± 10.9 and 133.9 ± 30.3 $U \cdot ml^{-1}$ in the morning and the afternoon, respectively. Although higher values were recorded the morning of the experimental session (102.3 ± 18.7 $U \cdot ml^{-1}$) with respect to the resting day, no significant difference emerged. A significant ($F_{(2, 36)}=8.14$, $P=0.0012$) sA-A response to the simulated intervention was shown (Figure 1), with peak values (279.7 ± 59.0 $U \cdot ml^{-1}$) registered 30 minutes after the accomplishment of the intervention. Post-hoc analysis showed differences only between peak values and the other two sA-A collections. On average, sA-A increased 174% from pre-exercise to 30 min after the completion of the simulated intervention, returning to pre-exercise levels at 90 min post-intervention. Baseline cortisol concentrations demonstrated a diurnal rhythm, with concentrations higher in the morning (16.7 ± 2.2 $nmol \cdot l^{-1}$), and reduced in the afternoon (2.7 ± 0.4 $nmol \cdot l^{-1}$). Although lower values were recorded the morning of the experimental session (11.3 ± 1.9 $nmol \cdot l^{-1}$) with respect to the resting day, no significant difference emerged. During the experimental setting (Figure 2), peak sC values (108.5% increase from pre-exercise) were always found at 30 min post-intervention, which differed ($F_{(1, 18)}=36.8$, $P<0.001$) from pre- and 90 min post-intervention. Small effect sizes were found between pre-exercise and peak sC ($r=0.5$) and sA-A ($r=0.04$) values. Low (range: 0.02-0.37) and non-significant correlations were observed between HR, [La], sA-A, sC, STAI and POMS values.

4. DISCUSSION

Firefighting is an occupation with a high variability and unpredictability of the working demands (Bos *et al.,* 2004, Kales *et al.,* 2007). The environmental, physical, and emotional stress encountered by firefighters might determine occupational hazards, job-related injuries and fatal events (Gledhill and Jamnik, 1992; VV. FF., 2004; Kales *et al.,* 2007). Although neuro-behavioural problems

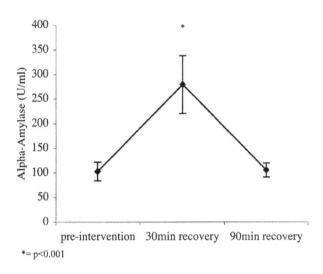

Figure 1. Means ± SEM of the salivary alpha-amylase values registered before, and after (30 min and 90 min) the simulated firefighting intervention.

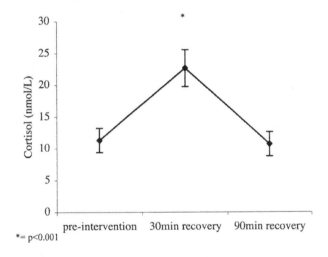

Figure 2. Means ± SEM of the salivary cortisol values registered before, and after (30 min and 90 min) the simulated firefighting intervention.

have been reported in firefighters (Ray *et al.,* 2006), the stress-related component of their work activity is relatively uninvestigated (Smith *et al.,* 2005). To study the interaction of emotional and physical stress of firefighting, in the present study a multivariate approach was applied, including the evaluation of physiological (i.e., cardiac load and blood lactate concentration), psychological (i.e., mood, anxiety, and subjective ratings of exertion) and hormonal (i.e., sC and sA-A) aspects of a firefighting intervention. Since researchers would be not allowed to interfere with real firefighting events, data were collected during a simulated intervention, which included climbing a firemen's ladder to a 3rd floor building to rescue a child, running 250 m to a dark chamber, finding an exit, and running 250 m to the finish line.

To help understand the intensities of efforts during the simulated intervention, the firefighters underwent a laboratory test in which their aerobic power was assessed while running on a treadmill wearing their own fire garments. In the literature (Gledhill and Jamnik, 1992; Sothmann *et al.,* 1992; Bos *et al.,* 2002), an aerobic power >45 $ml \cdot kg^{-1} \cdot min^{-1}$ has been suggested necessary to complete firefighting interventions successfully, also considering that the use of heavy, thick, multi-layered and bulky self-protective clothing causes an increase in energy expenditure and reductions of working tolerance (Louhevaara *et al.,* 1995). The aerobic assessment results showed that the firefighters in the present study had adequate fitness levels, despite by their own admission of not engaging in any additional fitness training programme.

According to the literature (Smith *et al.,* 1997, 2005) the simulated intervention imposed a heavy load on the firefighters, reflected by both physiological measures and subjective ratings. Peak [La] has often been used to represent the product of anaerobic glycolysis during exercise. The high post-intervention [La] peak values and the percentage of time spent (63%) working at a HR >85% of the individual HR_{max} during the firefighting intervention indicate that the anaerobic metabolism of the firefighters was highly taxed. Although the intensity of activity required depends on the type of intervention and heat stress (Wilmore and Costill, 1994), it is not possible to exclude that the progressive elevation of HR could be due in part to the thermoregulatory challenges imposed by the protective clothing (Smith *et al.,* 2005) rather than ambient temperature (i.e., 13 °C) and humidity (63%). In fact, although the firefighters mostly walked to find the exit in the dark room, their heart rates remained elevated. Furthermore, after 30 minutes of post-intervention passive recovery, HR was significantly higher than baseline values, indicating that thermoregulatory effects were still present.

The non-invasive nature of salivary measures provides new opportunities to study how biological and social processes interact. In the present study the acute reponses of sC and sA-A to the simulated firefighting intervention have been investigated to reflect hypothalamus pituitary adrenal axis and the sympatho-adrenal medullar system activities, respectively. In particular, Granger *et al.* (2007) reported that these measures are not redundant and their inclusion in biosocial studies has been suggested to highlight individual differences in stress-related vulnerability and resilience. Smith *et al.* (2005) observed elevated post-intervention sC values although a discrepancy emerged regarding the recovery period. In fact, in our study sC values returned to baseline level after 90 minutes of recovery post-intervention. Considering that the present study and that of Smith *et al.* (2005) investigated different firefighting interventions using different measurement approaches (i.e., blood and salivary), further research is needed to identify the physiological significance of these results. Given the interest in sA-A as a marker of sympathetic activity in relation to stress (for a review see Granger *et al.*, 2007) it is surprising that there have not yet been studies of the bio-behavioural differences in the context of high risk occupations. The present study addressed this potentially important issue. The present results showed the typical sA-A response with a rapid increase and recovery, in line with the response of the sympathetic nervous system to stress. Granger *et al.* (2007) suggested that 30 minutes after the stress stimulus sC peak values can be observed while sA-A might already decrease towards resting values. However, this suggestion is based on data collected after a stress stimulus different from exercise. In fact, Kiviglan and Granger (2006) showed that 40 minutes after 6-8 minutes of indoor rowing competition sA-A values remained significantly elevated from baseline, indicating that metabolic stress might require a longer recovery period. It could be also suggested that sA-A increases in relation to exercise reflect increases in plasma catecholamines and thermal stress (Chatterton *et al.*, 1996). The observed higher increase of sA-A (174%) with respect to sC (109%) following 30 minutes of passive recovery (i.e., sitting on a chair) is in agreement with the literature on their response to exercise (Chatterton *et al.*, 1996; Kivlighan and Granger, 2006), indicating a higher reactivity in the sympathetic nervous system (i.e., sA-A) than in the hypothalamic-pituitary-adrenal axis (i.e., sC). However, 90 minutes following the end of the simulated firefighting intervention sA-A also returned to baseline values. Despite the fact that the sympathetic nervous system and the hypothalamic-pituitary-adrenal axis react in coordination to generate a stress response, the present results confirmed the lack of significant correlation between the levels of sA-A and sC (Nater *et al.*, 2006; Granger *et al.*, 2007). The joint or dissociated activation of the hypothalamic-pituitary-adrenal and the sympathetic nervous system is still unresolved (Kiviglan and Granger, 2006; Granger *et al.*, 2007). The exact nature of the relationship between the sC and sA-A activities was beyond the aim of this research and the present results do not contribute to the solution of the ongoing debate. In fact, the sample collection strategy of the present study was designed specifically for sC and it might be possible that early sA-A peaks were missed (Gordis *et al.*, 2006). Examining individual differences in sC and sA-A activation in response to a rowing ergometer competition, Kivlighan and Granger (2006) claimed that different patterns might emerge in relation to the level of experience of the individuals and their performances. Given the limited number of participants in this study, it was not possible to investigate this particular issue and further research is highly recommended.

Recently, sA-A was reported to correlate significantly with the STAI scores following mathematically induced mental stress (Noto *et al.*, 2005) and video viewing of a surgery (Takai *et al.*, 2004), while no correlation was reported between cortisol and STAI (Takai *et al.*, 2004). On the other hand, fatigue and tension-anxiety in POMS reflected cortisol increases to episodic stress in young college students (Izawa *et al.*, 2007). In this study, the psychological responses to the simulated interventions did not mirror the physiological responses, indicating that the salivary hormones reflected changes as a result of intense exercise load rather than psychological stress. The STAI and POMS scores of the firefighters were lower than those reported for young Italian individuals (Farnè *et al.*, 1991, Battisti *et al.*, 2004) and in agreement with those reported by American colleagues (Smith *et al.*, 1997) and soldiers (Liebermann *et al.*, 2006). Also the lack of difference between before and after the intervention in STAI and POMS scores indicates that in the controlled environment of a firefighting training centre the firefighters feel in no danger and perceived the firefighting intervention as non-threatening. On the other hand, following a simulated firefighting intervention Smith *et al.* (1997) reported significant increases in state anxiety only in hot conditions, while no difference emerged during neutral condition (Smith *et al.*, 2001). Therefore, a completely different picture might emerge during real firefighting and a research design with a higher ecological validity (i.e., during real firefighting settings) is highly recommended.

Although the major limitation of this study is that the observations were obtained during a simulated condition and a quite different picture could emerge following real interventions, the paradigm used can provide a useful model for examining interactive effects of various stressors on physiology and behaviour of firefighters where a complex combination of stressors might be expected.

Acknowledgements

The authors would like to express their gratitude to the firefighters of the Italian Fire Fighting Corp for their support when carrying out the experimental sessions.

References

Baker, S.J., Grice, J., Roby, L. and Matthews, C., 2000, Cardiorespiratory and thermoregulatory response of working in fire-fighter protective clothing in a temperate environment. *Ergonomics*, **43**, pp.1350-1358.

Battisti, F. , De Franciscis, A., Tarsitani, L., Di Clemente, L., Di Stani, F., Calabresi, M., Bruti, G., Di Piero, V. and Biondi, M., 2004, Differenze di genere nelle dimensioni rabbia, depressione e ansia e possibile screening psicometrico di condizioni psicopatologiche in un campione di studenti (Gender differences in anger, depression and anxiety dimensions and psychometric screening of psychopathology in a sample of students). *Rivista di Psichiatria*, **39**, pp. 184-188.

Bishop, D., 2001, Evaluation of the Accusport lactate analyser. *International Journal of Sports Medicine*, **22**, pp. 525-30.

Borg, G., 1998, *Borg's Perceived Exertion and Pain Scales*. (Champaign, IL: Human Kinetics), pp. 2-16.

Bos, J., Kuijer, P.P. and Frings-Dresen, M.H.W., 2002, Definition and assessment of specific occupational demands concerning lifting, pushing, and pulling based on a systematic literature search. *Occupational and Environmental Medicine*, **59**, pp. 800 – 806.

Bos, J., Mol, E., Visser, B. and Frings-Dresen, M., 2004, The physical demands upon (Dutch) fire-fighters in relation to the maximum acceptable energetic workload. *Ergonomics*, **47**, pp. 446-460.

Chatterton, R.T., Vogelsong, K.M., Lu, Y.C., Ellman, A.B. and Hudgens, G.A., 1996, Salivary alpha-amylase as a measure of endogenous adrenergic activity. *Clinical Physiology*, **16**, pp. 433–448.

Farnè, M., Sebellico, A., Gnugnoli, D. and Corallo, A., 1991. *POMS: Profile of Mood States: Manuale Adattamento Italiano (Italian Manual)*. Firenze: O.S. Organizzazioni Speciali.

Gledhill, N. and Jamnik, V.K., 1992, Characterization of the physical demands of firefightings. *Canadian Journal of Sport Science*, **17**, pp. 207-213.

Gordis, E.B., Granger, D.A., Susman, E.J. and Trickett, P.K., 2006, Asymmetry between salivary cortisol and alpha-amylase reactivity to stress: relation to aggressive behavior in adolescents. *Psychoneuroendocrinology*, **31**, pp. 976-987.

Granger, D.A., Kivlighan, K.T., El-Sheikh, M., Gordis, E.B. and Stroud, L.R., 2007, Salivary alpha-amylase in biobehavioral research: recent developments and applications. *Annals of New York Academy of Sciences,* **1098**, pp. 122-144.

Impellizzeri, F. M. Rampinini, E., Coutts, A.J., Sassi, A. and Marcora, S. M., 2004, Use of RPE-Based Training Load in Soccer. *Medicine & Sciences in Sports & Exercise*, **36**, pp. 1042-1047.

Izawa, S., Sugaya, N., Ogawa, N., Nagano, Y., Nakano, M., Nakase, E., Shirotsuki, K., Yamada, K.C, Machida, K., Kodama, M. and Nomura, S., 2007, Episodic stress associated with writing a graduation thesis and free cortisol secretion after awakening. *International Journal of Psychophysiology,* **64**, pp. 141-145.

Kales, S.N., Soteriades, E.S., Christophi, C.A. and Christiani, D.C., 2007, Emergency duties and deaths from heart disease among firefighters in the United States. *New England Journal of Medicine,* **356**, pp. 1207-1215.

Kivlighan, K.T. and Granger, D.A., 2006, Salivary a-amylase response to competition: Relation to gender, previous experience, and attitudes. *Psychoneuroendocrinology*, **31**, pp. 703–714.

Li, T.L. and Gleeson, M., 2004, The effect of single and repeated bouts of prolonged cycling and circadian variation on saliva flow rate, immunoglobulin A and alpha-amylase responses. *Journal of Sports Sciences*, **22**, pp.1015-1024.

Lieberman, H.R., Niro, P., Tharion, W.J., Nindl, B.C., Castellani, J.W. and Montain, S.J., 2006, Cognition during sustained operations: comparison of a laboratory simulation to field studies. *Aviation, Space and Environmental Medicine*, **77**, pp. 929-935.

Louhevaara, V., Ilmarinen, R., Griefahn, B., Künemund, C. and Mäkinen, H., 1995, Maximal physical work performance with European standard based fire-protective clothing system and equipment in relation to individual characteristics. *European Journal of Applied Physiology and Occupational Physiology*, **71**, pp. 223-229.

McLellan, T.M. and Selkirk, G. A., 2004, Heat stress while wearing long pants or shorts under firefighting protective clothing. *Ergonomics*, **47**, pp. 75–90.

McNair, D., Lorr, M. and Droppleman, L.F., 1991, *POMS Profile of Mood States* (Italy: O.S. Publ.).

Nater, U.M, Rohleder, N., Gaab, J., Berger, S., Jud, A., Kirschbaum, C. and Ehlert, U., 2005, Human salivary alpha-amylase reactivity in a psychosocial stress paradigm. *International Journal of Psychophysiology*, **55**, pp. 333-342.

Nater, U.M., La Marca, R., Florin, L., Moses, A., Langhans, W., Koller, M.M. and Ehlert, U., 2006, Stress-induced changes in human salivary alpha-amylase activity-associations with adrenergic activity. *Psychoneuroendocrinology*, **31**, pp. 49–58.

Noto, Y., Sato, T., Kudo, M., Kurata, K. and Hirota, K., 2005. The relationship between salivary biomarkers and state-trait anxiety inventory score under mental arithmetic stress: a pilot study. *Anesthesia and Analgesia*, **101**, pp. 1873-1876.

Ray, M.R., Basu, C., Roychoudhury, S., Banik, S. and Lahiri, T., 2006, Plasma catecholamine levels and neurobehavioral problems in Indian firefighters. *Journal of Occupational Health*, **48**, pp. 210-215.

Smith, D. L., Petruzzello, S. J., Chludzinski, M. A., Reed, J. J. and Woods, J. A., 2005, Selected hormonal and immunological responses to strenuous live-fire firefighting drills. *Ergonomics*, **48**, pp. 55-65.

Smith, D. L., Petruzzello, S. J., Kramer, J. M., and Misner, J. E., 1997, The effects of different thermal environments on the physiological and psychological responses of firefighters to a training drill. *Ergonomics*, **40**, pp. 500-510.

Smith, D. L., Manning, T. S. and Petruzzello, S. J., 2001, Effect of strenuous live-fire drills on cardiovascular and psychological responses of recruit firefighters. *Ergonomics*, **44**, pp. 244-254.

Smith, D. L. and Petruzzello, S. J., 1998, Selected physiological and psychological responses to live-fire drills in different configurations of firefighting gear. *Ergonomics*, **41**, pp. 1141-1154.

Spielberger, C.D., 1983, *State-Trait Anxiety Inventory, Form Y Manual*, (CA: Consulting Psychologists Press).

Sothmann, M. S., Saupe, K., Jasenof, D. and Blaney, J., 1992, Heart rate response of firefighters to actual emergencies. Implications for cardiorespiratory fitness. *Journal of Occupational Medicine*, **34**, pp. 797-800.

Skosnik, P.D., Chatterton, R.T., Swisher, T. and Park, S., 2000, Modulation of attentional inhibition by norepinephrine and cortisol after psychological stress. *International Journal of Psychophysiology*, **36**, pp. 59-68.

Takai, N., Yamaguchi, M., Aragaki, T., Eto, K., Uchihashi, K. and Nishikawa, Y., 2004. Effect of psychological stress on the salivary cortisol and amylase levels in healthy young adults. *Archives of Oral Biology*, **49**, pp. 963-968.

Van Stegeren, A., Rohlederb, N., Everaerda, W. and Wolf, O., 2006, Salivary alpha amylase as marker for adrenergic activity during stress: Effect of betablockade. *Psychoneuroendocrinology*, **31**, pp. 137-141.

Viru, A. and Viru, M., 2004, Cortisol-essential adaptation hormone in exercise. *International Journal of Sports Medicine*, **25**, pp. 461-464.

VV.FF., 2004, *Statistica del Corpo Nazionale Vigili del Fuoco*. Italy: Ministero dell'Interno Dipartimento dei Vigili del Fuoco del Soccorso Pubblico e della Difesa Civile.

Walsh, N.P., Blannin, A.K., Clark, A.M., Cook, L., Robson, P.J. and Gleeson, M., 1999, The effects of high-intensity intermittent exercise on saliva IgA, total protein and alpha-amylase. *Journal of Sports Sciences*, **17**, pp. 129-134.

White, M. K. and Hodous, T. K., 1987, Reduced work tolerance associated with wearing protective clothing and respirators. *American Industrial Hygiene Association Journal*, **48**, pp. 304-310.

Wilmore, J. H. and Costill, D. L., 1994, *Physiology of Sport and Exercise*, (Champaign, IL: Human Kinetics), pp. 248-257.

CHAPTER TEN

Relationships between leisure-time energy expenditure and individual coping strategies for shift-work

S. Fullick, C. Grindey, B. Edwards, C. Morris, T. Reilly,
D. Richardson, J. Waterhouse and G. Atkinson

Research Institute for Sport and Exercise Sciences,
Liverpool John Moores University, Liverpool, UK

1. INTRODUCTION

For many years, shift-work has been required to provide emergency cover and essential services at all hours of the day and night, as well as for maintaining long-term industrial processes. Nevertheless, shift-work is no longer restricted to these types of occupations, but is increasingly found in modern 'call centres', where employees deliver financial and retail services around the clock to meet the demands of a '24-hour' society, in shops, and so on. It is not surprising, therefore, that approximately 13-14% of the European and South American workforce is now involved in a shift-work schedule that includes some time spent working at night (Spelten *et al.*, 1999; Harrington, 2001; Rajaratnam and Arendt, 2001; Costa, 2003). Many employees can also be found working 'unusual hours'; outside of the 'normal' 9:00 – 17:00 hour period, but not necessarily involving night work, *e.g.* the permanent early morning shifts worked by postal delivery personnel or the shorter morning and evening 'split-shifts' worked by public transport staff or office cleaners (Taylor *et al.*, 1997).

Shift-work schedules differ markedly in terms of organization, timing and duration of each shift, as well as in the speed of shift rotation of the shifts. Whilst there might be benefits to working 'unusual hours' and shifts, such as increased wages, shift-work, and in particular that including night work, has been associated with greater health problems in comparison to "normal" day work (Waterhouse *et al.*, 1992; Harrington, 2001). The health effects of shift-work can include a reduction in quality and quantity of sleep, insomnia, chronic fatigue, anxiety and depression, adverse cardiovascular and gastrointestinal effects and reproductive effects in women. More recently links between shift-work and an increased risk of obesity have been proposed (Lasfargues *et al.*, 1996; Karlsson *et al.*, 2001; Di Lorenzo *et al.*, 2003). The accumulative sleep deprivation that is associated with shift-work is also thought to have long-term effects in the form of "allostatic load", which refers to the cumulative wear and tear on body systems (McEwen, 2006).

Such loading has been forwarded as a contributory factor to hypertension, reduced parasympathetic tone, increased proinflammatory cytokines, increase oxidative stress, increased evening cortisol and insulin, as well as an overall increased risk of obesity (McEwen, 2007). The exact explanation for the detrimental effects of shift-work on health is complicated. Nevertheless, most general reviews have suggested that the health inequalities associated with shift-work are biological and behavioural in nature (Waterhouse *et al.*, 1992; Harrington, 2001; Costa, 2004; Knutsson, 2004). Harrington (2001) identified improvements in recreational facilities as a factor which could potentially ameliorate shift-work problems in the short-term. Furthermore, Harrington (2001) and Harma *et al.* (1982, 1988) highlighted the importance of physical fitness and activity in helping workers reduce the problems associated with shift-work. Whilst various studies have highlighted the problems associated with shift-work and have sought to develop recreational/leisure/physical activity recommendations to help alleviate such problems, few have addressed the implementation of practical coping strategies within an 'actual' working environment.

The extent to which individuals cope with shift-work is very heterogenous (Lasfargues *et al.*, 1996; Karlsson *et al.*, 2001; Di Lorenzo *et al.*, 2003). As such, most researchers would agree that it is imperative to consider how individuals cope with working shifts, and how they deal with the possible health problems that they might experience. Whilst a number of reviews, booklets and guides on how to cope with irregular working hours have been produced (Monk and Folkard, 1992; Harrington, 2001; Costa, 2003), little attention has focused on how individuals actually cope with shift-work and examining how effective are the strategies they employ at sustaining health and well-being. A more systematic approach to exploring individual coping strategies may help us understand why some individuals seem to be more successful than others (Spelten *et al.*, 1993).

Coping refers to individuals' behavioural and cognitive efforts to manage situations that are viewed as taxing personal resources (Carver *et al.*, 1989 ; Soderstrom *et al.*, 2000). Generally, researchers distinguish between two broad types of coping strategy: approach/engagement-oriented strategies (involving active attempts to confront and resolve the problem) and avoidance/disengagement strategies (reducing the associated emotional distress or evading the problem) (Tobin *et al.*, 1989; Klag and Bradley, 2004). Some study findings indicate that engaging or approaching problems is more beneficial and will prevent burnout as opposed to avoiding or disengaging from the problem (Ceslowitz, 1989; Chang *et al.*, 2006). Nevertheless, Lazarus (1993) suggested that there are no universally good or bad coping processes, merely those that might often be better or worse than others in a particular individual. Indeed, it has been suggested that individuals use both disengagement and engagement strategies to deal with shift-work-related problems, regardless of shift-schedule or job-type (Spelten *et al.*, 1999). Since coping is a dynamic process, the strategies employed may also evolve with time and experience. For example Spelten *et al.* (1999) found that, regardless of the shift schedule, shift-working nurses with inflexible sleeping habits tended to avoid or disengage from the problems whilst permanent night-working nurses utilised both engagement and disengagement strategies when dealing with sleep and social/

family disturbances. It should be noted that the vast majority of participants were female; coping is thought to show gender differences (Tamres *et al.*, 2002).

Shift-workers may become desynchronised from their family's habits and routines and, therefore, become dissatisfied with the amount of time spent with them. It is feasible that participation in leisure-time physical activity may not only have stress-reducing effects but also increase time spent with the family if leisure activities can be pursued as a group (Beermann and Nachreiner, 1995; Presser, 2000; Nomaguchi and Bianchi, 2004). Mechanisms for the stress-reducing effects of physical activity may involve increases in self-esteem, self-efficacy and energy (Wijndaele *et al.*, 2007). These states might evoke feelings of competence, through which individuals may be able to appraise or perceive a stressor as less harmful or threatening. An increase in self-efficacy, energy levels and social support through increased leisure time physical activity and, therefore, a decrease in stress and an increase in the ability to cope would seem to be beneficial to shift-workers in theory. Yet, no previous research work has established that the degree of shift-work coping is even related to participation in physical activity.

2. METHOD

2.1 Participants

From a sampling frame of approximately 200, 95 participants (24 females and 71 males) volunteered to complete a modified version of the SSI, together with the leisure-time physical activity questionnaire validated by Lamb and Brodie (1990). Some characteristics of the sample are shown in Table 1.

2.2 Research Design

Copies of an adapted version of the SSI were distributed by the research team to the various organisations that participated in the study, along with pre-paid envelops in which the SSIs were to be returned to the research team. This distribution process ensured that participants were allowed enough time to complete the questionnaire as well as maintaining anonymity. The SSI represents a well-established and validated (Barton *et al.*, 1990) battery of questions that have been used frequently on shift-workers to measure perceived problems and issues. The SSI covers items referring to: biographical and demographic information, chronotype, major difficulties caused by working shifts (adaptation to shift-work, fitness to undertake job content, social life, fatigue, daytime sleepiness, shift system advantages, psychological well-being), and problems associated with each shift (sleep disturbance, alertness on the job, workload, and items specific to the night-shift), health and well-being, and the ability to cope with night work (Takahashi *et al.*, 2005). Likert scales are used throughout the SSI.

Table 1. Characteristics of shift-workers studied.

Characteristics of Shift-workers Studied	
Number Studied	95
Number of Females	24
Number of Males	71
Mean (SD) Age	37.2 (8.9) years
Age Range	22-59 years
Mean (SD) Experience working shift patterns	120.62 (103.66) months
Mean Height (SD)	1.75 (0.08) m
Mean Female Height (SD)	1.68 (0.07) m
Mean Male Height (SD)	1.78(0.003) m
Mean Weight (SD)	81.0(11.29) kg
Mean Female Weight (SD)	73.74(13.30) kg
Mean Male Weight (SD)	79.78(1.087) kg
Mean BMI (SD)	26.2(3.42)%
Mean Female BMI (SD)	25.84(3.87)%
Mean Male BMI (SD)	26.27(3.21)%
Shift Rotation (Direction & Speed)	Forward/Backward, Fast/Slow, All incorporated a period of nights
Marital Status	
Married/Living With Partner	76%
Single	14%
Divorced	10%
Percentage of Participants with dependents	40.85% (1-3 dependents age range 0-70+)
Marital Status of Those With Dependents:	
Married/Living With Partner	39.85%
Single Parent	1%
Job Type/Title & Number of Participants:	
Police Service	Group 1 –22
FireFighters/Watch Managers/Control Operators/Station Officer	Group 2 –66
Bus drivers/Bus Engineers	Group 3 –6
Flight Attendant	Group 4 –1

To date, the SSI has not included sections designated to explore diet and physical activity during leisure time. Therefore, a leisure time physical activity (LTPA) questionnaire was added to the SSI. The LTPA questionnaire was an adapted version of Lamb and Brodie's (1990) LTPA questionnaire, which allows recording of physical activities that are participated in during leisure time over a 14-day period. The LTPA was complemented with additional questions regarding time spent watching television, transportation, adherence to exercise regimens, availability/accessibility to exercise facilities and barriers to participating in LTPA. Whilst combining the LTPA questionnaire with the social and domestic component of the SSI helped shorten the questionnaire, it was felt that, at 40 pages, the questionnaire was too long to expect a reasonable response rate. Therefore, modified versions of specific sections in the original SSI were used. The original Composite Morning Questionnaire (CS) within the SSI was changed for the validated shortened version of the preference scale questionnaire (PS). Diaz-Morales and Sanchez-Lopez (2004) found that the relationship between the CS and the PS is high ($r = 0.76$), which indicates adequate convergent validity. Smith *et al.* (2002) suggested that whilst both CS and PS are quite adequate psychometrically, the PS is preferable as it is simpler to use and is not influenced by the respondent's sleep-wake schedule.

2.3 Data analysis

According to previous research on factors which influence shift-work tolerance, the most important predictors were delimited to be age, time spent in shift-work (experience), gender, marital status and the overall shift-work engagement and disengagement coping scores. These were entered into an initial exploratory step-wise multiple regression model, with leisure-time energy expenditure over a 14-day period entered as the dependent (outcome) variable. Following this initial analysis, a second exploratory step-wise multiple regression model was implemented to consider the predictive value of the various subscales for individual coping indices. There were four individual subscales of coping mechanisms related to social, domestic, sleep and work-dependent problems, which were entered into the step-wise multiple regression model, with leisure-time energy expenditure over a 14-day period entered as the dependent variable.

3. RESULTS

The distribution of the outcome variable of energy expenditure in leisure-time was found to be slightly skewed. Therefore, these data were analysed before and after logarithmic transformation. Results of the multivariate regression analyses did not differ substantially between logged and non-logged data. Therefore, the Beta coefficients and associated P-values presented below are for the non-logged data.

Time spent in shift-work ($\beta = 26.36$ kJ.week^{-1}, *P=0.051;* Figure 1) and gender ($\beta = 7168.9$ kJ.week^{-1}, *P=0.023;* Figure 2) were found to be predictors of leisure-time energy expenditure, with the most experienced, male shift-workers expending the most energy during leisure-time activities. The overall 'disengagement' coping score (the overall score is the sum of all four sub-scales related to sleep, domestic life, work performance and social life) was found to be a positive predictor of leisure-time energy expenditure ($\beta = 956.27$ kJ.week^{-1}, *P=0.054;* Figure 3). There was no relationship between physical activity and overall 'engagement' coping scores (*P=0.756).*

In males, the individual disengagement subscale of sleep disturbances (β = -1078.1 kJ.week^{-1}, $P=0.086$) was found to be negatively correlated to energy expenditure, whereas disengagement of domestic-related disturbances was found to be positively related to leisure-time energy expenditure (β = 1961.92 kJ.week^{-1}, $P=0.001$). Nevertheless, the r-squared statistic for both these predictors in combination was quite low (14%). These disengagement indices were not found to relate to the energy expenditure of female shift-workers ($P=0.762$).

4. DISCUSSION

These data suggest that experienced male shift-workers participate in the most leisure-time physical activity. Some indices related to an individual's coping strategy were also found to correlate with leisure-time energy expenditure; the male shift-workers with higher levels of physical activity in leisure-time 'disengaged' more from their domestic-related problems, but less from their sleep-related problems. These findings have important implications for the design of physical activity interventions, especially in targeting the least active shift-workers who cope less well with certain stressors related to working at unusual hours.

The most experienced shift-workers were found to participate in more leisure-time physical activity. Obviously, an individual's experience of shift-work will with age (Baker et al., 2004). Whilst a number of researchers have suggested that older shift-workers have more pronounced difficulties and health issues in comparison with younger shift-workers (Harma, 1996; Nachreiner, 1998; Furnham and Hughes, 1999; Seo et al., 2000; Pati et al., 2001; Baker et al., 2004; Rouch et al., 2005), an alternative view is that time spent doing shift-work could be a moderating factor and promote adaptation (Oginska et al., 1993; Bohle and Tiley, 1998; Baker et al., 2004; Bonnefond et al., 2006). Rutenfranz et al. (1985) suggested that a number of phases are lived through if an individual remains in shift-work. The first 1-5 years of shift-work (the first phase) comprise an adaptation phase, whereby workers attempt to adapt and adjust to new working schedules and to deal with social, family, domestic and leisure activities and obligations. Kundi et al. (1979) indicated that the first five years of shift-work also have the strongest effect on subjective health and well-being, highlighting the process of self-selection, which can also hamper interpretation of study findings. That is, those with a greater tolerance and ability to cope with shift-work opt to stay on a shift schedule but those who are unable to tolerate or cope with shift-work leave. Therefore, those individuals that continue to work a shift schedule (i.e. who have a greater experience of working shifts) have been able to adapt their livestyles; they accept the various forms of disruption and desynchronisation associated with shift-work as opposed to 'new-comers', who must go through the adaptive process and find suitable strategies to help them tolerate and cope with shift-work. Shift-workers have been shown to value time similarly to day workers. (Herbert, 1983; Hornberger and Knauth, 1993; Knutson, 2003; Baker et al., 2004; Lipvocan et al., 2004); therefore, their attempts and ability to adhere to societal norms and diurnal activities such as leisure time physical activity may influence

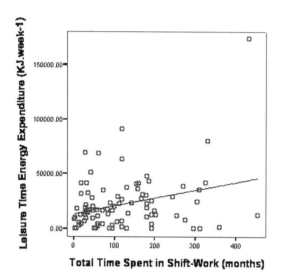

Figure 1 Relationship between Energy Expenditure (kJ.week^{-1}) and time Spent in Shift-work. Data are total energy expenditure for 14-days (kJ.week^{-1}), with time spent in shift-work being total number of months an individual had spent in shift-work.

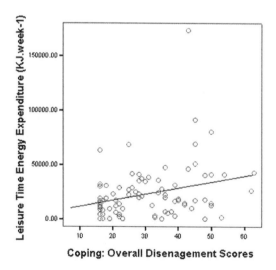

Figure 2 Relationship between Energy Expenditure and Gender. Data are mean ± SE.

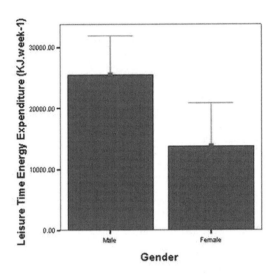

Figure 3. Relationship between Energy Expenditure (kJ.week^{-1}) and overall Coping Disengagement Scores. Data are total energy expenditure for 14-days (kJ.week^{-1}) with Coping Disengagement Scores being individual total scores from the four subscales of the SSI.

the amount of work and life satisfaction they experience. Since the more experienced shift-worker is thought to have adapted to a 'non-diurnal' existence, and deals better with the problems associated with 'free/spare' time and time/activity/obligation management in comparison with a less experienced worker, his/her ability to participate in leisure time physical activity and to schedule it may be somewhat easier. Therefore, promotion of an active lifestyle may be particularly important within the first 5 years of a shift-workers career.

Higher levels of leisure-time physical activity were found to be associated with a disengagement style of coping strategy. It is clear that the very nature of shift-work requires individuals to have a dynamic and flexible adaptation/coping process which allows for the variable circumstances a rotating shift system brings. Such coping could be related to the relatively new personality construct associated with shift-work tolerance - 'hardiness'. Hardiness has been described as an amalgamation of attitudes that enhance health and mood despite stressful circumstances (Maddi *et al.*, 2006). This attitude construct consist of commitment, control and challenge. Since the shift-worker often lacks 'control' over his/her shift pattern, both commitment and challenge would seem to be the primary factors influencing shift-workers' hardiness and therefore the ability to cope with shift systems. Challenge may refer to the requirements of the job, and these might contribute to a shift-worker's tolerance to shift-work due to shifts being part of the job they enjoy. Commitment is highly individualised, especially in relation to shift-work. However, this commitment could relate to issues such as higher rates of pay, and the fact that shift-work might allow for second jobs. Workers who are 'hardy' may in fact be able to adapt more favourably to shift-work, allowing for domestic, social and leisure time activities such as physical activity to be scheduled within their day. This begs the question as to whether: experienced shift-workers are better able to adapt, and therefore more able to schedule activities such as physical activity into their time; 'hardier' shift-workers are better able to cope with shift-work, due to their commitment to their work schedule (for whatever reason), and, therefore, more willing/able to schedule physical activity within their day; or experienced shift-workers exhibit both 'hardiness' and 'adaptive' coping strategies, so allowing for participation leisure activities and the fulfilment of social and domestic tasks. Such theories and questions are relatively new and require further research but may be a useful tool in understanding the differences between those who can and cannot tolerate or cope with shift schedules (Wedderburn and Scholarios, 1993; Harma, 1996; Nachreiner, 1998; Maddi *et al.*, 2006).

Females were found to be less active than males. Gender has been found to influence the ability to cope with the pressures of shift-work, with women showing more signs and symptoms of intolerance than men, until the age of about 50 (Oginska *et al.*, 1993; Spelten *et al.*, 1993; Nachreiner, 1998; Winwood *et al.*, 2006). The divergence in coping with shift schedules can be attributed mainly to social rather than biological factors. The roles and activities which female workers engage in tend to differ from those of their male counterparts. Female shift-workers generally engage more in domestic and household obligations.

Furthermore, those with children or dependents experience an increase in such obligations and often feel as though they work a 'double-shift' (working their shift at work and coming home to fulfil domestic duties). Female shift-workers often cite 'a lack of time' or their 'shift system' as a factor affecting their participation in leisure time physical activity. There could also be gender differences in the quality of leisure time activity as well as the quantity. Males are less likely to combine their leisure time activities with other activities such as housework or childcare, or time spent with family and friends. All of these variables are affected by working shift schedules, and so this increases the likelihood of female workers having to combine their 'spare/leisure' time with domestic and social obligations. Clearly, this influences female shift-workers' participation in leisure time physical activity (Bird and Fremont, 1991; Nomaguchi and Bianchi, 2004; Baker *et al.*, 2004, Demerouti *et al.*, 2004; Lipovcan *et al.*, 2004; Karlsen *et al.*, 2006). The combination of effects of gender and experience clearly highlights some key areas which require further study, particularly with regard to why more experienced male shift-workers appear to be able to expend more energy in leisure time physical activity.

In support of the above points, male shift-workers with higher levels of leisure time physical activity 'disengaged' more from their domestic-related problems, but less from their sleep-related problems, in comparison with their female counterparts. Coping mechanisms are, at best, complex with no definitive 'right' or 'wrong' strategy as to which shift-workers should employ. There may well be more advantageous general methods of coping, ones that are often cited as more favourable than others for those working shifts. However, the effects of shift schedules on workers are individual in nature; the coping strategy employed must be individually chosen by the person concerned, therefore. Moreover, the transient nature of coping strategies must be appreciated, i.e. as people grow and evolve, or as the situation changes, the coping strategy also will evolve to meet the varying needs of the individuals and their environment (Lazarus, 1993). Many perceive disengagement or avoidance as a negative form of coping with regard to shift-work, with some studies suggesting that increased use of disengagement strategies by the shift-worker actually increased the number of problems experienced (Spelten *et al.*, 1993). However, the results of the present study, in conjunction with previous research, have highlighted the need to employ a mixed copying strategy. This will allow an individual to construct a method that matched the changing difficulties of shift-work and the shift-worker. This study suggests that males who expend more energy through leisure-time physical activity tend to disengage more from domestic problems. It appears plausible to suggest that physical activity is used by male shift-workers as a strategy to disengage from domestic issues or problems.

Avoidance/disengagement has often been cited as a more favourable coping strategy than approach/engagement in situations that cannot be controlled by the individual (Roth and Cohen, 1986; Lazarus, 1993; Spelten *et al.*, 1999; Karlsen *et al.*, 2006; Winwood *et al.*, 2006). For example, a shift-worker's ability to control for domestic problems is somewhat hampered by the shift system and the 'non-diurnal' lifestyle. Since the shift system is to a certain extent 'out' of their control, it is reasonable to assume that approach/engagement strategies (trying to change or solve the problem) are not feasible (unless a worker chooses to change his job). Therefore, avoidance/disengagement strategies (avoiding/distancing from/escaping from the problems) may be the only workable option. Clearly, it is possible that male shift-workers in this study used leisure time physical activity as a tool to distance themselves from domestic issues that they were unable to influence due to the shift-work. Conversely, the same male shift-workers disengaged far less from their sleep-problems. Sleep and sleep hygiene may be issues that can be controlled more by the individual. Whilst there is little a shift-worker can do to prevent sleep at undesirable times of the day (as this is dictated by the shift roster), there are certain activities and routines that can aid sleep. However, the male shift-workers appeared to use instead a more engaging strategy via the medium of leisure time physical activity. Whilst such a statement is highly presumptuous, it is reasonable after all that those male shift-workers utilized other disengaging strategies and mechanisms or indeed used a combination of engaging and disengaging strategies in an attempt to deal with issues surrounding sleep. Nonetheless, physical activity has been linked with increased sleep quality, but the links between physical activity, sleep and shift-work are unclear, and further investigation is needed. It is also plausible that male shift-workers may have been more aware of the links between physical activity and sleep in general, and therefore more inclined to participate in greater levels of leisure time physical activity in comparison to females.

The question remains as to whether being more physically active allows a shift-worker to cope better with shift patterns or whether coping better with shift patterns allows more time for a worker to participate in physical activity. It is feasible that leisure-time physical activity, whether it be used as an engaging or disengaging strategy, is a tool that can be utilized to increase a shift-workers' tolerance and therefore their ability to cope with unusual working hours (Roth and Cohen, 1986; Lazarus, 1993; Spelten *et al.*, 1993; Spelten *et al.*, 1999; Karlsen *et al.*, 2006; Winwood *et al.*, 2006). This view is in agreement with previous research conducted by Winwood *et al.*, (2007) and Eriksen and Bruusgaard (2003) who reported that individuals with more active leisure pursuits reported significantly better sleep, recovery between periods of work and less likely to develop persistent fatigue. These authors attributed their findings to the down-regulation of stress-induced brain arousal, and stimulation of the pleasure-reward brain neurophysiology associated with physical leisure time activities (Eriksen and Bruusgaard, 2003; Winwood *et al.*, 2007). To allow for a thorough exploration of coping mechanism, shift-work and physical activity it would appear a new more extensive measurement tool needs to be designed. We recommend that physical activity interventions for shift-workers should be designed with careful consideration of individual domestic responsibilities and perceived disruption to sleep.

References

Baker, A., Roach, G., Ferguson, S, Dawson, D., 2004, Shift-work experience and the value of time. *Ergonomics,* **47**, pp. 307-317.

Barton, J., Folkard, S., Smith, L.R., Selton E.R., and Totterdell, P.A., 1990, *Standard Shift-work Index Manual*, SAPU Memo No: 1159 MRC/ESRC
Social and Applied Psychology Unit, Department of Psychology, University of Sheffield, Sheffield, S10 2TN.

Beermann, B. and Nachreiner, F., 1995, Working shifts – different effects for women and men? *Work and Stress,* **9**, pp. 289-297.

Bird, C.E. and Fremont, A.M., 2006, Gender, time use, and health. *Journal of Health and Social Behaviour,* **32**, pp.114-129. CTA

Bohle, P. and Tiley, A., 1989, The impact of night work on psychological well-being. *Ergonomics,* **32**, pp.1089-1099.

Bonnefond A, Harma M, Hakola T, Sallinen M, Kandolin I, Virkkala J, 2006, Interaction of age with shift-related sleep-wakefulness, sleepiness, performance, and social life. *Experimental Aging Research,* **32**, pp. 185-208.

Carver, C.S, Scheier, M.F., and Weintraub, J.K., 1989, Assessing coping strategies: A theoretically based approach. *Journal of Personality and Social Psychology,* **56**, pp. 267-283.CTA

Ceslowitz, S., 1989, Burnout and coping strategies among hospital staff nurses. *Journal of Advanced Nursing,* **14**, pp. 553-558.

Chang, E.M., Daly. J., Hancock, K.M., Bidewell, J.W., Johnson, A., Lambert, V.A., and Lambert, C.E., 2006, The relationships among workplace stressors, coping methods, demographic characteristics, and health in Australian nurses. *Journal of Professional Nursing,* **22**, pp. 30-38.

Costa G., 2003, Shift-work and occupational medicine: an overview. *Occupational Medicine - Oxford,* **53**, pp. 83-88.

Costa G., 2004, Multidimensional aspects related to shift-workers health and well being. *Rev Saude Publica*, **38** (Supplement), pp. 86-91.

Demerouti, E., Geurts, S.A.E., Bakker, A.B, and Euwema, M., 2004, The impact of shift-work on work-home conflict, job attitudes and health. *Ergonomics*, **47**, pp. 987-1002.

Diaz - Morales, J. and Sanchez-Lopez, M., 2004, Composite and preference scales of mornings and reliability and factor invariance in an adult university sample. *Spanish Journal of Psychology*, **7**, pp. 93-100.

Di Lorenzo, L., G. De Pergola, C. Zocchetti, N. L'Abbate, N. Pannacciulli, M. Cignarelli, R. Giorgino and L. Soleo, 2003, Effect of shift-work on body mass index: results of a study performed in 319 glucose-tolerant men working in a Southern Italian industry. *International Journal of Obesity*, **27**, pp. 1353-1358.

Eriksen, W., and Bruusgaard, D., 2003, Do physical leisure time activities prevent fatigue? A 15 month prospective study of nurses' aides. *British Journal of Sports Medicine*, **38**, pp. 331-336.

Furnham, A., and Hughes, K., 1999, Individual difference correlates of night work and shift-work rotation. *Personality and Individual Differences*, **26**, pp. 941-959.

Härmä, M., 1996, Ageing, physical fitness and shift-work tolerance. *Applied Ergonomics*, **27**, pp. 25-29.

Harma, M. I., J. Ilmarinen, et al., 1988, Physical-training intervention in female shift-workers 2. The effects of intervention on the circadian-rhythms of alertness, short-term-memory, and body-temperature. *Ergonomics*, **31**, pp. 51-63. CTA

Harma, M. I., llmarinen, J., Knauth, P., Rutenfranz, J., and Hanninen, P., 1988, Physical training intervention in shift-workers. 1. The effects of intervention on fitness, fatigue, sleep, and psychomotor symptoms. *Ergonomics*, **31**, pp. 39-50.

Harrington JM., 2001, Health effects of shift-work and extended hours of work. *Occupational and Environmental Medicine*, **58**, pp. 68-72.

Herbert, A., 1983, The influence of shift-work on leisure activities. A study with repeated measurement. *Ergonomics*, **26**, pp. 565-574.

Hornberger, S. and Knauth P., 1993, Interindividual differences in the subjective valuation of leisure time utility. *Ergonomics*, **36**, pp. 255-264.

Karlsen, E., Dybdahl, R., and Vitterso, J., 2006, The possible benefits of difficulty: How stress can increase and decrease subjective well-being. *Scandinavian Journal of Psychology*, **47**, pp. 411-417.

Karlsson, B., A. Knutsson and Lindahl, B., 2001, Is there an association between shift-work and having a metabolic syndrome? Results from a population based study of 27,485 people. *Journal of Occupation and Environmental Medicine*, **58**, pp. 747-752.

Klag, S. and Bradley, G., 2004, The role of hardiness in stress and illness: An exploration of the negative affectivity and gender. *British Journal of Health Psychology*, **9**, pp. 137-161.

Knutsson, A., 2003, Health disorders of shift-workers. *Occupational Medicine*, **53**, pp. 103-108.

Knutsson, A., 2004, Methodological aspects of shift-work research. *Chronobiology*

International, **21**, pp. 1037-1047.

Kundi, M., Koller, M., Cervinka, R. and Haider, M., 1979, Consequences of shiftwork as a function of age and years on shift. *Chronobiologia*, **6**, pp. 123.

Lamb, K.L. and Brodie, D.A., 1990, The assessment of physical activity by leisure-time physical activity questionnaires. *Sports Medicine,* **10**, pp. 159-180.

Lasfargues, G., Vol, S., Caces, E., LeClesiau, H., Lecomte, P., Tichet, J., 1996, Relations among night work, dietary habits, biological measures, and health status. *International Journal of Behavioural Medicine,* **3**, pp. 123-134.

Lazarus, R.S., 1993, Coping theory & research: past, present, and future. *Psychosomatic Medicine,* **55**, pp. 234-247.

Lipovcan, K., Larsen, P., Zganec., 2004, Quality of life, life satisfaction and happiness in shift and non-shift-workers. *Revista De Saude Publica*, **38**, pp. 3-10.

McEwen, B.S., 2007, Sleep deprivation as a neurobiologic and physiologic stressor: allostasis and allostatic load. *Metabolism Clinical and Experimental,* **55**(suppl 2), S20-S23.

Maddi, S.R., Harvey, R.H., Khoshaba, D.M., Lu, J.L., Persico, M., and Brow, M., 2006, The personality construct of hardiness, III: Relationships with repression, innovativeness, authoritarianism, and performance. *Journal of Personality,* **72**, pp. 575-598.

Monk, T.H. and Folkard, S., 1992, *Making Shift-Work Tolerable,* (Basingstoke: Taylor and Francis).

Nachreiner, F., 1998, Individual and social determinants of shift-work tolerance. *Scandinavian Journal of Work and Environmental Health,* **24**, pp. 35-42.

Nomaguchi, K.M, and Bianchi, S.M., 2004, Exercise time: Gender differences in the effects of marriage, parenthood, and employment. *Journal of Marriage and Family,* **66**, pp.413.

Oginska, H., Pokorski, J., Oginski, A., 1993, Gender, aging and shift-work intolerance. *Ergonomics,* **36**, pp.161-168.

Pati, A.K., Chandrawanshi, A., and Reinberg, A., 2001, Shift-work: consequences and management. *Current Science*, **81**, pp. 32-52.

Presser, H.B., 2000, Nonstandard work schedules and marital instability. *Journal of Marriage and Family,* **62, pp.** 93-110.

Rajaratnam S.M.W and Arendt, J., 2001, Health in a 24-hr society. *Lancet,* **358**, pp. 999-1005.

Roth, S. and Cohen, L.J., 1986, Approach, avoidance, and coping with stress. *American Psychologist,* **41**, pp. 813-819.

Rouch, I., Wild, P., Ansiau, D and Marquie, J.C., 2005, Shift-work experience, age and cognitive performance. *Ergonomics,* **48**, pp. 1282-1293.

Rutenfranz, J., Haider, M. and Koller, M., 1985, Occupational health measures for night workers and shiftworkers. In *Hours of Work: Temporal Factors in Work Scheduling*, edited by Folkard, S. and Monk, T., (New York: Wiley).

Seo, Y.J., Matsumoto, K., Park, Y.M., Shinkoda, H., and Noh, T.J., 2000, The relationship between sleep and shift system, age and chronotype in shift-workers. *Biological Rhythm Research,* **31**, pp. 559-579.

Smith, C.S., Folkard, S., Schmieder, A., Parra, L.F., Spelten, E., Almiral, Helena,

Sen, R.N., Sahu, S., Perez, L.M and Tisak, J., 2002, Investigation of morning-evening orientation in six countries using the preferences scale. *Personality and Individual Differences,* **32,** pp. 949-968.

Soderstorm, M., Dolbier, C., Leiferman, J., and Steinhardt, M., 2000, The relationship of hardiness, coping strategies, and perceived stress to symptoms of illness. *Journal of Behavioural Medicine,* **23**, pp. 311-327.

Spelten, E., Totterdell, P., Costa, G., 1999, A process model of shiftwok and health. *Journal of Occupational Health Psychology,* **4,** pp. 207-218.

Spelten, E; Smith, L; Totterdell, P; Barton, J; Folkard, S, and Bohle, P., 1993, The relationship between coping strategies and GH scores in nurses. *Ergonomics,* **36**, pp. 227-232.

Takahashi, M., Tanigawa, T., Tachibana, N., Mutou, K., Kage, Y., Smith, L., Iso, H., 2005, Modifying effects of perceived adaptation to shift work on health, wellbeing, and alertness on the job among nuclear power plant operators. *Industrial Health,* **43**, pp. 171-178.

Tamres, L.K., Janicki, D., and Helgeson, V.S., 2002, Sex differences in coping behaviour: A meta-analytic review and an examination of relative coping. *Personality and Social Psychology Review,* **6**, pp. 2-30.

Taylor, E., Briner, R.B. and Folkard, S., 1997, Models of shift-work and health: An examination of the influence of stress on shift-work theory. *Human Factors,* **39**, pp. 67-82.

Tobin, D.L., Holroyd, K.A., Reynolds, R.V. and Wigal, J.K., 1989, The hierarchical factor structure of the coping strategies inventory. *Cognitive Therapy and Research,* **13**, pp. 343-361.

Waterhouse J, Folkard S. and Minors D., 1992, Shift-work, Health and Safety. An overview of the scientific literature 1978-1990. *HSE contract research report.* HMSO: London.

Wedderburn A. and Scholarios, D., 1993, Guidelines for shift-workers: trials and errors? *Ergonomics,* **36**, pp. 211-218.

Wijndaele, K., Matton, L., Duvigneaud, N., Lefevre, J., Bourdeaudhuij, I.D., Duquet, W., Thomis, M. and Philippaerts, R.M., 2007, Association between leisure time physical activity and stress, social support and coping: A cluster-analytical approach. *Psychology of Sport and Exercise,* **8,** pp. 425-440.

Winwood, P.C., Winefield, A.H., Lushington, K., 2006, Work-related fatigue and recovery: the contribution of age, domestic responsibilities and shift-work. *Journal of Advanced Nursing,* **56**, pp. 438-449.

Winwood, C.P., Bakker, A.B., and Winefield, A.H., 2007, An investigation of the role of non-work-time behavior in buffering the effects of work strain. *Journal of Occupational and Environmental Medicine,* **49**, pp. 862-871.

Part III
Sports

CHAPTER ELEVEN

Effects of blinded differences in ambient conditions on performance and thermoregulatory responses during a 4-km cycling time trial

G. Atkinson[1], C. Jackson[1], B. Drust[1], W. Gregson[1] and A. St Clair Gibson[2]

[1]Research Institute for Sport and Exercise Sciences, Liverpool John Moores University, Liverpool, UK
[2]School of Psychology and Sport Sciences, Northumbria University, UK

1. INTRODUCTION

Many applied physiologists have investigated the effects of high ambient temperatures on exercise performance. During prolonged sub-maximal exercise, it is known that time to exhaustion is significantly reduced in hot compared to cooler conditions (Galloway and Maughan, 1997; Parkin et al., 1999; Arngrimsson et al., 2004). The termination of sub-maximal exercise in hot conditions has been reported to coincide with a core body temperature of ~40°C in trained participants (Nielsen et al., 1993; Nielsen et al., 1997; Gonzalez-Alonso et al., 1999). This relationship between exercise capacity in the heat and a 'critical' core body temperature seems quite robust, since time to exhaustion has been found to decrease when pre-exercise core temperature is increased by 0.1-0.2°C (Gonzalez et al., 2000; Gregson et al., 2002) and vice versa (Lee and Haymes, 1995; Booth et al., 1997).

Although offering an insight into human responses to exercise at high ambient temperatures, the above studies, involving time to exhaustion protocols, are more relevant to exercise 'capacity' than exercise performance in a 'real world' context. In most athletic competitions and occupational tasks, individuals are aware of an end point to the physical activity, i.e. they can estimate how long they are likely to exercise for and are, consequently, able to self-select a work-rate that is deemed suitable for the predicted duration of exercise (Atkinson et al., 2007). Moreover, this work-rate selection is a dynamic process; work-rate might be altered during exercise according to changes in perception of effort.

Much research effort (reviewed recently by Atkinson et al., 2007) has been devoted to ascertaining how the work-rate selection (pacing strategy) process is controlled. Pacing during exercise is thought to be regulated in a complex anticipatory system which monitors afferent feedback from various physiological systems, and then regulates the work-rate so that potentially limiting changes do not occur before the end-point of exercise is reached (St Clair Gibson and Noakes, 2004). The pacing strategy 'algorithm', sited in the brain, is thought to have afferent input from interoceptors such as heart rate and respiratory rate, as well as exteroceptors providing information on local environmental conditions. Knowledge of time, modulated by the cerebellum, basal ganglia and primary somatosensory cortex, is also thought to provide input to the pacing algorithm as would information stored in memory about similar exercise bouts completed in the past (Lambert *et al.*, 2005; Noakes *et al.*, 2005). Although this model of pacing is consistent with classical homeostatic control theories, how all this information is assimilated by the different regions of the brain is not known at present.

Pacing strategies and associated mechanisms can be studied using a time-trial protocol. The results of studies which have involved time trials suggest that overall performance (i.e. total time or average power output) is, like time to exhaustion, reduced at high ambient temperatures (Tatterson *et al.*, 2000; Marino *et al.*, 2004; Tucker *et al.*, 2004). In the most recent studies (Slater *et al.*, 2005; Altareki *et al.*, 2006), moderate duration (6-10 min) exercise performance was found to be reduced at higher temperatures but core temperature was not found to have reached the critical level (~40°C) associated with the point of exhaustion. Therefore, in the context of a moderate duration time trial, it can be postulated that power output is reduced during exercise in the heat in an anticipatory manner in relation to rate of heat production and storage in order to allow the exercise bout to be completed without causing T_C to reach a critical level and disturb thermal homeostasis (Tatterson *et al.*, 2000; Marino *et al.*, 2004; Tucker *et al.*, 2004; Slater *et al.*, 2005).

Bearing in mind that participants were not blinded to each ambient condition in all the previous relevant studies, our interest lies in whether the relationship between the thermoregulatory responses to moderate duration (6-10 min) exercise and pacing strategy is robust during self-selected exercise, or is it prone to human reactivity effects? Such influences would provide greater insight into the mechanisms of human fatigue during externally-valid bouts of exercise. Moreover, there have been recent calls for applied physiologists to consider generally the extent to which human reactivity might affect study conclusions about human performance (Grossman *et al.*, 2005). Therefore, we aimed to examine the relationship between thermoregulatory responses to exercise and human performance when participants are deceived about true ambient temperature. We hypothesised that 4-km cycling time trial performance is unaffected by a blinded five-degree change in ambient temperature, even though the magnitude of this change is sufficient to alter core and skin temperatures.

2. METHODS

The study comprised two phases. First, a pilot study was designed to examine the extent to which participants could subjectively detect, at rest, different ambient temperatures. We aimed to identify two ambient temperatures which would be predicted to elicit different body temperatures, according to past research (Lind,

1963), but not necessarily perceived to be different by blinded participants. Second, an experiment was designed to examine the thermoregulatory responses to exercise and human performance at the two identified ambient temperatures and which participants were led to believe were the same.

2.1 Pilot study

Seven males, with mean±SD age of 24±2 years, height of 1.74±0.04 m and body mass of 72±9 kg, provided informed written consent to participate in the pilot study. The study was approved by the University Ethics Committee. Participants visited the laboratory on three occasions. Without knowing true ambient temperature, participants entered an environmental chamber on three occasions and sat at rest for 15 minutes in temperatures of 13°, 30° and 35°C. These temperatures were selected according to the relationship between core temperature responses during exercise and environmental conditions decribed by Lind (1963). We predicted that the difference between 13 and 30-35 °C ambient temperatures would be detected by participants, but that the difference between 30 and 35 °C would not be recognised, even though these ambient temperatures should mediate different core temperatures, especially at high metabolic rates (Lind, 1963). Humidity was controlled at 50% during all three visits. Thermal discomfort scores (Toner *et al.*, 1986) were recorded 30 s after entry into the environmental chamber and then after 5, 10 and 15 minutes. The participants were exposed to the three temperatures in a single-blind counterbalanced sequence in order to control for sequential effects of learning (Keppel *et al.*, 1992).

The mean±SD thermal discomfort scores during the 15-min exposure to ambient temperatures of 13°, 30° and 35°C were 2.6±0.6, 5.3±0.7 and 5.8±0.7 units, respectively. Mean thermal discomfort scores were found to be significantly different between the 13°C and both the 30°C (*P=0.001*) and 35°C (*P<0.0005*) conditions. Nevertheless, thermal comfort scores were not significantly different between 30°C and 35°C temperatures (*P=0.145*). Based on these data, we predicted that participants were sensitive to relatively large differences in ambient temperature (13 vs 30 °C) but would not detect the difference between ambient temperatures of 30°C and 35°C. Therefore, these latter two ambient temperatures were selected in our main deception-based experiment.

2.2 Main Experiment

2.2.1 Participants

Following an estimation of required statistical power (see *Experimental design*), 8 males, with mean± SD age of 29±7 years, height of 1.80±0.05 cm, body mass of 74±9 kg and maximum power output of 296±23 W, provided informed written consent to participate in this study, which was approved by the University's ethics committee. All participants were physically active and were familiar with cycling exercise. Four of the participants in this main study were also involved in the prior pilot investigation.

2.2.2 Experimental design

Participants visited the laboratory on four occasions. During the first visit, the height and body mass of each participant were recorded, and all participants were familiarised with the equipment to be used during the study. Participants then performed an incremental cycle ergometer test (Padilla *et al.*, 1999; Mujika and Padilla, 2002). This test allowed for the determination of the lactate threshold and maximal power output (W_{max}), as well as the prediction of maximal oxygen uptake ($\dot{V}O_{2max}$).

During the second visit, participants performed a maximal self-paced 4-km time trial for familiarisation purposes. In our laboratory, the 4-km time trial has been found to show a standard error of measurement of 6.1 s and 8 W (coefficients of variation: 1.6% and 3.2%) for performance time and mean power output, respectively (Altareki *et al.*, 2006). These small coefficients of variation allowed us to estimate that differences between trials of 5% in power output and 2% in 4-km time would be detected with a sample size of 8 participants, with statistical power set at 80% (Atkinson and Nevill, 2006). The delimited smallest worthwhile difference in power output is similar to the 6.4% and 6.25% differences observed between temperature conditions studied by Tatterson *et al.* (2000) and Tucker *et al.* (2004) respectively.

On the two subsequent visits, participants performed a maximal self-paced 4-km cycle time trial at ambient temperatures of 30°C (TT_{30}) and 35°C (TT_{35}). The decision to use these temperatures was based on the results of the pilot study. Participants were informed, falsely, that they would be exercising at an ambient temperature of 32.5°C on both occasions.

2.2.3 Lactate threshold and maximal power test

A Computrainer Pro ergometer (Computrainer Pro, Racer Mate Inc., Seattle, USA) was used for the incremental cycle test and for all time trials. This ergometer allowed participants to use their own bicycles. The Computrainer was individually calibrated for each participant prior to testing. A discontinuous 4-min staged protocol was used as described by Padilla *et al.* (1999). We started each participant at 145 W instead of the recommended 110 W, since our participants were well-trained. Blood lactate was measured on the completion of each stage via a finger prick using the Lactate Pro (LT 1710, Kyoto, Daninchi, Kagaki, Japan). This device has been shown to be sensitive to changes during exercise and reliable for lactate analysis (McNaughton *et al.*, 2002). The lactate threshold was defined as the point at which blood lactate concentration increased in a non-linear fashion (Denis et al., 1982; Jones and Doust, 1998; Svedahl and MacIntosh, 2003). The protocol also allowed for the calculation of W_{max} and estimation of $\dot{V}O_{2max}$ using the following equations: -

$$W_{max} = W_f + ((t/240) \times 35) \qquad \text{(Kuipers et al., 1985)}$$

$$\dot{V}O_{2max} = (0.01141 \times W_{max}) + 0.435 \qquad \text{(Hawley and Noakes, 1992)}$$

Where W_f is the power (in W) maintained for the last complete stage and t is the time the last stage is maintained for (in s).

2.2.4 Performance time trials

Within 1 week of the incremental and familiarisation sessions, participants reported to the laboratory for the first of their two performance trials. The participants undertook the two trials in a counterbalanced order with four participants performing TT_{35} first and the remaining four performing TT_{30} first. All trials took place in an environmental chamber with relative humidity controlled at 50-55%. Participants were required to refrain from consuming alcohol and caffeine during the 24-h period prior to each trial and not to eat during the 3 h prior to a test.

Participants were issued with disposable ingestible temperature sensors (Cortemp, HQInc., USA), which they were instructed to take at least five hours prior to their arrival at the laboratory. The reliability of these sensors has been found to be acceptable in repeated trials and they have also been found to agree with measurements of rectal temperature (Gant et al., 2006), although the exercise-mediated increase in intestinal temperature might lag behind that of oesophageal temperature (Lee et al., 2000). Participants were also requested to consume 400-600 ml of water 2 hours before they were due to arrive at the laboratory to ensure adequate levels of hydration (Convertino et al., 1996). However, no further fluids were ingested prior to and during the test so that intestinal temperature recordings were not compromised (Wilkinson et al., 2008). Participants recorded their nutritional and fluid intake for the 24-hour period prior to their first time trial thus allowing them to replicate their habits for their second trial.

On arrival at the laboratory, nude body mass was recorded (Seca, Germany) and all equipment to record physiological data was attached. Participants then entered the environmental chamber and sat for 20 minutes to allow for the recording of resting values. After the Computrainer was calibrated participants completed a 10-min warm up at 40% of their W_{max}. Within 1 min of completing the warm up participants began the 4-km time trial. The only feedback given to the participants during time trials was distance completed at each kilometre. Participants were towel-dried on completion of the time trial and nude body mass was again recorded. The two performance trials were separated by 4-7 days to allow for sufficient recovery. Each participant's time trials were conducted at the same time of day to minimize the effects of circadian rhythms (Reilly and Brooks, 1986). Participants were also requested to maintain the same level of training throughout their involvement in the study and to refrain from heavy exercise the day before each trial.

2.2.5 Physiological measurements

Heart rate was recorded at 5-s intervals using a heart rate monitor (Polar, Finland). Core temperature (T_C) was measured via ingested temperature sensors (Cortemp, HQInc.) and recorded using a portable data logger attached to the participant's waist (Cortemp, 2000, HQInc., Palmetto, Florida, USA). Skin temperature was measured using a Squirrel data logger 1250 (Grant Instruments, Cambridge, UK) via skin thermistors attached to the chest (T_{chest}), forearm ($T_{forearm}$), thigh (T_{thigh}) and calf (T_{calf}). All thermistors were attached at the same relative anatomical location in both trials. Mean skin temperature (T_{SK}) was calculated using the following equation: -

$$T_{SK} = T_{chest} \times 0.3 + T_{forearm} \times 0.3 + T_{thigh} \times 0.2 + T_{calf} \times 0.2 \quad \text{(Ramanathan, 1964)}.$$

2.2.6 Subjective measurements

Ratings of perceived exertion (RPE), measured using the Borg scale (Borg, 1998), and thermal discomfort (TCS), measured using the Toner scale (Toner *et al.*, 1986), were recorded on the completion of each kilometre during the time trials. The participants' TCS were also recorded at rest and after the warm up.

2.2.7 Time trial measurements

Time, speed and power output were recorded every second by the Computrainer throughout each trial. Averages for the complete trial and for each kilometre were calculated from these data.

2.2.8 Data analysis

Data analysis was carried out using the Statistical Package for Social Sciences (SPSS). A two-factor general linear model was used to compare power output, speed, split times, T_C, T_{SK}, HR, RPE and TCS over the time trial distance and between the two heat conditions. Statistically significant interactions between condition and time-trial distance were followed-up with Bonferroni-corrected paired t-tests. The assumption of sphericity was explored and controlled for according to the methods outlined by Atkinson (2001). Paired sample t-tests and associated 90% confidence intervals were used to examine the differences in performance time and body mass loss between the two conditions. The level of statistical significance was set at $P<0.05$. All values presented are mean±SD.

3. RESULTS

3.1 Incremental cycle ergometer test

The mean±SD maximal power output (W_{max}) in the incremental lactate threshold test was 296±23 W. The mean lactate threshold (Denis *et al.*, 1982; Jones and Doust, 1998; Svedahl and MacIntosh, 2003) occurred at 69.3±9.7% of this W_{max}. Predicted mean $\dot{V}O_{2max}$ was 52.4±5.7 ml·kg⁻¹·min⁻¹.

3.2 Time trial performance

Power output ($F_{1,7}$ = 0.92, P=0.37) and cycling speed ($F_{1,7}$ = 1.25, P=0.30), averaged over the 4-km time trial, were not significantly different between TT_{30} and TT_{35} (Table 1). Accordingly, the time taken to complete the time trial was also not significantly different between TT_{30} and TT_{35} (t_7 = 1.13, P=0.30). Nevertheless, a significant interaction between condition and distance was found for power output ($F_{3,21}$ = 3.15, P=0.047). Mean power output was 14.0 W lower in TT_{35} compared to TT_{30} during the first kilometre, but only 0.1, 3.5, 1.3 W lower during the remaining three kilometres (Figure 1). The interaction between condition and distance for speed ($F_{3,21}$ = 2.54, P=0.08) and 1-km split times ($F_{3,21}$ = 2.84, P=0.06) approached significance (Figure 1).

Table 1. Time, average power output and average speed during maximal self-paced 4-km cycling time trials in 30°C (TT_{30}) and 35°C (TT_{35}) conditions. Data are means±SD. No significant differences in all three outcome variables were found between the two ambient temperatures (P>0.25).

	30°C	*35°C*
Time (Seconds)	376.9 ± 17.8	380.8 ± 22.2
Average power (W)	301 ± 38	296 ± 44
Average speed (kph)	38.3 ± 1.8	37.9 ± 2.2

3.3 Thermoregulatory responses

Core temperature increased significantly during both time trials ($F_{1.3,9.2}$ = 38.1, P<0.0005). Nevertheless, T_C was significantly higher throughout TT_{35} compared to TT_{30} ($F_{1,7}$ = 9.39, P=0.018). Mean T_C was 37.5±0.2°C compared to 37.3±0.2°C during TT_{35} and TT_{30} respectively. No significant interaction between condition and time was found ($F_{1.5,10.8}$ = 0.11, P=0.84, Figure 2). Accordingly, the rates of rise in T_C during the time trials were similar; 0.04±0.02°C·min⁻¹ in TT_{30} and

$0.05\pm0.02°C\ min^{-1}$ in TT_{35} (Figure 2). Mean skin temperature was also significantly higher throughout TT_{35} compared to TT_{30} ($F_{1,7}= 219.9$, P<0.0005). Mean T_{SK} was $35.3 \pm 0.4°C$ compared to $33.8 \pm 0.5°C$ during TT_{35} and TT_{30} respectively. A significant interaction between time and condition was also found for this variable($F_{2.5,17.6}= 3.25$, P=0.05); T_{SK} increased at a rate of $0.0086\pm0.02°C\ min^{-1}$ in TT_{35} compared to a smaller rate of $0.0003\pm0.01°C\ min^{-1}$ in TT_{30} (Figure 2).

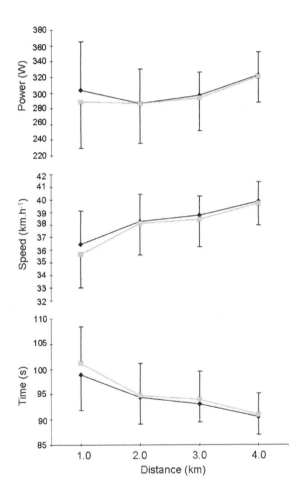

Figure 1. (A) Power output, (B) speed and (C) split time during maximal self-paced 4-km cycling time trials under 30°C (TT30) and 35°C (TT35) conditions. A significant interaction was observed between condition and distance for power output (p=0.047).

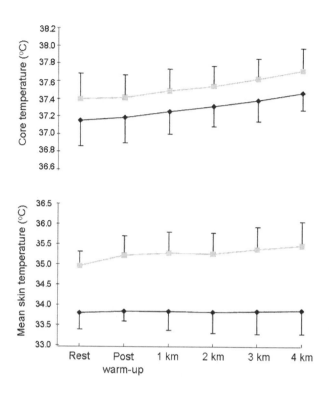

Figure 2. (A) Core temperature (TC) and (B) mean skin temperature (TSK) during maximal self-paced 4-km cycling time trials under 30°C (TT30) and 35°C (TT35) conditions. A significant interaction was observed between condition and time for TSK (p=0.05).

3.4 Subjective responses

Ratings of perceived exertion increased significantly during TT_{30} and TT_{35} ($F_{1.4,9.5}$ = 62.8, P<0.0005), but were not significantly different between trials ($F_{1,7}$ = 0.67, P=0.44). Mean RPE was 16.9±1.4 and 16.7±1.9 in TT_{30} and TT_{35} respectively. There was no significant interaction between condition and time for RPE ($F_{2.2,15.1}$ = 0.16, P=0.87). Thermal discomfort increased significantly throughout both time trials ($F_{2.2,15.6}$ = 63.36, P<0.0005), but was significantly higher throughout TT_{35} ($F_{1,7}$ = 6.90, P=0.034). Mean thermal discomfort was 6.0±0.4 compared to 5.5±0.5 units during TT_{35} and TT_{30} respectively. There was no significant interaction between condition and time ($F_{2.6,18}$ = 0.66, P=0.56) (Figure 3).

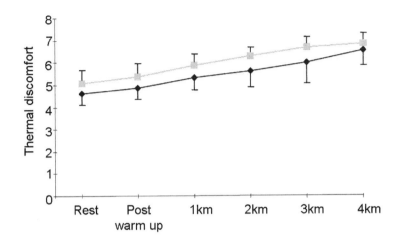

Figure 3. Rating of perceived exertion (RPE) during maximal self-paced 4-km cycling time trials under 30°C (TT30) and 35°C (TT35) conditions.

3.5 Heart rate and body mass

Heart rate increased significantly throughout TT_{30} and TT_{35} ($F_{2.2,15.4} = 605.3$, $P<0.0005$), but was not significantly different between conditions ($F_{1,7} = 0.54$, $P=0.49$). Heart rate was 139 ± 3 and 140 ± 5 averaged over TT_{30} and TT_{35} respectively. There was no significant interaction between condition and time ($F_{2,13.8} = 0.12$, $P=0.89$). The change in body mass was not statistically significantly different between conditions ($t_7 = 2.22$, $P=0.06$), although body mass loss was 0.34 ± 0.14 kg compared to 0.54 ± 0.20 kg in TT_{30} and TT_{35} respectively.

4. DISCUSSION

To the authors' knowledge, the present study is unique in being designed to deceive human participants about ambient conditions during self-selected exercise. Some previous researchers have deceived participants about the true distance or duration of the exercise bout, which is thought to be an important factor that influences the selection of a sustainable power output (Nikopoulos et al., 2001; Ansley et al., 2004). Our focus on ambient conditions was based on the reported importance humans place on this variable when selecting a pacing strategy, even during moderate- rather than long-duration exercise (Altereki et al., 2006). The focus on an externally-valid moderate-duration time-trial protocol has resulted in new knowledge about how humans process information relevant to pacing strategy. It is apparent that humans can 'detect' higher thermal discomfort when blinded ambient, core and skin temperatures are 5°C, 0.2°C and 1.5°C higher, respectively, throughout moderate-duration exercise. Nevertheless, a concomitant reduction in performance occurs only transiently during the first few minutes of exercise before the highest body temperatures, thermal discomfort and perceived exertion have been attained.

The finding that power output was initially lower in the hotter condition is comparable to the findings of Marino *et al.* (2004) who observed the fact that runners ran more slowly right from the start of a self-paced 8-km run in 35°C compared to 15°C. They concluded that the reduction in pace, hence the rate of heat production, was a result of an anticipatory mechanism to allow the completion of the exercise bout before a critical limiting temperature was reached. In contrast to their study, where running speed was slower in the hotter condition at all time points throughout the 8-km trial, power output in the present study, having been lower in TT_{35} during the first kilometre, became almost identical to that in TT_{30} during the remaining kilometres. The findings of the present study also contrast with those of Tatterson *et al.* (2000), who found power output only became significantly lower in the hotter condition during the last 20 minutes of his 30-minute cycle time trial, and those of Tucker *et al.* (2004) who found that although power output was lower in the hot condition compared to the cool condition after 30% of his 20-km time trial, it only became significantly lower during the last 20% of the time trial. The exact reasons for these conflicting findings are difficult to pinpoint. Obviously, exercise duration differed between these past studies as well as the present study. Nevertheless, because the present study is the only one of these to involve blinded ambient conditions, it would be interesting to repeat previous longer-duration studies under blinded conditions of ambient temperature.

In order to explain the initially lower power output seen in TT_{35}, it is also relevant to consider the interaction of physiological and subjective responses in our blinded conditions. It is speculated that the higher T_{SK} in TT_{35} after the warm up period (35.2 ± 0.3°C in TT_{35} versus 33.8 ± 0.5°C in TT_{30}) provided afferent input to pace-setting mechanisms in the brain and so led participants to believe there would be a reduced capacity for heat diffusion due to the higher environmental temperature. These differences in skin temperature also appear to have been translated into higher subjective feelings of heat as is evidenced by the significantly higher thermal discomfort scores in TT_{35}. As a result, it is proposed that these cues led participants to set an initially lower work rate in an anticipatory manner in order to ensure the bout of exercise was completed without causing an excessive level of heat storage and ultimately a disturbance in thermal homeostasis.

After being initially lower, power output in TT_{35} became almost identical to that seen in TT_{30} for the remainder of the time trial, rather than dropping off and mirroring the 'u' shaped profile seen in TT_{30}, as would be anticipated of a power-output profile in a cycling time trial (Atkinson and Brunskill, 2000; Nikolopoulos *et al.*, 2001; Ansley *et al.*, 2004; Tucker *et al.*, 2004; Altareki *et al.*, 2006). It can be postulated that, after a few minutes of the exercise, participants have responded to afferent feedback, which indicated that the rate of heat storage was below the originally anticipated level and still within acceptable limits (considering the anticipated duration of the exercise bout) and facilitated an increase in power output. Nevertheless, such a notion is difficult to confirm without further experimental control and/or measurement of other variables. For example, an experiment could be designed in which participants are deceived with respect to both the distance of the time trial and the ambient conditions. Electroencephalographic measurements of changes in regional brain activity might also offer some insight into how pace is initially set and/or adapted during an exercise period.

One limitation of the present study is that the attempted blinding may have been affected by the fact that four participants who had taken part in the pilot study were also included in the main experiment. This was due to problems encountered with recruiting suitable participants for the performance study. These participants were therefore aware of the fact, that the study may involve two temperatures, because they had been informed on completion of the pilot study that they had been subjected to two different hot temperatures. Therefore, although they were not informed of the temperatures, and the fact there were two different temperatures during the performance trial, they may have subconsciously been anticipating a difference between the conditions.

In conclusion, it appears that individuals can 'detect' higher thermal discomfort when blinded ambient and core temperatures are 5°C and 0.2°C higher, respectively, throughout moderate-duration exercise. Nevertheless, a concomitant reduction in performance occurred only during the first few minutes of the 4-km time trial. This subjective sensitivity to a difference in ambient temperature seemed to be lower when participants were at rest and not expecting exercise at all.

References

Altareki, N.A., Drust, B., Cable, N.T., Atkinson, G., Gregson, W.A. and George, K., 2006, The effects of environmental heat stress (35°C) on simulated 4 km cycling time-trial performance. *Medicine and Science in Sports and Exercise*, **38**, S354.

Ansley, L., Schabort, E., St Clair Gibson, A., Lambert, M.I. and Noakes, T.D., 2004, Regulation of pacing strategies during successive 4-km time trials. *Medicine and Science in Sports and Exercise*, **36**, pp. 1819-1825.

Arngrimsson, S.A., Petitt, D.A., Borrani, F., Skinner, K.A. and Cureton, K.J., 2004, Hyperthermia and maximal oxygen uptake in men and women. *European Journal of Applied Physiology*, **92**, pp. 524-532.

Atkinson, G., 2001, Analysis of repeated measurements in physical therapy research. *Physical Therapy In Sport*, **2**, pp. 194-208.

Atkinson, G. and Brunskill, A., 2000, Pacing strategies during a cycling time trial with simulated headwinds and tailwinds. *Ergonomics*, **43**, pp. 1449-1460.

Atkinson, G. and Nevill, A.M., 2006, Method agreement and measurements error in exercise physiology. In: *Physiology Testing Standards: The British Association of Sports and Exercise Sciences Guide, Volume 1, Sports Testing*, edited by Winter, E., Jones, A.M., Davison, R.C.R., Bromley, P. and Mercer, T.

Atkinson, G., Peacock, O. and Passfield, L., 2007, Variable versus constant power strategies during cycling time-trials: prediction of time savings using an up-to-date mathematical model. *Journal of Sports Sciences*, **25**, pp. 1001-1019.

Booth, J., Marino, F. and Ward, J.J., 1997, Improved running performance in hot humid conditions following whole body pre-cooling. *Medicine and Science in Sports and Exercise*, **29**, pp. 943-949.

Borg, G.A.V., 1998, *Borg's Perceived Exertion and Pain Scales,* (Champaign, IL: Human Kinetics).

Convertino, V.A., Armstrong, L.E., Coyle, E.F., Mack, G.W., Sawka, M.N., Senay, L.C. Jr. and Sherman, W.M., 1996, American College of Sports Medicine position stand. Exercise and fluid replacement. *Medicine and Science in Sports and Exercise*, **28**, i-vii.

Denis, C., Fouquet, R., Poty, P., Geyssant, A. and Lacour, J.R., 1982, Effect of 40 weeks of endurance training on the anaerobic threshold. *International Journal of Sports Medicine*, **3**, pp. 208-214.

Galloway, S.D.R. and Maughan, R.J., 1997, Effects of ambient temperature on the capacity to perform prolonged cycle exercise in man. *Medicine and Science in Sports and Exercise*, **29**, pp. 1240-1249.

Gant, N., Atkinson, G. and Williams, C., 2006, The validity and reliability of intestinal temperature during intermittent running. *Medicine and Science in Sports and Exercise,* **38**, pp. 1926-1931.

Gonzalez-Alonso, J., Teller, C., Andersen, S.L., Jensen, F.B., Hyldig, T. and Nielsen, B., 1999, Influence of body temperature on the development of fatigue during prolonged exercise in the heat. *Journal of Applied Physiology*, **86**, pp. 1032-1039.

Gonzalez-Alonso, J., Mora-Rodriguez, R. and Coyle, E.F., 2000, Stroke volume during exercise: interaction of environment and hydration. *American Journal of Physiology, Heart and Circulatory Physiology*, **278**, pp. 321-330.

Gregson, W.A., Drust, B., Batterham, A. and Cable, N.T., 2002, The effects of pre-warming on the metabolic and thermoregulatory responses to prolonged submaximal exercise in moderate ambient temperatures. *European Journal of Applied Physiology*, **86**, pp. 526-533.

Grossman, P., Gibala, M.J., Burgomaster, K.A. and Heigenhauser, G.H.F., 2005, A comment on Burgomaster et al. and a general plea to consider behavioral influences in human physiology studies. *Journal of Applied Physiology*, **99**, pp. 2473-2475.

Hawley, J.A. and Noakes, T.D., 1992, Peak power output predicts maximal oxygen uptake and performance in trained cyclists. *European Journal of Applied Physiology*, **65**, pp. 79-83.

Jones, A.M. and Doust, J.H., 1998, The validity of the lactate minimum test for determination of the maximal lactate steady state. *Medicine and Science in Sports and Exercise*, **30**, pp. 1304-1313.

Keppel, G., Saufley, W.H. and Tokunaga, H., 1992, *Introduction to Design and Analysis. A Students Handbook,* (New York: W.H. Freeman and Company).

Kuipers, H., Verstappen, H.A., Keizer, H.A. and Geurten, P., 1985, Variability of aerobic performance in the laboratory and its physiological correlates.

International Journal of Sports Medicine, **6**, pp. 197-201

Lambert, E.V., St Clair Gibson, A. and Noakes, T.D., 2005, Complex systems model of fatigue: integrative homeostatic control of peripheral physiological systems during exercise in humans. *British Journal of Sports Medicine*, **39**, pp. 52-62.

Lee, D.T. and Haymes, E.M., 1995, Exercise duration and thermoregulatory responses after whole body pre-cooling. *Journal of Applied Physiology*, **79**, pp. 1971-1976.

Lee, S., Williams, W., and Schneider, S., 2000, Core temperature measurement during supine exercise: esophageal, rectal, and intestinal temperatures. *Aviation, Space and Environmental Medicine*, **71**, pp. 939-945.

Lind, A.R., 1963. A physiological criterion for setting thermal environmental limits for everyday work. *Journal of Applied Physiology*, **18**, pp. 51-56.

Marino, F. E., Lambert, M.I. and Noakes, T.D., 2004, Superior performance of African runners in warm humid but not cool environmental conditions. *Journal of Applied Physiology*, **96**, pp. 124-130.

McNaughton, L.R, Thompson, D., Phillips, G., Backx, K. and Crickmore, L., 2002, A comparison of the Lactate Pro, Accusport, Analox GM7 and Kodak Ektachem lactate analysers in normal, hot and humid conditions. *International Journal of Sports Medicine*, **23**, pp. 130-135.

Mujika, I. and Padilla, S., 2002, Event selection. In *High Performance Cycling*, edited by Jeukendrup, A.E., (Champaign, IL: Human Kinetics), pp. 79-90.

Nielsen, B., Hales, J.R., Strange, S., Christensen, N.J., Warberg, J. and Saltin, B., 1993, Human circulatory and thermoregulatory adaptations with heat acclimation and exercise in a hot, dry environment. *Journal of Physiology*, **460**, pp. 467-485.

Nielsen, B., Strange, S., Christensen, N, J., Warberg, J. and Saltin, B., 1997, Acute and adaptive responses in humans to exercise in a warm, humid environment. *Pflugers Archive*, **434**, pp. 49-56.

Nikolopoulos, V., Arkinstall, M.J. and Hawley, J.A., 2001, Pacing strategy in simulated cycle time trials is based on perceived rather than actual distance. *Journal of Science and Medicine in Sport*, **4**, pp. 212-219.

Noakes, T.D., St Clair Gibson, A. and Lambert, E.V., 2005, From catastrophe to complexity: a novel model of integrative central neural regulation of effort and fatigue during exercise in humans: summary and conclusions. *British Journal of Sports Medicine*, **39**, pp. 120-124.

Padilla, S., Mujika, I., Cuesta, G. and Goiriena, J.J., 1999, Level ground and uphill cycling ability in professional road cycling. *Medicine and Science in Sports and Exercise*, **31**, pp. 878-885.

Parkin, J.M., Carey, M.F., Zhao, S. and Febbraio, M.A., 1999, Effect of ambient temperature on human skeletal muscle metabolism during fatiguing submaximal exercise. *Journal of Applied Physiology*, **86**, pp. 902-908.

Ramanathan, N.L., 1964, A new weighting system for mean surface temperature of the body. *Journal of Applied Physiology*, **19**, pp. 531-533.

Reilly, T. and Brooks, G.A., 1986, Exercise and the circadian variation in body temperature measures. *International Journal of Sports Medicine*, **7**, pp. 358-

362.

Slater, G.J., Rice, A.J., Sharpe, K., Tanner, R., Jenkins, D., Gore, C.J. and Hahn, A.G., 2005, Impact of acute weight loss and/or thermal stress on rowing ergometer performance. *Medicine and Science in Sports and Exercise*, **37**, pp. 1387-1394.

St Clair Gibson, A. and Noakes, T.D., 2004, Evidence for complex system integration and dynamic neural regulation of skeletal muscle recruitment during exercise in humans. *British Journal of Sports Medicine*, **38**, pp. 797-806.

Svedahl, K. and MacIntosh, B.R., 2003, Anaerobic threshold: The concepts and methods of measurement. *Canadian Journal of Applied Physiology*, **28**, pp. 299-323.

Tatterson, A.J., Hahn, A.G., Martin, D.T. and Febbraio, M.A., 2000, Effects of heat stress on physiological responses and exercise performance in elite cyclists. *Journal of Science and Medicine in Sport*, **3**, pp. 186-193.

Toner, M.M., Drolet, L.L. and Pandolf, K.B., 1986, Perceptual and physiological responses during exercise in cool and cold water. *Perceptual and Motor Skills*, **62**, pp. 211-220.

Tucker, R., Rauch, L., Harley, Y.X.R. and Noakes, T.D., 2004, Impaired exercise performance in the heat is associated with an anticipatory reduction in skeletal muscle recruitment. *Pflugers Archive*, **448**, pp. 422-430.

Wilkinson, D.M., Carter, J.M., Richmond, V.L., Blacker, S.D. and Rayson, M.P., 2008, The effect of cool water ingestion on gastrointestinal pill temperature. *Medicine and Science in Sports and Exercise*, **40**, pp. 523-528.

CHAPTER TWELVE

Lower limb function in the maximal instep kick in soccer

Adrian Lees[1], Ian Steward[1], Nader Rahnama[1,2] and Gabor Barton[1]

[1]Research Institute for Sport and Exercise Sciences, Liverpool John Moores University, Liverpool, UK
[2]Faculty of Physical Education & Sport Sciences, Isfahan University, Isfahan, Iran

1. INTRODUCTION

The worldwide popularity of soccer has stimulated an extensive interest in the scientific analysis of the skills of the game. The soccer kick is the defining skill of soccer and although there are many variants of this skill, it is the maximal soccer instep kick of a stationary ball that has received most attention in the biomechanical literature (e.g. Lees and Nolan, 1998). The soccer kick is three dimensional (3D) in nature but it is only recently that studies involving 3D analysis have been undertaken. A kinematic description of the kicking leg for the maximal instep and side-foot kicks has been reported by Levanon and Dapena (1998) and a kinetic analysis by Nunome et al. (2002). This latter group also investigated the effects of foot-ball impact (Nunome et al., 2006a) and kicking foot preference on selected kinematics and kinetics of the instep kick (Nunome et al., 2006b). Kellis et al. (2004) investigated the influence of approach angle on the 3D kinematics and kinetics of the support leg.

A limitation of these studies is their selected focus. Most previous researchers have focused on the function of the kicking leg and a good deal is now known about the dynamics of this aspect. Ground reaction force data have been reported only by Kellis et al. (2004) who are also the only researchers to have described the 3D characteristics of the support leg. Kinematics of the pelvis has only been reported by Levanon and Dapena (1998). In particular, no research group has attempted to investigate the whole of the lower-body function during the kicking action.

While the studies referred to above have involved 3D analysis to compare conditions (instep vs side step kick; preferred vs non-preferred; approach angles), in none of these studies have researchers attempted to explain the observable features in the kicking action. Observable features in a kicking skill are the angled approach to the kick, an orientation of the body away from the ball at impact, a consistently flexed knee of the support leg during stance, cocking of the kicking leg, and flexion at the knee of the support and kicking legs at impact. These features not only define kicking technique but suggest that the skill of kicking benefits from being performed in this way. Therefore, the aim of this study was, through a comprehensive analysis of lower-body function, to quantify the kinematics and kinetics of the lower limb during the maximal soccer instep kick and to explain observable features of the kick.

2. METHODS

2.1 Participants

Ten soccer players participated in this investigation (mean ± s: age 23.4±2.5 years; mass 71.2 ±6.8 kg; height 1.781±0.0048 m). Each gave their informed consent following approval from the institution's Ethics Committee. All participants were skilled at kicking and played in top amateur or semi-professional clubs. All were right footed and kicked the ball with the right foot with participants wearing a commonly available indoor soccer boots (boot mass=380 g). Retro-reflective markers (25 mm diameter) were placed on the participant's lower limbs. Marker clusters, each made up of four markers attached to a rigid thermoplastic shell, were firmly taped on the right and left shank and thigh. Separate markers were attached to the pelvis (left and right anterior superior iliac spine, and sacrum half way between the posterior superior iliac spines) and to each foot (left and right lateral and medial malleolus, heel and 5th metatarsal head). Before data collection, eight calibration markers were added at the left and right greater trochanter (2), iliac crest (2) and left and right medial (2) and lateral (2) knee epicondyle. These markers were removed before kicking commenced.

2.2 Data collection

All kicks were made in an indoor environment with a soccer ball of standard size and pressure. Participants were given an opportunity to warm up and practice maximal instep kicks before data recording began. They were asked to approach the ball as they would normally do when taking a penalty kick, which was an angled approach using two to four steps. The ball was placed beside a force platform and kicked into a target located in a goal mouth with net. The target was a sheet of wood (1.5 m * 1.5 m) and was used both to ensure the kick was reasonably directed and to create a sound on impact. A microphone was placed close to the target to record impact and a second microphone was placed beside the ball to record foot impact with the ball. The time between the events triggered by the microphones was used together with the distance between ball and target to calculate the average ball speed. The ball was placed so that the support foot landed on a force platform (Kistler, model 9287B, Winterthur, Switzerland) and force data were sampled at 960 Hz. The 3D position of each marker was simultaneously recorded at 240 Hz using an 8-camera opto-electronic motion capture system (Proreflex, Qualysis, Gothenburg, Sweden). Ten successful trials were conducted and analysed.

2.3 Data analysis

The 3D motion data were tracked (QTM, Qualysis, Gothenburg, Sweden) and exported to a 3D motion analysis package that computed joint kinematics and kinetics (Visual3D, C-Motion, Rockville, USA). A 12 Hz fourth order Butterworth low-pass filter was used to smooth all displacement data. The laboratory and segment local co-ordinate systems were defined as illustrated in Figure 1. The local coordinate system was defined at the proximal joint centre for each segment. For the foot, the proximal joint centre was located midway between the medial and lateral malleolus markers and its distal joint centre was located 0.05 m from the 5th metatarsal towards the middle of the foot in the plane of the three markers defining the foot (medial and lateral malleolus, and 5th metatarsal). For the shank, the proximal joint centre was located mid-way between the medial and lateral knee markers, while the distal joint centre was at the ankle as defined for the foot. For the thigh, the proximal joint centre was located at a distance of 0.1 m from the hip marker towards the middle of the thigh in the plane of the three markers defining the thigh (medial and lateral knee, and hip). The distal joint centre was at the knee as defined for the shank. For the pelvis, the proximal joint centre was located mid-way between the iliac crest markers and the distal joint centre was located mid-way between the thigh proximal joint centres. For all segments the positive Z (internal/ external rotation) axis was defined in the direction of distal to proximal joint centres. The positive Y (abduction/adduction) was defined as perpendicular to the Z axis and the plane of the segment (as determined by the three or four markers defining the segment), while the X (flexion-extension) axis was defined as the vector cross product of Y and Z.

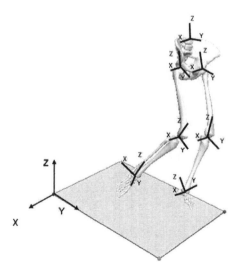

Figure 1. Laboratory and segment coordinate systems. Segmental coordinate systems are located at the proximal end of each segment.

Variables chosen to represent the kick were support (left) and kicking (right) hip, knee and ankle angles, angular velocities, moments and powers. In addition, the orientation of the pelvis relative to the laboratory, and kicking leg hip, knee and ankle joint centre vector velocities, all relative to the laboratory, were computed. Joint angles were obtained from an X-Y-Z Cardan rotation sequence, as in pilot work this was shown to be the most appropriate rotation sequence. Joint angular velocities, internal moments and powers were referenced to the distal segment as recommended by O'Connor and Hamill (2005) and Schache *et al.* (2007). Variables are presented numerically as mean ± SD, and graphically as mean ± SE.

2.4 Data reduction

Each trial was defined by the instances of right foot take-off (RFTO), left foot touch-down (LFTD), ball contact (CONTACT) and left foot take-off (LFTO). The RFTO was determined when the 5th metatarsal marker on the right foot was 0.1 m above the ground; LFTD was determined from the force platform data once the vertical force had exceeded 20 N; CONTACT was determined by the binary signal recorded by the microphone at ball contact. The LFTO was determined when the vertical force decreased below 20 N. Some subjects dragged their foot making this event indistinct. In such cases LFTO was defined as CONTACT+36 data points (equal to 0.15 s) which was a mean value taken from other subjects' data. The period from RFTO to LFTO was termed the event duration. The period RFTO to LFTD was termed the flight phase; the period LFTD to CONTACT the pre-contact phase; and the period CONTACT to LFTO the post-contact phase. For graphical presentations, the event duration was normalised to 101 points and averaged across trials and participants (a total of 100 kicks).

3. RESULTS

The mean values for the instances of LFTD and CONTACT occurred at 26% and 58% of the mean event duration of the kick and are highlighted as vertical lines on relevant graphs. The mean ±SD ball velocity achieved was $25.7 ± 2.4$ m.s^{-1}, with a range of $22.2 – 28.9$ m.s^{-1}.

The ground reaction force generated as the support foot made contact with the ground (Figure 2) was characterised by a wholly lateral Fx force component and a wholly posterior (or braking) Fy force component. Force components showed their mean peak (Fx= -4.1 N.kg^{-1}; Fy=-5.7 N.kg^{-1} and Fz (vertical) = 14.3 N.kg^{-1}) at 37%, 43% and 37%, respectively of the event duration, which was approximately the middle of the pre-contact phase. The vertical force appeared to be maintained at an elevated level during the pre-contact phase.

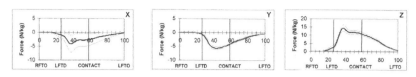

Figure 2. Component ground reaction forces. Variables are presented as a % of event duration.

For the support leg, the ankle joint (Figure 3 – Ankle) showed a small plantar-flexion prior to and after LFTD, followed by dorsi-flexion as the shank moved over the fixed foot until CONTACT and then plantar-flexed again during the follow through. In the pre-contact phase there was a substantial plantar-flexion moment and a corresponding power absorption. Power generation occurred only in the post-contact phase which also produced the greatest plantar-flexion angular velocity. The knee joint of the support leg (Figure 3 – Knee) extended prior to LFTD, but flexed after LFTD, and although extended during the later part of the pre-contact phase, remained flexed to 42 ±7° (full extension = 0°) at CONTACT. These actions are also reflected in the knee angular velocity. Knee extension moment during the pre-contact phase supported the body and power was absorbed. However, both knee extension and power generation began before CONTACT. The support leg hip joint (Figure 3 – Hip) was in a position of marked flexion at RFTO and subsequently only extended, most rapidly prior to ball contact. Hip extension was associated with an extension moment until CONTACT and a flexion moment thereafter. Consequently, the hip generated most of its power during the flight and pre-contact phase and little once impact had taken place. The joint longitudinal rotational angles about the support leg (Figure 4) suggest that during the pre-contact phase the foot everts (7°) and then inverts (2°) relative to the shank, the shank internally rotates (22°) relative to the thigh, and the thigh externally rotates (7°) and then internally rotates (4°) relative to the pelvis. The large internal rotation of the shank relative to the thigh is in part due to the contrary direction of rotation of the thigh.

The pelvis was tilted forward in the sagittal plane (Figure 5 – X) at RFTO and progressively tilted backward with a suggestion of a more rapid backward tilt prior to CONTACT. Subsequently, the pelvis stabilised at CONTACT and then produced a small forward tilt which reflected the flexion of the hip during the follow through. With regard to frontal plane motions, the pelvis was level at RFTO (Figure 5 – Y) and raised toward the kicking leg side as the motion progressed, but it remained very stable (mean change 2.0°) during the pre-contact phase. With regard to transverse plane motions, the kicking side of the pelvis was retracted at RFTO (Figure 5 – Z) and progressively rotated forwards, more rapidly just before ball contact.

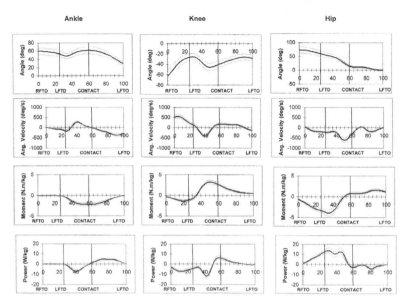

Figure 3. Support leg flexion-extension joint angle, angular velocity, moment and power (ankle and hip:- flexion (+), extension (-); knee:- extension (+), flexion (-); power generation (+), power absorption (-)). Variables are presented as a % of event duration.
Moment and power are relative to body mass.

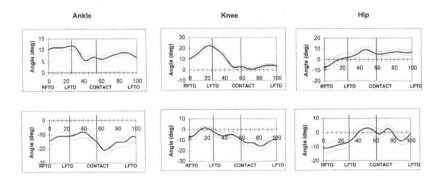

Figure 4. Support leg (top row) and kicking leg (bottom row). Z axis rotation (foot:- inversion (+), eversion (-); shank and thigh:- external rotation (+), internal rotation (-)).
Variables are presented as a % of event duration.

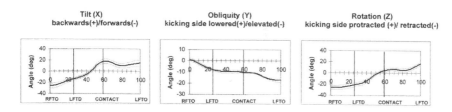

Figure 5. Pelvic angles relative to laboratory coordinate system.
Variables are presented as a % of event duration.

For the kicking leg, the ankle joint (Figure 6 – Ankle) showed rather small changes in mean ranges (angle = 9.2°, angular velocity = 330 °.s^{-1} (5.75 rad.s^{-1}), moment = 0.24 N.m.kg^{-1} and power = 0.56 W.kg^{-1}). The kicking leg knee joint (Figure 6 – Knee) flexed markedly and then extended but remained flexed at CONTACT (57±8°) compared to full extension (0°) attained during the follow through. The knee angular velocity reached its peak at contact as has been widely reported elsewhere. Maximum knee joint extension moment was reached at 37% of the event duration coinciding with the greatest ground reaction forces. This extension moment with the flexing motion of the knee resulted in power absorption. The joint moment reduced towards ball contact and became a flexion moment. This was associated with very high levels of power absorption (58 W.kg^{-1}). The precise magnitude of these variables around impact are known to be influenced by the data processing procedures (see Nunome *et al.*, 2006a) and need to be taken into account when interpreting data at impact. The kicking leg hip joint (Figure 6 – Hip) extended from RFTO but began to flex at LFTD. The hip flexed through ball contact but showed a marked reduction in flexion velocity before CONTACT. This coincided with an increase in knee joint angular velocity as has been widely reported elsewhere (and which can be seen in Figure 6 – Knee Angular Velocity) as CONTACT is approached. The hip joint produced mainly a flexion moment that reached its peak at LFTD. There was little power absorbed (2.0 W.kg^{-1}) by the hip, but substantial power generated (20.9 W.kg^{-1}) in the pre-contact phase. The rotations of the kicking leg shank and thigh about their longitudinal axes (Figure. 4) suggest that during the pre-contact phase the foot inverts (3°) and then everts (10°) relative to the shank, the shank internally rotates (8°) in two stages relative to the thigh, and the thigh externally rotates (10°) and then internally rotates (2°) relative to the pelvis. The internal rotation of the shank and eversion of the foot suggests that the kicking foot is being orientated to control the impact with the ball.

The fast rotational motion of the kicking foot and shank about the knee will require a large centripetal force whose reaction will act on the thigh and pelvis and may influence the orientation of the body in the frontal plane. The net joint force at the kicking knee, acting along the longitudinal axis of the thigh, is presented in Figure 7. This became substantial as the foot velocity increased to its peak at ball contact and is balanced by the lateral lean of the body away from the ball.

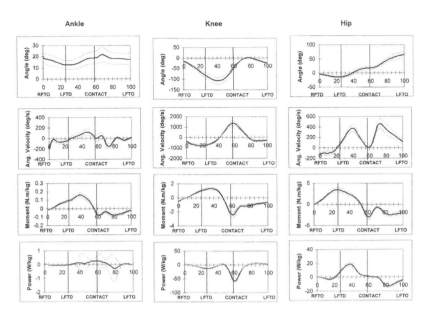

Figure 6. Kicking leg flexion-extension joint angle, angular velocity, moment and power (ankle and hip:- flexion (+), extension (-); knee:- extension (+), flexion (-); power generation (+), power absorption (-)). Variables are presented as a % of event duration. Moment and power are relative to body mass.

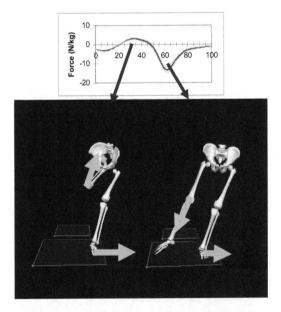

Figure 7. The net Z-axis force acting at the knee along the thigh of the kicking leg as a percentage of event duration together with frontal plane orientations after LFTD and at CONTACT showing the lateral ground reaction force and the net Z axis force.

4. DISCUSSION

4.1 The kicking leg

It is commonly thought that knee flexion at impact is a protective mechanism as the joint is prevented from reaching full extension at full velocity. At impact, the knee is flexed to 57° and during the impact which lasts approximately 9 ms (Nunome *et al.*, 2006a), typically extends a further 12°. Leg extension continues during the follow-through to maximal extension (0°). So, after contact with the ball the knee still has around 45° of extension available to it, which is rather a large margin of safety, so there may be additional causes for this. One possible cause is that the flexed leg enables the foot to clear the ground as it makes contact with the ball. At impact the support leg is also flexed (42°), the pelvis is lowered on the kicking side, and the kicking foot is plantar-flexed, all of which require compensatory knee flexion of the kicking leg for the foot to clear the ground. A second possible cause is that energy could be fed into the shank and foot system during the impact period due to active knee extension. The knee extension moment is negative just before impact (this is confirmed by the more precise analysis of Nunome *et al.*, 2006a), so this does not seem likely. A third possibility is that when the knee is flexed, the shank has a greater range of internal/external rotation relative to the thigh (Blankevoort *et al.*, 1988) which would be of some value in orientating the foot for impact with the ball in order to achieve a more precise impact position.

A second observable feature of the maximal instep soccer kick is the cocking of the kicking leg. The term refers to the flexion of the knee as the thigh is brought forwards during the pre-contact phase (Nunome *et al.*, 2002). The cocking of the leg reduces the moment of inertia of the leg and allows the work done by muscles around the hip (Figure. 6 – Hip Power) to enhance kicking leg angular velocity. This is a well known mechanical principle but, in addition, cocking of the kicking leg produces a stretch-shorten cycle, responsible for maximising the extension moment and work done at the knee joint (Dorge *et al.*, 2002). Knee extension peak moment (Figure 6) occurred at 37% of the event duration and coincided with other force maxima, but how these variables are related is not obvious.

4.2 The pelvis

The pelvis rotates about a vertical (Z) axis from backward retraction of the kicking side at RFTO to forward retraction at LFTO (Figure 5 – Z) and about a medio-lateral axis (X) from backward tilt to forward tilt (Figure 5 – X). Evident in these graphs is a rapid change in rotation during the pre-contact phase. These rapid changes could be due to an interaction between the pelvis and support and kicking legs. It has been shown in Figure 4 that the support leg thigh initially externally rotates relative to the pelvis, and then internally rotates. The internal rotation implies rotation of the pelvis towards the thigh thus enhancing the forward retraction of the kicking leg side of the pelvis. The thigh of the kicking leg may also influence the pelvic tilt during the pre-contact phase as the reduction in thigh angular velocity may allow enhanced backward tilt at the pelvis at the proximal (hip) joint (Dorge *et al.*, 2002; Nunome *et al.*, 2002). A more detailed analysis than conducted here would be necessary to identify the forces and torques acting at the hip of the kicking leg and possible contribution to pelvic tilt.

The frontal plane orientation of the pelvis remains very stable during the pre-contact support phase, moving no more than 2°. This would serve to provide a stable base for judging kicking foot height and position relative to the ball, and consequently may be a factor influencing movement consistency.

4.3 The support leg

Soccer players prefer to make an angled approach to the ball which results in an orientation of the body in its frontal plane away from the ball. Whether this has any benefits to performance, directly in terms of ball speed, is equivocal (Isokawa and Lees, 1988; Kellis *et al.*, 2004) but there may be other benefits, for example movement control and fluency of performance. The fact that even sub-maximal kicks are performed with angled approaches suggests that any benefits may not be directly related to ball speed.

It has been noted that the horizontal ground reaction forces are wholly negative in both the X and Y directions (Figure 2). These data are in contrast to those reported by Kellis *et al.* (2004) who found a small positive force at the early stages of contact in the Y (anterior-posterior) direction. There is no obvious reason why that should occur but may be due to the one step approach they used, in contrast to the multiple step approach used in this study. The wholly negative forces indicate that the body is being continually slowed down during the kick. The approach speed generated from the few strides of the approach is not high, so it is suggested that the approach speed itself is not a factor in performance (in terms of high ball speed) but a means for achieving stability in a dynamic performance.

The negative X axis force (Figure 2) results in a braking action to stabilise frontal plane motion. The angled approach enables the support leg to be planted with a frontal plane orientation away from the ball. Even when the axial force acting on the kicking leg (Figure 10) reaches its peak at around 60% of event duration, there is little disturbance to the orientation of the pelvis as noted above. This control of the dynamics of motion would occur through internal muscular tensions, the reflection of which appear in the ground reaction force as elevated force values at around 60% of event duration (Figure 2X, 2Z).

The flexion of the support leg knee at contact (42°) is similar to the peak flexion that occurs in running (Milliron and Cavanagh, 1990) but this angle is held for longer suggesting that knee flexion in the support leg has a role to play other than absorbing impact with the ground. One benefit of a flexed knee is that it has the ability to extend and to contribute to performance (a small extension has been identified just prior to ball contact, Figure 3 – Knee). This would serve to lift the body and add to the vertical velocity of the foot at impact. It may also help to redirect the velocity vector of the foot to influence precision of foot-ball contact. Knee extension during the pre-contact phase may also be the means of controlling the perturbing effect of the axial force acting on the kicking leg. The extension moment around the knee joint (Figure 3 – Knee Moment) is maximal prior to ball contact and although declining as the axial force acting on the kicking leg reaches its peak, is nevertheless in a good state of muscular tension to resist any unbalancing that this force may cause. Thus, a second benefit of the flexed support leg knee is that it provides the means for creating dynamic stability.

4.4 Performance issues

Performance is determined not only by sound technique, but also by the work done around the joints. The work done around joints can be obtained by integrating the net joint power over the pre-contact phase and can be evaluated by inspecting the power graphs in Figures 3 and 6. The hip joints of both the support and kicking legs perform substantial positive work prior to ball contact. The work done by the support leg hip joint maintains body speed during the early pre-contact phase. The work done by the kicking leg hip serves to increase the rotational velocity of the kicking leg, subsequently enhancing its end-point speed. The knee joints of both the support and kicking legs initially do negative work but then positive work. The support leg has to perform the role of shock absorber at contact and it is caused to flex, doing negative work, but can then play an active role in extension as noted above. The kicking leg is involved in the cocking motion, essentially a stretch-shorten cycle, first doing negative work and then positive work as also noted above. The ankle of the support leg only does negative work during the pre-contact phase and so makes no active contribution to performance. The work done by the kicking leg ankle is variable and small, and so has little contribution to make to performance, although its role in accuracy may be considerable. It appears that in the maximal instep soccer kick, the hip muscles are the main workhorses, the knee muscles have a complex role to play but can contribute to performance while the ankle muscles have little contribution to performance. These findings may be of value in the physical training of soccer players.

5. CONCLUSIONS

This analysis has attempted, through the provision of comprehensive 3D data, to increase understanding of lower -limb function in the maximal instep soccer kick by identifying and explaining several of the observable features of the kick in terms of their benefits and risks. A complex sequence of events is apparent which are related to both control and performance. Control is achieved by creating dynamic stability – a balancing of the dynamic forces so as to perturb the motion as little as possible, providing a stable platform for ball contact to take place. The angled approach also allows the body to tilt away from the ball, raising the kicking leg. This raising of the kicking leg in turn allows the foot to clear the ground and the knee to extend through impact with the ball while also maximising the distance of the foot to increase end-point speed. The hip muscles are the major contributors to performance, with some extra benefit coming from the knee extensors of both the support and kicking legs. This greater understanding of lower-limb function in the maximal instep soccer kick will be of value to all those with an interest in kicking and related skills.

References

Blankevoort, L., Huiskes, R. and de Lange, A.,1988, The envelope of passive knee joint motion. *Journal of Biomechanics*, **21**, pp. 705-720.

Dorge, H. C., Bull Anderson, T., Sørensen, H. and Simonsen, E. B., 2002, Biomechanical differences in soccer kicking with the preferred and the non-preferred leg. *Journal of Sports Sciences*, **20**, pp. 293-299.

Isokawa, M. and Lees, A., 1988, A biomechanical analysis of the instep kick motion in soccer. In *First World Congress of Science and Football*, edited by Reilly, T., Lees, A., Davids, K. and Murphy, W., (London: E & F.N. Spon), pp. 449-455.

Kellis, E., Katis, A. and Gissis, I., 2004, Knee biomechanics of the support leg in soccer kicks from three angles of approach. *Medicine and Science in Sports and Exercise*, **36**, pp. 1017-1028.

Lees, A. and Nolan, L , 1998, Biomechanics of soccer - A review. *Journal of Sports Sciences*, **16**, pp. 211-234.

Levanon, J. and Dapena, J., 1998, Comparison of the kinematics of the full-instep and pass kicks in soccer. *Medicine and Science in Sports and Exercise*, **30**, pp. 917-927.

Milliron, J. M. and Cavanagh, P. R., 1990, Sagittal plane kinematics of the lower extremity during distance running. In *Biomechanics of Distance Running*, edited by Cavanagh, P.R., (Champaign, IL: Human Kinetics).

Nunome, H., Asai, T., Ikegami, Y. and Sakurai, S., 2002, Three dimensional kinetic analysis of side-foot and instep kicks. *Medicine and Science in Sports and Exercise*, **34**, pp. 2028-2036.

Nunome, H., Lake, M., Georgakis, A. and Stergioulas, L. K., 2006a, Impact phase kinematics of the instep kick in soccer. *Journal of Sports Sciences*, **24**, pp. 11-22.

Nunome, H., Ikegami, Y., Kozakai, R., Apriantono, T. and Sano, S., 2006b, Segmental dynamics of soccer instep kick with the preferred and non-preferred leg. *Journal of Sports Sciences*, **24**, pp. 529-541.

O'Connor, K. M. and Hamill, J., 2005, Frontal plane movements do not accurately reflect ankle dynamics during running. *Journal of Applied Biomechanics*, **21**, pp. 85-95.

Schache, A. G., Baker, R. and Vaughan, C. L., 2007, Differences in lower limb transverse plane joint moments during gait when expressed in two alternative reference frames. *Journal of Biomechanics*, **40**, pp. 9-19.

CHAPTER THIRTEEN

Polymorphisms of the angiotensin-converting enzyme (ACE/ID) and their distribution in developing young adult Rugby Union players

W. Bell[1], J.P. Colley[2], J.R. Gwynne[1], L. Llewellyn[1], S.E. Darlington[3]
and W.D. Evans[3]

[1]University of Wales Institute Cardiff, Cyncoed, Cardiff, Wales
[2]Wales Gene Park, Institute of Medical Genetics, Cardiff University, Wales
[3]Department of Medical Physics and Clinical Engineering, University Hospital of Wales, Cardiff, Wales

1. INTRODUCTION

Rugby Union football is a team game involving skills of high and low intensity exercise. It is played between two opposing teams, each containing 15 players. The duration of play is two periods of 40 minutes. Broadly speaking, players are classified into two main playing units; forwards (n=8) and backs (n=7). The major function of the forwards is to secure and maintain possession of the ball during open (rucks and mauls) and set-pieces (scrummage and line-out) of play. The ball is then released to the backs whose objective is, by running and kicking, to ground the ball in the opposing in-goal area.

Rugby Union differs from other team games because it encourages vigorous, dynamic and sustained physical contact to support and execute a variety of playing skills. The phenotypes of players are varied, and to be successful, need to be extremely robust in terms of somatic (height, weight, physique, body composition) and physiological (aerobic and anaerobic fitness, power, strength) characteristics. Generally speaking, players in each unit are given specific requirements which they are expected to fulfil. Nevertheless, despite these demands, *all* players are expected to demonstrate high levels of basic skills in the various phases of play, thus enabling the team to perform in a unified and coordinated manner (Duthie *et al.*, 2003).

The human endocrine renin-angiotensin system plays an important role in cardiovascular function (Rosendorff, 1996). The pathway is initiated by the release of renin from the kidney, primarily in response to salt depletion, volume loss, or sympathetic nervous system activation, to produce the decapeptide angiotensin I from its precursor angiotensinogen. In turn, angiotensin I is catalyzed by the angiotensin-converting enzyme (ACE) which converts angiotensin I into the octapeptide angiotensin II, and at the same time inactivates bradykinin, a powerful vasodilator.

Angiotensin II, which is mediated largely through hormone receptors AT_1 and AT_2, (Stroth and Unger, 1999), has multiple effects on the cardiovascular system. It is a potent vasoconstrictor, increases blood pressure, and regulates salt and water homeostasis; it also functions as a muscle growth factor, which may explain the increased frequency of the D allele in power athletes. Bradykinin, apart from inducing vasodilation via receptors BK1 and BK2, has favourable effects on endothelium function and substrate utilization. In addition to its general circulatory function, renin-angiotensin systems operate locally via autocrine and paracrine systems, in adipose (Jonsson *et al.*, 1994), heart and lung (Pieruzzie *et al.*, 1995), and skeletal muscle (Dragovic *et al.*, 1996) tissues.

A polymorphism of the human gene encoding ACE (17q23, Hubert *et al.*, 1991), consists of the presence (insertion, I allele) or absence (deletion, D allele) of a 287 amino acid base-pair fragment found within intron 16 (Rigat *et al.*, 1990; Tiret *et al.*, 1992). Individuals carry two versions of the ACE allele, therefore three ACE genotypes are identified: individuals homozygous for the insertion (II) and deletion (DD) alleles, and individuals heterozygous for the ID allele. The distribution of the frequency of alleles in Caucasians is approximately Gaussian (0.25, 0.50, 0.25 for II, ID and DD genotypes respectively). The presence of the I allele is associated with low levels of ACE activity in serum and tissues, whereas the D allele is associated with increased levels of ACE activity (Danser *et al.*, 1995).

A number of studies have suggested that the I allele is associated positively with enhanced endurance performance. The majority of these studies have focussed on *senior elite athletes* participating in individual sports requiring prolonged effort and high metabolic cost; these sports include mountaineering (Montgomery *et al.*, 1998), long-distance swimming (Tsianos *et al.*, 2004), cycling (Lucia *et al.*, 2004), triathlons (Collins *et al.*, 2004), distance running (Myerson *et al.*, 1999; Nazarov *et al.*, 2001) and rowing (Gayagay *et al.*, 1998). Investigators who have used admixtures of athletes from different endurance events have similarly found an over-representation of the I allele (Alvarez *et al.*, 2000; Nazarov *et al.*, 2001). On the other hand, some studies have failed to identify a definitive relationship between the I allele and endurance performance (Taylor *et al.*,1999; Rankinen *et al.*, 2000; Scott *et al.*, 2005).

The D allele is thought to be associated anaerobically with short-distance sporting events such as sprinting (Myerson *et al.*, 1999; Nazarov *et al.*, 2001), and swimming (Woods *et al.*, 2001).

In general terms, the mechanisms responsible for an association between the I allele and endurance performance are thought to be the result of an increase in local cellular growth, with improvements in muscle efficiency and metabolism, rather than any direct central cardio-respiratory effect. The relationship between the D allele and enhanced anaerobic performance is believed to arise as a consequence of an angiotensin II - mediated increase in muscle growth and strength (Woods *et al.*, 2000; Jones *et al.*, 2002; Payne and Montgomery, 2003).

A further consideration of the ACE I/D genotype is its sensitivity to systematic athletic training. Individuals exercising with appropriately designed training programmes have, in general, acquired beneficial adaptations to functional variables such as aerobic endurance (Montgomery *et al.*, 1999; Woods *et al.*, 2002; Williams *et al.*, 2004), maximal oxygen uptake ($VO_{2\,max}$) (Hagberg *et al.*, 1998), left ventricular mass (Montgomery *et al.*, 1997; Diet *et al.*, 2001), and muscle strength (Folland *et al.*, 2000; Colakoglu *et al.*, 2005; Pescatello *et al.*, 2006). However, other authors (Sonna *et al.*, 2001; Thomis *et al.*, 2004) have found little influence of the ACE I/D gene on variables such as, muscular size, endurance and strength.

We are not aware of any studies which have detailed the genotypes of Rugby Union football players. In fact, the number of investigations characterising the status of games players of any code is surprisingly limited. Myerson *et al.* (1999) reported the genotypes of UK Olympic standard field-hockey players (n=53) and ice-hockey players (n=34), Taylor *et al.* (1999) a group of Australian national field-hockey players (n=26), and Alvarez *et al.* (2000) professional handball players (n=15). All these sports, however, have been part of larger hetrogeneous groups of athletes.

The purpose of the present study, therefore, was to compare differences in genotype and allele frequency between: (i) *developing* male Rugby Union football players and control subjects; (ii) forwards and backs, the two major playing units in Rugby Union football, and (iii) between individual players classified on the basis of functional playing requirement.

2. METHODS

2.1 Study design, subjects and active control group

A cross-sectional case-control study design was used employing a candidate gene approach. A university-based group consisting of 109 senior squad and centre of excellence young adult Rugby Union players (cases) and 108 active controls (Booth and Lees, 2006), all Caucasian, were recruited to the study over a period of three years. To maintain group homogeneity, subjects were selected from a larger cohort of players by the Director of Rugby on the basis of current playing performance and the potential to play at senior level. Many of the players had previously competed at their respective age-group international level (U16, U18, U19, U20 and U21). Players competed in Division 1 of the Welsh National League.

The active control group (n = 108) comprised subjects from a wide variety of sporting disciplines other than rugby, who participated either at recreational or competitive levels in activities such as association football, athletics, cricket, hockey, basketball, swimming, tennis, rowing, gymnastics and squash.

Since physical activity has been a selective pressure on genes for survival, the premise for using a physically active, rather than a purely sedentary group as control subjects, is based on the assumption that where studies involve exercise, physically active subjects are considered more suitable as controls than sedentary subjects (Booth *et al.*, 2002; Booth and Lees, 2006). The study was approved by the University Ethics Committee and written informed consent obtained from individuals.

2.2 ACE genotyping

Individuals provided a 5-ml saliva sample into a sterile container. Specimens were stored collectively at 4°C until buccal cell DNA extraction was carried out (Qiagen, QIAamp DNA Micro Kit). Three-primer polymerase chain reaction (PCR) was used to assay the region of interests for I and D variants of the ACE gene. Since established primers failed to achieve good amplification, and forward and reverse oligonucleotides (oligos) did not provide a perfect match to the genomic sequence NC_00017.9, oligos were manually designed using the sequence NC_00017.9 obtained from the National Centre for Biotechnology Information database (http://www.ncbi.nlm.nih.gov). Sequencing this region, with oligos spanning the original primers, showed NC_00017.9 to be satisfactory. The size of the expected PCR products were Forward > Reverse I – 540 bp; Forward > Reverse D – 252 bp; Internal > Reverse I – 335 bp. Therefore, the primers used were as follows:

Forward   5′ - GACTCTGTAAGCCACTGCTGGAG - 3′
Reverse   5′ - TCGCCAGCCCTCCCATGCCCATAA - 3′
Internal Forward   5′ - TGGGACCACAGCGCCCGCCACTAC - 3′

Twenty-five ng of template DNA, 25 pmoles of each oligonucleotide, 0.2 mM of each dNTP, 0.5 units of Ampli Taq Gold DNA Polymerase and 1X PCR Buffer (15 mM MgCl$_2$) were mixed in a total volume of 25 µl. Cycling conditions were 94°C for 12 minutes (heat activation step), followed by 35 cycles of 94°C denaturing for 30 s, annealing at 60°C for 30 s, extension at 72°C for 30 s, with a final extension period of 10 min at 72°C. A 5-µl sample of PCR product was mixed with 5 µl of Gel Loading Buffer and loaded into the well of a 2% agarose gel and electrophoresis run at 100 V. Allele identification was visualised using ethidium bromide staining. ACE (I/D) genotypes are displayed in Figure 1; lanes 1 and 6 are DNA ladders, lanes 2 and 5 illustrate the ID genotype, lane 3 the I genotype and lane 4 the D genotype.

2.3 Statistics

Independent t-tests were employed to identify differences between demographic characteristics. Pearson's chi-square test was used to compare genotype and allele frequency between players and control subjects, between forwards and backs, and between individual positions (SPSS version 12.0). The alpha level for significance was set at $P<0.05$. Effect size and power values were also calculated (Cohen, 1988).

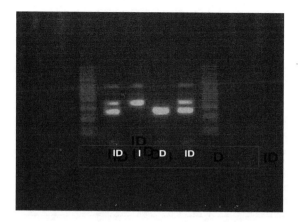

Figure 1. An image of the ACE genotype ID is displayed in lanes 2 and 5.
Lane 3 is the I genotype and lane 4 the D genotype.

3. RESULTS

The demographic data for rugby players and control subjects are given in Table 1. Players were significantly older (19.7±2.0 vs 18.4±0.8 years), taller (1.819±0.069 vs 1.792±0.065 m) and heavier (90.3±12.6 vs 72.9±9.1 kg) than active controls (p=0.003).

Table 1. Descriptive data (mean ± SD) for age, stature and body mass of
rugby players (n=109) and active control (n=108) subjects.

	Age (years)	Stature (m)	Body mass (kg)
Rugby players (n=104)*	19.7 ± 2.0	1.819 ± 0.069	90.3 ± 12.6
Active controls (n=108)	18.4 ± 0.8	1.792 ± 0.065	72.9 ± 9.1
	***	***	***

*** = P<0.003
* Five values unavailable for age, stature and body mass

The results showed no significant differences in genotype or allele frequency between rugby players and active controls. The proportion of genotypes for rugby players (Table 2) was 0.21, 0.49 and 0.30 for II, ID and DD respectively. In comparison, active controls had similar proportions for II (0.20), fewer ID (0.40), but more DD (0.40) ($c^2 = 2.38$, 2df, P = 0.31); effect size = 0.15 and power = 0.49. There was a tendency for the proportion of the D allele to be slightly greater than the I allele in both rugby players (0.55 vs 0.45 respectively) and active controls (0.60 vs 0.40 respectively), but differences were not significant (c^2 = 1.168, 1df, P=0.28); effect size = 0.1, power = 0.55.

Table 2. Genotype and allele frequency of rugby players (n=109) and active (n=108) control groups.

	Genotype			Allele frequency	
	II	ID	DD	I	D
Players	23 (0.21)	53 (0.49)	33 (0.30)	99 (0.45)	119 (0.55)
Controls	22 (0.20)	43 (0.40)	43 (0.40)	87 (0.40)	129 (0.60)

Relative frequency of genotypes and allele frequency given in brackets
Genotype chi-square (players vs controls) = 2.38, 2df, P=0.31. Effect size = 0.15, power = 0.49
Allele frequency chi-square (players vs controls) = 1.168, 1df, P = 0.28. Effect size = 0.1, power = 0.55

Descriptively, there was a contrast in the II genotype distribution of forwards (0.27) compared to backs (0.15) (Table 3). Backs had a greater proportion of ID (0.55 vs 0.43 respectively) and the same proportion of DD (0.30 vs 0.30 respectively); however, none of these differences were significant (c^2 = 2.55, 2df, P = 0.279); effect size = 0.217 and power = 0.51. The I and D allele frequency within forwards (0.48 vs 0.52) and backs (0.42 vs 0.58) and between forwards and backs (0.48 vs 0.42 and 0.52 vs 0.58 respectively) were similar in magnitude and not significant (c^2 = 0.729, 1df, P= 0.393); effect size = 0.1 and power = 0.31. The greater frequency of the D allele in backs, although not significant, suggests a greater capacity for power activities. The overall result indicates that there are no significant differences in genotype or allele frequency when players are classified on the basis of forwards or backs.

Table 4 presents genotypes and allele frequency between functional playing positions. There are various ways in which individual positions can be assembled depending on availability of players and strategy of play employed; thus the position or function of players is not a static one. On the basis of the categories chosen as being generally applicable in the modern game, there were no significant differences between functional position and genotype (c^2 = 4.52, 6df, P= 0.607); effect size = 0.44 and power = 0.93, or allele frequency (c^2 = 1.454, 3df, P= 0.693); effect size = 0.165, power = 0.31. It is noticeable that there is a larger frequency of the D allele in all positional groups except the front row (P/H). Unfortunately, this group analysis resulted in the size of some of the individual/positional categories being small.

Table 3. Genotype and allele frequency of rugby forwards (n=56) and backs (n=53).

	Genotype			Allele frequency	
	II	ID	DD	I	D
Forward	15 (0.27)	24 (0.43)	17 (0.30)	54 (0.48)	58 (0.52)
Backs	8 (0.15)	29 (0.55)	16 (0.30)	45 (0.42)	129 (0.60)

Relative frequency of genotypes and allele frequency given in brackets
Genotype chi-square (forwards vs backs) = 2.55, 2df, P=0.279. Effect size = 0.217, power = 0.51
Allele frequency chi-square (forwards vs backs) = 0.729, 1df , P=0.393. Effect size 0.1, power = 0.31

Table 4. Genotype and allele frequency of rugby players (n = 109) classified by playing position.

	Genotype			Allele frequency	
	II	ID	DD	I	D
P/H	8 (0.33)	9 (0.38)	7 (0.29)	25(0.52)	23 (0.48)
L/BR	7 (0.22)	15 (0.47)	10 (0.31)	29 (0.45)	35 (0.55)
SH/OH	3 (0.14)	13 (0.62)	5 (0.24)	19 (0.45)	23 (0.55)
C/W/FB	5 (0.16)	16 (0.50)	11 (0.34)	26 (0.41)	38 (0.59)

P/H = Props and hookers, L/BR = locks and back row, SH/OH = scrum half and outside half, C/W/FB = centres, wings and full backs
P/H and L/BR = Forwards. SH/OH and C/W/FB = Backs
Relative frequency of genotypes and allele frequency given in brackets
Genotype chi-square (positions) = 4.52, 6df, P=0.607. Effect size = 0.44, power = 0.93
Allele frequency chi-square (positions) = 1.454, 3df , P = 0.693. Effect size = 0.165, power = 0.31

4. DISCUSSION

The aim of the present study was to determine the distribution of the ACE (I/D) genotype in *developing* young adult Rugby Union players. It was not designed to identify the classification of the ACE (I/D) genotype in *senior elite* players. Nor was it concerned with the genotype and allele frequencies of somatic characteristics such as body size, proportion, composition and physique, which are clearly very important in player selection and playing position. Our immediate concern was to identify the distribution of the genotype and allele frequency in young adult players, and consider the appropriateness of the ACE (I/D) polymorphic characteristics in furthering their potential to senior playing status.

A number of principles outlining an acceptable gene-environmental model have been proposed by Jones *et al.* (2002). To fulfil these requirements the present study examined a homogeneous group of developing young adult Rugby Union players (ability, gender, age, ethnicity) participating in similar training programmes (nature, duration, frequency, intensity) and having comparable lifestyle conditions (eating, sleeping, working).

The majority of studies that have been carried out thus far have used the word *elite* in describing their study group. The term is more pertinent to individual sports than team games since individual sports are able to provide a definitive measure of what is regarded as *elite* performance. When applied to team sports we have found the use of the term unhelpful. The situation is more complicated in team games because of the variety of activities carried out, and the need for multiple player co-operation.

In an environment requiring high levels of performance involving multiple skills (as in Rugby Union), increased performance in one skill may well interfere with another, so there needs to be trade-off between specialist and generalist phenotypes (Van Damme *et al.*, 2002). As a consequence it is important that morphological (size, shape, and body composition) and physiological (aerobic and anaerobic metabolism, strength and power) phenotypes are compatible with playing position and requirement (Duthie *et al.*, 2003; Bell *et al.*, 2005).

The genotype and allele frequency of rugby players (Table 2) was not significantly different (P=0.31 and 0.28 respectively) from those of active controls. The power values of both these analyses (0.49 and 0.55 respectively) were large (Cohen, 1988), so the result seems fairly definitive in suggesting that the distribution of the ACE (I/D) genotypes in *developing* rugby players is similar to those of active controls. This result was confirmed by comparing rugby players with a control group (n = 1906) free of cardiovascular disease (Myerson *et al.*, 1999). The genotype chi-square was P = 1.107 and power = 0.8.

Table 3 provides the genotype distribution of forwards (n=56) compared with backs (n=53). The distinction between these two groups arises largely because of their contrasting functional role. Although there are marked morphological and physiological differences between the two groups, there are no differences in the distribution of the ACE I/D polymorphism (P=0.279) or allele frequency (P=0.393) between groups. Forwards, where absolute power is a prerequisite still retain a fairly well-balanced genotype (Table 3, II = 0.27, ID = 0.43, DD = 0.30: I = 0.48, D = 0.52).

We took the opportunity to compare the genotypes of the present group of young adult rugby players with those of three other team games identified in the literature (field-hockey, ice-hockey, handball). There were no significant differences between the four team games (c^2 = 3.42, 6df, P = 0.754); effect size = 0.39 and power = 0.92. This would suggest that team games which employ a broad variety of skills and utilize a wider range of the aerobic-anaerobic spectrum, have a more balanced genotype than extreme metabolic sports such as sprinting and long-distance running.

A further question which arises is whether differences occur in the ACE (I/D) genotype when individuals are classified on the basis of individual/ functional playing requirement. In the present study (Table 4) these classifications were based on current playing philosophy, although lack of numbers prohibited a more refined breakdown of positions. Table 4 illustrates the greater proportion of the II genotype in forwards (PH/LBR) (0.33 and 0.22 respectively), compared with backs (SH/OH and C/W/FB) (0.14 and 0.16 respectively); that is, forwards had a genotype supporting endurance rather than power, a result which might be viewed as being contrary to expectation. However, differences between these individual/positional categories were not significant (c^2 = 4.52, 6df, P=0.607); the power value for this analysis was large (0.93). The frequency of the D allele was marginally greater than the I allele in most playing positions, but was not significant (c^2 = 1.454, 3df, P=0.693); power = 0.31.

It seems, therefore, that the ACE (I/D) genotype and allele frequency in *developing* players does not differ between forwards or backs, or between individual/functional playing positions. As such, therefore, it is unlikely to serve as a useful genetic marker for the identification of *developing* young adult rugby players. Given the multifactorial nature of sporting performance, particularly team games, this may not be surprising. In such circumstances, maturity of performance will need to be attained by exposure to relevant and more demanding training regimes, and participation at a superior level of competition.

The ACE (I/D) gene is known to affect inter-individual variation in both performance and trainability (e.g. Montgomery *et al.*, 1998; Montgomery *et al.*, 1999; Colakoglu *et al.*, 2005), nonetheless Pitsiladis and Scott (2005) argued that the current state of genetic evidence does not lend itself to the prediction of sporting talent or preparation for training. Considering the positive response of the ACE (I/D) gene to functional overload (e.g. Folland *et al.*, 2000) it may be worthwhile to examine the relationship between the ACE (I/D) polymorphism and position-specific training.

In their study of healthy Japanese students Zhang *et al.* (2003) found that ACE II subjects had a significantly greater % of type I fibres (50.1±13.9 vs 30.5±13.3 %, P<0.01) and a corresponding lower % of fast-twitch type IIb fibres (16.2±6.6 vs 32.9±7.4 %, P<0.01) than DD subjects. In South African rugby players (Jardine *et al.*, 1988) the muscle fibre distribution in type IIb fibres between forwards and backs was essentially the same (52.6±10.9 vs 56.8±12.1 %, P>0.05), which tends to lend some support to the finding that rugby players have a reasonably well balanced genotype. Apart from the obvious morphological and physiological differences between Rugby Union players, and given the physical nature of the game, genotypes with a high proportion of the D allele, and a corresponding increase in type IIb fibres, might well be advantageous.

In summary, the genotype and allele frequencies of the ACE I/D polymorphisms required to meet the metabolic demands of *developing* young adult rugby players are more balanced than in *elite* individual sports demanding extreme endurance or power. One of the likely reasons for this is the employment of a wider range of skills which utilize a more extensive range of the aerobic-anaerobic spectrum, compared with extreme *elite* sports which tend to use a disproportionate frequency of a specific allele.

Acknowledgements: We should like to express our appreciation for the technical expertise of J. Maynard and V. Humphries (Wales Gene Park) and C. Davey (UWIC).

References

Alvarez, R., Terrados, N., Ortolano, R., Iglesias-Cubero, G., Reguero, J.R., Batallo, A., Cortina, A., Fernandez-Garcia, B., Rodriguez, C., Braga, S., Alvarez, V. and Coto, E., 2000, Genetic variation in the renin-angiotensin system and athletic performance. *European Journal of Applied Physiology,* **82**, pp. 117-120.

Bell, W., Evans, W. D., Cobner, D.M. and Eston, R.G., 2005, The regional placement of bone mineral mass, fat mass, and lean soft tissue mass in young adult Rugby Union players. *Ergonomics,* **48**, pp. 1462-1472.

Booth, F.W. and Lees, S.J., 2006, Physically active subjects should be the control group. *Medicine and Science in Sports and Exercise,* **38**, pp. 405-406.

Booth, F.W, Chakravarthy, M.V. and Spangenburg, E.E., 2002, Exercise and gene expression: physiological regulation of the human genome through physical activity. *The Journal of Physiology,* **543**, pp. 399-411.

Cohen, J., 1988, *Statistical Power Analysis,* 2nd ed., (New Jersey, NY: Lawrence Erlbaum Associates).

Colakoglu, M., Cam, F.S., Kayitken, B., Cetinoz, F., Colakoglu, S., Turkmen, M. and Sayin, M., 2005, ACE genotype may have an affect on single versus multiple set preferences in strength training. *European Journal of Applied Physiology,* **95**, pp. 20-26.

Collins, M., Xenophontos, S. L., Cariolou, M.A., Mackone, G.G., Hudson, D.E., Anastasiades, L. and Noakes, T.D., 2004, The ACE gene and endurance performance during the South African Ironman Triathlons. *Medicine and Science in Sports and Exercise,* **36**, pp. 1314-1320.

Danser, A.H.J., Schalekamp, M.A.D.H.., Bax, W.A., Van den Brink, A.M., Saxena, P.R., Riegger, G. A. J. and Schunkert, H., 1995, Angiotensin-converting enzyme in the human heart. Effects of the deletion/insertion polymorphism. *Circulation,* **92**, pp. 1387-1388

Diet, F., Graf, C., Mahnke, N., Wassmer, G., Predel, H.G., Palma-Hohmann, I., Rost, R. and Bohm, M., 2001, ACE and angiotensinogin genotypes and left ventricular mass in athletes. *European Journal of Clinical Investigation,* **31**, pp. 836-842.

Dragovic, T., Minshall, R., Jackman, H.L., Wang, L-X. and Erdos, E.G., 1996, Kininase II-type enzymes. Their putative role in muscle energy metabolism. *Diabetes,* **45**, Suppl. 1, S34-S37.

Duthie, G., Pyne, D. and Hooper, S., 2003, Applied physiology and game analysis of Rugby Union. *Sports Medicine,* **33**, pp. 973-991.

Folland, J., Leach, B., Little, T., Hawker, K., Myerson, S., Montgomery, H. and Jones. D., 2000, Angiotensin-converting enzyme genotype affects the response of human skeletal muscle to functional overload. *Experimental Physiology*, **85**, pp. 575-579.

Gayagay, G., Yu, B., Hambly, B., Boston, T., Hahn, A., Celermajer, D.S. and Trent, R.J., 1998, Elite athletes and the ACE I allele – the role of genes in athletic performance. *Human Genetics,* **103**, pp. 48-50.

Hagberg, J.M, Ferrell, R.E., McCole, S.D., Wilund, K.R. and Moore, G.E., 1998, V0$_2$max is associated with the ACE genotype in postmenopausal women. *Journal of Applied Physiology,* **85**, pp. 1842-1846.

Hubert, C., Houot, A-M., Corvol, P. and Soubrier, F., 1991, Structure of the angiotensin I-converting enzyme gene. Two alternate promoters correspond to evolutionary steps of a duplicated gene. *Journal of Biological Chemistry*, **266**, pp. 15377-15383.

Jardine, M.A., Wiggins, T.M., Myberg, K.H. and Noakes, T.D.,1988, Physiological characteristics of rugby players including muscle glycogen content and muscle fibre composition. *South African Medical Journal,* **73**, pp. 529-532.

Jones, A., Montgomery, H.E. and Woods, D.R., 2002, Human performance: a role for the ACE genotype. *Exercise and Sport Sciences Reviews*, **30**, pp. 184-190.

Jonsson, J.R., Game, P.A., Head, R.J. and Frewin, D.B.,1994, The expression and localisation of the angiotensin-converting enzyme mRNA in human adipose tissue. *Blood Pressure*, **3**, pp. 72-75.

Lucia, A., Gomez-Gallego, F., Chicharro, J.L., Hoyos, J., Celeya, K., Cordova, A., Villa, G., Alonso, J.M., Barriopedro, M., Perez, M. and Earnest, C.P., 2004, Is there an association between ACE and CKMM polymorphisms and cycling performance status during 3-week races? *International Journal of Sports Medicine*, **25**, pp. 442-447.

Montgomery, H.E., Clarkson, P., Dollery, C. M., Prasad, K., Losi, M-A., Hemingway, H., Statters, D., Jubb, M., Girvain, M., Varnava, A., World, M., Deanfield, J., Talmud, P., McEwan, J. R., Mckenna, W. J. and Humphries, S., 1997, Association of angiotensin converting enzyme gene I/D polymorphism with change in left ventricular mass in response to physical training. *Circulation*, **96**, pp. 741-747.

Montgomery, H.E., Marshall, R., Hemingway, H., Myerson, S., Clarkson, P., Dollery, C., Hayward, M., Holliman, D. E., Jubb, M., World, M., Thomas, E. L., Brynes, A. E., Saeed, N., Barnard, M., Bell, J. D., Prasad, K., Rayson, M., Talmud, P.J. and Humphries, S.E., 1998, Human gene for physical performance. *Nature*, **393**, pp. 221.

Montgomery, H., Clarkson, P., Barnard, M., Bell, J., Brynes, A., Dollery, C., Hajnal, J., Hemingway, H., Mercer, D., Jarman, P., Marshall, R., Prasad, K., Rayson, M., Saeed, N., Talmud, P., Thomas, L., Jubb, M., World, M. and Humphries, S., 1999, Angiotensin-converting enzyme gene insertion/deletion polymorphism and response to physical training. *Lancet*, **353**, pp. 541-545.

Myerson, S., Hemingway, H., Budget, R., Martin, J., Humphries, S. and Montgomery, H., 1999, Human angiotensin I-converting enzyme gene and endurance performance. *Journal of Applied Physiology,* **87**, pp. 1313-1316.

Nazarov, I. B., Woods, D. R., Montgomery, H.E., Shneider, O. V., Kazakov, V. I.,

Tomilin, N. V. and Rogozkin, V.A., 2001, The angiotensin converting enzyme I/D polymorphism in Russian athletes. *European Journal of Human Genetics*, **9**, pp. 797-801.

Payne, J. and Montgomery, H., 2003, The renin-angiotensin system and physical performance. *Biochemical Society Transactions*, **31**, pp. 1286-1289.

Pescatello, L.S., Kostek, M.A., Gordish-Dressman, H., Thompson, P.D., Seip, R.L., Price T.B., Angelopoulos, T.J., Clarkson, P.M., Gordon, P.M., Moyna, N.M., Visich, P.S,. Zoeller, R.F., Devaney, J.M. and Hoffman, E.P., 2006, ACE ID genotype and the muscle strength and size response to unilateral resistance training. *Medicine and Science in Sports and Exercise*, **38**, pp. 1074-1081.

Pieruzzi, F., Abassi, Z.A. and Keiser, H.R., 1995, Expression of renin-angiotensin system components in the heart, kidneys, and lungs of rats with experimental heart failure. *Circulation*, **92**, pp. 3105-3112.

Pitsiladis, Y.P. and Scott, R., 2005, Essay: The makings of the perfect athlete. *Lancet*, **366**, S16-S17.

Rankinen, T., Wolfarth, B., Simoneau, J.A., Maier-Lenz, D., Rauramaa, R., Rivera, M.A., Bouley, M.R., Chagnon, Y.C., Perusse, L., Keul, J. and Bouchard, C., 2000, No association between angiotensin-converting enzyme ID polymorphism and elite endurance athlete status. *Journal of Applied Physiology*, **88**, pp. 1571-1575.

Rigat, B., Hubert, C., Alhenc-Gelas, F., Cambien, F., Corvol, P. and Soubrier, F., 1990, An insertion/deletion polymorphism in the angiotensin 1-converting enzyme gene accounting for half of the variance of serum enzyme levels. *Journal of Clinical Investigation*, **86**, pp. 1343-1346.

Rosendorff, C., 1996, The renin-angiotensin system and vascular hypertrophy. *Journal American College of Cardiology*, **28**, pp. 803-812.

Scott, R.A., Moran, C., Wilson, R.H., Onywera, V., Boit, M.K., Goodwin, W.H., Gohlke, P., Payne, J., Montgomery, H. and Pitsiladis, Y.P., 2005, No association between angiotensin converting enzyme (ACE) gene variation and endurance status in Keyans. *Comparative Biochemistry and Physiology*, Part A, **141**, pp. 169-175.

Sonna, L.A., Sharp, M.A., Knapik, J.J., Cullivan, M., Angel, K.C., Patton, J.F. and Lilly C.M., 2001, Angiotensin-convertying enzyme genotype and physical performance during US army basic training. *Journal of Applied Physiology*, **91**, pp. 1355-1363.

Stroth, U. and Unger, T., 1999, The renin-angiotensin system and its receptors. *Journal of Cardiovascular Pharmacology*, **33**, S21-S28.

Taylor, R.R., Mamotte, C.D.S., Fallon, K. and van Bockxmeer, FM., 1999, Elite athletes and the gene for angiotesin-converting enzyme. *Journal of Applied Physiology*, **87**, pp. 1035-1037.

Thomis, M.A.I., Huygens, W., Heuninckx, S., Chagnon, M., Maes, H.H.M., Claessens A.L., Vlietnick, R., Bouchard, C. and Beunen, G.P., 2004, Exploration of myostatin polymorphisms and the angiotensin-converting enzyme insertion/deletion genotype in responses of human muscle to strength training. *European Journal of Applied Physiology*, **92**, pp. 267-274.

Tiret, L., Rigat, B., Visvikis, S., Breda, C., Corvol, P., Cambien, F., and Soubrier, F., 1992, Evidence from the combined segregation and linkage analysis, that a variant of the angiotensin I-converting enzyme (ACE) gene controls plasma ACE levels. *American Journal of Human Genetics,* **51**, pp. 197-205.

Tsianos, G., Sanders, J., Dhamrait, S., Humphries, S., Grant, S. and Montgomery, H., 2004, The ACE gene insertion/deletion polymorphism and elite endurance swimming. *European Journal of Applied Physiology*, **92**, pp. 360-362.

Van Damme, R., Wilson, R.S., Vanhooydonk, B. and Aerts, P., 2002, Performance constraints in decathletes. *Nature*, **415**, pp. 755-756.

Williams, A.G., Dhamrait, S.S., Wooton, P.T.E., Day, S.H., Hawe, E., Payne, J.R., Myerson, S.G., World, M., Budgett, R., Humphries, S.E. and Montgomery, H.E., 2004, Bradykinin receptor gene variant and human physical performance. *Journal of Applied Physiology*, **96**, pp. 938-942.

Woods, D.R., Brull, D. and Montgomery, H.E., 2000, Endurance and the ACE (I/D) polymorphism. *Science Progress*, **83**, pp. 317-336.

Woods, D.R., Hickman, M., Jamshidi, Y., Brull, D., Vassiliou, V., Jones, A., Humphries, S. and Montgomery, H., 2001, Elite swimmers and the D allele of the ACE I/D polymorphism. *Human Genetics,* **108**, pp. 230-232.

Woods, D.R., World, M., Rayson, M.P., Williams, A.G., Jubb, M., Jamshidi, Y., Hayward, M., Mary, D.A.S.G., Humphries, S.E. and Montgomery, H.E., 2002, Endurance enhancement related to the human angiotensin I-converting enzyme I-D polymorphism is not due to differences in the cardiorespiratory response to training. *European Journal of Applied Physiology,* **86**, pp. 240-244.

Zhang, B., Tanaka, H., Shono, N., Miura, S., Kiyonaga, A., Shindo, M. and Saku, K., 2003, The I allele of the angiotensin-converting enzyme gene is associated with an increased percentage of slow-twitch I fibres in human skeletal muscle. *Clinical Genetics*, **63**, pp. 139-144.

CHAPTER FOURTEEN

The stress and attraction of cross-country running

T. Reilly, N. Chester and L. Sutton

Research Institute for Sport and Exercise Sciences, Liverpool John Moores University, Liverpool, UK

1. INTRODUCTION

Recreational and competitive running events are held on road, synthetic tracks, parkland, and on more rugged off-road terrain (Creagh *et al.*, 1998a). Over the last few decades, road running has prospered with the 'marathon boom' of the early 1980s, and the increasing popularity of shorter road races such as half-marathon and 10-km events. There has not been a corresponding rise in cross-country running as a means of improving health and fitness, despite some efforts to promote variations of this sport such as the 'HellRunner' series (TrailPlus, 2008).

Studies in which cross-country running has been the focus have largely been concerned with physical (biomechanical and physiological) responses to the varying resistance, terrain and gradient associated with the 'off-road' environment. Davies (1980) showed that air resistance increased the energy cost of running, implying that exercise on road, cross-country and outdoor track surfaces incurs greater energy expenditure than running on a treadmill or indoor track. When considering outdoor running, the surface terrain also affects energy expenditure. Creagh and Reilly (1997) found an increased energy cost of running over forest terrain compared to road running. The increased demand was mainly attributed to the biomechanical differences in stride pattern associated with the varying undergrowth and gradient of rough terrain. The same research group has also reported greater heart rate responses during cross-country running compared to fell-running and orienteering (Creagh *et al.*, 1998b). Jensen *et al.* (1999) stated that runners utilising 'heavy' terrain develop a better running economy (manifest as reduced oxygen uptake for a given speed) than do track runners. Changes in gradient tend to tax the anaerobic capacity more than running on the level (Olesen, 1992), and cause changes in skeletal muscle activation that include greater engagement of muscles in the lower body (Sloniger *et al.*, 1997a, 1997b).

Further to the established physiological adaptations, running can also have psychological benefits. Whilst the psychological benefits of exercise in general are acknowledged, there is little specific information about the unique influence of individual sports. There is a variety of subjective responses to exercise, all operating simultaneously, and there may be aspects that are particular to running and the environment in which it takes place. Most investigations in which the potential psychological effects of running are described have focused on road running, which has been suggested to improve mental well-being (Wilson *et al.*, 1980; Chan and Lai, 1990; Raglin, 2007), without necessarily increasing endurance capacity (Suter *et al.*, 1991). Harte and Eifert (1995) compared attention levels and hormone responses of subjects running outdoors and indoors on a treadmill. They found that both responses were altered in such a way that a normally pleasant task such as running may become tedious if the setting is not right. Krenichyn (2006) investigated outdoor environments in the broad context of physical activity promotion, but specific forms of running such as road, track or cross-country were not considered. The author noted that nature and the physical features of parkland were regarded by the participants in a positive manner, and that participation in activities such as brisk walking and running in such environments had a beneficial effect on mental well-being.

The prevalence and avoidance of injuries may also be a factor in choice of terrain for runners. Reilly and Foreman (1983) found running surface to be the most frequently ascribed cause of injury by road runners, and advised carrying out parts of training on more compliant surfaces such as parkland. This finding supported recommendations from a previous review, in which hard surfaces and cambered roads were deemed to be causal in 'shin splints', ankle and knee joint problems (McDermott and Reilly, 1981).

The aims of the present study were to establish motives, practices and preferences of runners participating on both terrains; to determine whether participants' subjective responses reflect current scientific findings; and to explore the potential of cross-country as a means of promoting health and well-being.

2. METHODS

2.1 Participants

The survey involved 91 runners of different levels of participation, ranging from recreational runners (n = 22) to club (n = 40), regional (n = 21) and national competitors (n = 8). Participants were recruited at training venues, the North-West and Northern road-relay championships and through personal approaches. Characteristics of the respondents are shown in Table 1.

2.2 Questionnaire

A questionnaire was administered to establish subjective responses to different aspects of running. Participants were required either to write a response; select the most appropriate answer; rank set responses in order of importance, 1 being the most important and 5 the least; or rate their agreement with given statements on a Likert-type scale, ranging from 0 (strongly disagree) to 10 (strongly agree). Topics included demographic information; participation in, and reasons for, different forms of running; physical fitness and injuries; psychological well-being; and environmental factors. When considering level of performance, runners were categorised as recreational or competitive (club, regional and national level performers).

Table 1. Characteristics of the participants (n = 91).

| | Male (n = 60) | | Female (n = 31) | |
	Mean ± SD	Range	Mean ± SD	Range
Age (years)	33.8 ± 13.6	15 - 64	30.8 ± 12.0	15 - 59
Current weekly training distance (km)	51.8 ± 29.4	0.0 – 112.7	34.2 ±19.1	0.0 – 80.5
Weekly training distance at peak fitness (km)	83.9 ± 43.1	19.3 – 281.8	51.3 ± 19.2	16.1 – 96.6

2.3 Statistical Analyses

Data were analysed using SPSS for Windows version 14. Statistical significance was set at $P \leq 0.05$. To determine if the data were from normally distributed populations, the residuals were analysed using the Shapiro-Wilk test. Each set of data analysed violated the relevant assumption of normality; therefore, nonparametric tests were used. Mann-Whitney U tests were used to determine if there were significant differences between competitive runners and recreational runners in the current average weekly distances run, in the distance covered when at peak fitness and in the distances covered on cross-country surfaces. Wilcoxon tests were used to compare the perceived cardiovascular and psychological benefits of road and cross-country running. Tied scores were reported when considered substantial. To identify if there were significant relationships between age and a number of aspects of running in the country (i.e., a lack of noise, improved air quality and the beauty of the countryside) a series of Spearman's correlations were used. Single classification Chi-square goodness-of-fit tests were used to determine if there was a significant difference between the frequencies of the responses for the most important reason for running (i.e., to maintain cardiovascular fitness, to enhance psychological well-being, to compete or to maintain/lose weight). Data are reported in the text as mean values ± standard deviation, except where nonparametric tests are reported, in which case data are expressed as median values ± interquartile range, unless otherwise stated.

3. RESULTS

All respondents confirmed their participation in running and 93% rated it as their main form of exercise. Indeed 12% of participants, of whom 64% were competitive runners, stated that running was their only form of physical exercise. As expected, competitive runners tended to cover greater distances than recreational runners in their current training programmes, running on average 48.3 ± 36.2 km, in comparison to 27.4 ± 20.5 km (U = 383.50, Z = -3.49, $P < 0.001$), and at peak fitness, covering 72.5 ± 48.3 km and 40.3 ± 18.9 km (U = 307.50, Z = -4.21, $P < 0.001$), respectively. Similarly, the cross-country running distance covered per week increased with competitive level (recreational 3.4 ± 7.9 km; competitive 11.3 ± 18.9 km; $U = 427.50$, Z = -3.08, $P = 0.002$). The competitive runners completed 12% more total training (and 7% more as a proportion of their training load) running cross-country compared to recreational runners, whilst the two groups spent similar amounts of training time running on the road (Figure 1). Across both sexes, road proved to be the most frequently used surface, comprising 59% of the runners' total training. Cross-country running was the second most commonly used training surface (27%).

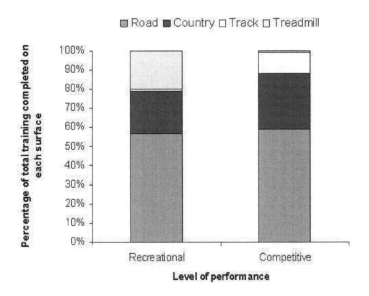

Figure 1. The proportion of running distance performed on different surfaces by recreational and competitive runners.

Improving or maintaining cardiovascular fitness was most commonly cited as the main reason for running, selected by 40% of respondents. Overall, 'to compete' was the second most frequently selected reason for running (32%), followed by 'to enhance psychological well-being or to de-stress' (14%) and 'to maintain or lose weight' (9%). There was a significant difference between the frequency of responses for the reasons for running ($\chi^2_3 = 24.23$, $P < 0.001$). Of the recreational runners, improving or maintaining cardiovascular fitness was the most frequently cited reason (60%), whereas the competitive runners repeatedly listed competing as their most important reason for running (43%). Differences between proportions were again significant (recreational $\chi^2_2 = 6.70$, $P < 0.035$; competitive $\chi^2_3 = 26.49$, $P < 0.001$).

The respondents reported a high enjoyment with regards the health and cardiovascular benefits of running (9.1 ± 1.1) and stated that running provided a very good cardiovascular workout compared to other forms of exercise (8.5 ± 2.0). Cross-country running was deemed to provide a greater (T = 306.50, Z = -2.94, $P = 0.003$) cardiovascular workout when compared to road running (Figure 2), although it should be noted that a substantial amount of tied ranks (n = 40) may bias the statistical results.

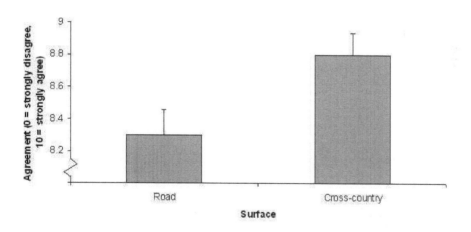

Figure 2. Respondents' agreement with the statement 'A very good cardiovascular workout is achieved through a) road running b) cross-country running' ($\bar{x} \pm$ SE).

Respondents strongly agreed that running was helpful in enhancing psychological well-being (8.5 ± 1.4). For the sample as a whole, cross-country running was believed to provide a greater psychological high than road running (this was reflected in the mean [cross-country 8.0 ± 1.7; road 7.6 ± 1.7], but not the median values [cross-country 8.0 ± 2.0; road 8.0 ± 2.3]; T = 554.00, ties = 33, Z = -2.20, $P = 0.028$ $t_{89} = 2.06$, $P < 0.001$). Furthermore, respondents stated that fewer injuries were experienced when running on country as opposed to road (mean [cross-country 6.5 ± 2.4; road 5.7 ± 2.7], median [cross-country 6.0 ± 4.0; road 6.0 ± 4.0]; t_{89} = 2.55, P = 0.012). When asked if injuries were a necessary burden of exercise and whether injuries were any more prevalent during running compared to other forms of exercise, responses were 4.7 ± 2.9 and 5.6 ± 2.5, respectively.

Participants highly agreed with the statement 'running helps me to de-stress' (8.5 ± 1.4). The respondents were in agreement that cross-country running and the associated environment were beneficial for mental relaxation and provided an escape from 'city life' (7.2 ± 2.7). The 'beauty of the countryside' was deemed to be the greatest motive for running cross-country, followed by 'improved air quality' (Figure 3). Whilst there were no significant differences between sexes, appreciation of these environmental factors increased progressively with age ('beauty' r_s = 0.38; 'air quality' r_s = 0.25; $P \leq 0.015$). Other responses to 'I enjoy running in the country as opposed to the city because of...', included enhanced feelings of safety and solitude.

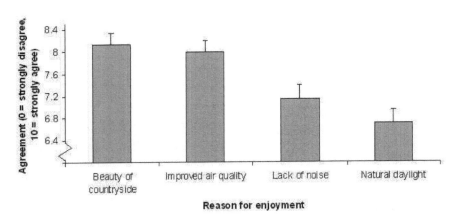

Figure 3. Average rating on a Likert-type scale (0 – 10) of participant responses to the statement 'I enjoy running in the country as opposed to the city because of...' ($\bar{x}$ ± SE).

4. DISCUSSION

The respondents in the current sample represented both sexes and a variety of ages and levels of performance. Overall, improving or maintaining cardiovascular fitness was most frequently cited as the most important reason for running, and it was believed that cross-country running provided a better cardiovascular workout than road running. There is undoubtedly a difference in cardiovascular response to the different forms of running, as established by Creagh *et al.* (1998b). The heart rate responses to cross-country running, fell running and orienteering were determined and compared to road running, the mean heart rates for each event varying with the differing demands of the terrain. Road and cross-country running elicited the greatest mean heart rates, and the cross-country runners showed the highest minimum heart rate of the four groups of participants. Cross-country runners spent 96% of the time exercising above 170 beats.min^{-1}, as opposed to 88% in the road runners. Percent time with heart rates above 170 beats.min^{-1} was 67% and 56% in fell running and orienteering, respectively. The cardiovascular response to exercise seems to be affected by changes in surface, vegetation and gradient. Physiological changes brought about by these environmental factors include altered musculoskeletal responses and greater overall energy expenditure when compared to horizontal running on a less varied terrain. Varying gradients are a common feature of cross-country courses. Sloniger *et al.* (1997a) reported a greater anaerobic capacity during uphill compared to horizontal running, and attributed the additional oxygen deficit to the increased mass of skeletal muscle activated in the lower extremities. This increased activation during exhaustive uphill running is achieved through an altered pattern of muscle activation that involves increased use of some muscles and less use of others (Sloniger et al., 1997b).

Creagh *et al.* (1998a) measured the effects of changing terrain on the biomechanical characteristics of running stride. With each progressively rougher terrain, velocity and step length decreased, whilst vertical hip displacement, knee lift and peak flexion angular velocity increased ($P < 0.01$). The authors concluded that running stride is significantly altered in response to changes in running surface. Further to changing gradient and surface, the wind resistance experienced during outdoor running also increases the energy cost of running (Davies, 1980).

Cross-country training has also been suggested to increase running economy. Jensen *et al.* (1999) found both orienteers and track runners to show greater oxygen consumption when running on cross-country terrain as opposed to a track. Whilst running on 'heavy terrain' reduced running economy, the impairment was lower in the orienteers compared to the track runners, which the authors attributed to their specific cross-country training. It is generally accepted that running off-road elicits a higher energy demand than road running. This increased demand is brought about by the altered physiological and biomechanical responses to the changing environment, and appears to be acknowledged by contemporary runners. Despite agreement with the greater cardiovascular benefits of cross-country running compared to road running, respondents in the current study also reported enjoyment of the health benefits associated with running in general, and rated it highly compared to other forms of exercise.

Respondents in the present study were generally neutral in their opinion as to whether injuries were a necessary affliction of regular exercise and whether injury prevalence was greater when comparing running to other forms of exercise. The participants stated that fewer injuries were experienced when running on country as opposed to road. This observation is in line with the consensus in the existing literature that cross-country surfaces provide greater compliance and result in fewer impact-related injuries in runners. Alongside inadequate footwear and excessive training load, hard surfaces are implicated in training-related injuries such as joint problems and 'shin splints' (McDermott and Reilly, 1981; Reilly and Foreman, 1983). Furthermore, the trauma associated with foot-strike has been found to be the main cause of red blood cell haemolysis during running, indicating a greater risk of red blood cell destruction and potential iron deficiency in athletes participating in foot-impact sports, the greatest risk being associated with high training volumes (Telford *et al.*, 2003). A reduction in impact forces, such as training on a 'softer', more compliant surface, as occurs in cross-country running, may reduce the mechanical trauma associated with running on harder surfaces.

Aside from the potential for injury prevention, cross-country terrain was also favoured by the respondents for its aesthetic qualities. The subjective responses of the runners mirrored previous findings in the literature. In the present study, the beauty of the countryside and improved air quality featured highly as motives for running cross-country. In a previous study by Krenichyn (2006), participants cited physical features such as hills, trails and scenery as positive aspects for the promotion of walking and running. Nature was described as stimulating the senses, restoring mental capacities, and parkland was deemed to be an important outdoor resource. Results of the present study also suggest that older runners are more agreeable to the virtues of cross-country in terms of the beauty of the surroundings. Overall, cross-country running was believed to elicit a significantly greater 'psychological high' than road running. Since the association of endorphins in the brain with uplifting mental states, the phenomenon of 'runners' high' has been alluded to (Szabo *et al.*, 2001), but has not been researched specifically in the context of cross-country running. The most common description of runners' high has been reported to be a general relaxation, with total euphoria being least used by runners as a description of their feelings (Masters, 1992). Following health and competition, psychological well-being was the next motive given by the current respondents for participation in running.

Road running was most common amongst respondents, despite cross-country running being the more frequently selected or highly ranked option in questions requiring a preference between the two. The competitive runners made greater use of cross-country surfaces during their training, possibly for protective reasons given their considerably greater training load, yet still performed the majority of their running on road surfaces. The overall lower participation in cross-country running may be due to greater availability of indoor or road surfaces for training, and the increased comfort and safety of these environments during poor weather conditions and darkness. Further reasons may be that fewer competitive events are held on cross-country surfaces compared to road, and there appears to be less publicity for cross-country running as a means of health promotion.

5. CONCLUSIONS

The balance of scientific evidence, together with the results from the present study, suggests that cross-country running is a relatively safe, mentally invigorating and physiologically beneficial form of exercise. The use of cross-country in the current sample increased, in both proportionate and absolute terms, the higher the level of running performance. Advantages of cross-country that were highlighted centred around the exposure to the natural environment and surface compliance. These advantages should not mask the benefits of other forms of running for health-related benefits. Despite a preference for cross-country running in terms of fitness, well-being and injury prevention, participation in both training and competition was lower than in road running. Greater promotion of cross-country running, in its traditional and modified contemporary versions, both as a means of improving health and well-being and as a competitive sport, is advocated to increase levels of participation.

Acknowledgements

This investigation was supported by a grant from Radox plc. The authors would like to acknowledge the assistance of Dr. Mark Scott with the statistical analyses.

References

Chan, D.W. and Lai, B., 1990, Psychological aspects of long-distance running among Chinese male runners in Hong Kong. *International Journal of Psychosomatics*, **37**, pp. 30-34.

Creagh, U. and Reilly, T., 1997. Physiological and biomechanical aspects of orienteering. *Sports Medicine*, **24**, pp. 409-418.

Creagh, U., Reilly, T. and Lees, A., 1998a, Kinematics of running on 'off-road' terrain. *Ergonomics*, **41**, pp. 1029-1033.

Creagh, U., Reilly, T. and Nevill, A., 1998b, Heart rate response to 'off-road' running events in female athletes. *British Journal of Sports Medicine*, **32**, pp. 34-38.

Davies, C.T.M., 1980, Effects of wind assistance and resistance on the forward motion of a runner. *Journal of Applied Physiology*, **48**, pp. 702-709.

Harte, J.L. and Eifert, G.H., 1995, The effects of running, environment, and attentional focus on athletes' catecholamine and cortisol levels and mood. *Psychophysiology*, **32**, pp. 49-54.

Jensen, K., Johansen, L. and Karkkainen, O.P., 1999, Economy in track runners and orienteers during path and terrain running. *Journal of Sports Sciences*, **17**, pp. 945-950.

Krenichyn, K., 2006, 'The only place to go and be in the city': women talk about exercise, being outdoors, and the meanings of a large urban park. *Health Place*, **12**, pp. 631-643.

Masters, K.S., 1992, Hypnotic susceptibility, cognitive dissociation and runners' high in a sample of marathon runners. *Clinical Hypnosis*, **34**, pp. 193-201.

McDermott, M. and Reilly, T., 1981, Common injuries in track and field athletes – 1. Racing and jumping. In *Sports Fitness and Sports Injuries*, edited by Reilly, T., (London: Faber and Faber), pp. 135-158.

Olesen, H.L., 1992, Accumulated oxygen deficit increases with inclination of uphill running. *Journal of Applied Physiology*, **73**, pp. 1130-1134.

Raglin, J.S., 2007, The psychology of the marathoner: of one mind and many. *Sports Medicine*, **37**, pp. 404-407.

Reilly, T. and Foreman, T.K., 1983, Recreational marathon running: some observations on injuries and correlates of performance. In *Proceedings of the Ergonomics Society's Conference*, edited by Coombes, K., (London: Taylor and Francis), pp. 187-192.

Sloniger, M.A., Cureton, K.J., Prior, B.M. and Evans, E.M., 1997a, Anaerobic capacity and muscle activation during horizontal and uphill running. *Journal of Applied Physiology*, **83**, pp. 262-269.

Sloniger, M.A., Cureton, K.J., Prior, B.M. and Evans, E.M., 1997b, Lower extremity muscle activation during horizontal and uphill running. *Journal of Applied Physiology*, **83**, pp. 2073-2079.

Suter, E., Marti, B., Tschopp, A. and Wanner, H.U., 1991, Effects of jogging on mental well-being and seasonal mood variations: a randomized study with healthy women and men. *Schweizerische Medizinische Wochenschrift*, **121**, pp. 1254-1263.

Szabo, A., Billet, E. and Turner, J., 2001, Phenylethylamine, a possible link to the antidepressant effects of exercise? *British Journal of Sports Medicine*, **35**, pp. 342-343.

Telford, R.D., Sly, G.J., Hahn, A.G., Cunningham, R.B., Bryant, C. and Smith, J.A., 2003, Footstrike is the major cause of hemolysis during running. *Journal of Applied Physiology*, **94**, pp. 38-42.

TrailPlus, 2008, *The 2008 Puma HellRunner Series* [online]. Glossop, UK. Available from: www.hellrunner.co.uk [accessed June 2008].

Wilson, V.E., Morley, N.C. and Bird, E.I., 1980, Mood profiles of marathon runners, joggers and non-exercisers. *Perceptual and Motor Skills*, **50**, pp. 117-118.

CHAPTER FIFTEEN

The effect of breast support on kinetics during overground running performance

J.L. White[1], J.C. Scurr[1] and N.A. Smith[2]

[1]Department of Sport and Exercise Science, University of Portsmouth, UK
[2]School of Sport, Exercise and Health Sciences, University of Chichester, UK

1. INTRODUCTION

The need for firm breast support, especially for larger breasted women, has been documented (Gehlsen and Albohm, 1980; Lorentzen and Lawson, 1987; Boschma et al., 1994; Mason et al., 1999). Breast pain is common during exercise. From a survey of 59 women, Lorentzen and Lawson (1987) reported that 56% had experienced sports-related breast pain. The majority of pain related to breast motion and occurred more frequently in larger-breasted women. A linear relationship between breast displacement and pain was reported by Mason et al. (1999). The use of a sports-bra was recommended by these authors, after they found a reduction in perceived breast pain when a sports-bra was worn. Excessive breast motion during physical activity not only causes discomfort, but embarrassment which may discourage females from participating in regular exercise (Shivitz, 2001). Therefore, a good sports-bra can help to increase confidence and consequently performance.

There are two basic styles of sports-bras: compression bras that flatten the breasts and evenly distribute their mass across the chest and encapsulation bras that support each breast separately in its own cup (Starr et al., 2005). The compression bra design is thought to be more effective for females with smaller breasts (cup size A or B), however, females with larger breasts (cup size C and above) may require more support and encapsulation bras are thought to be more effective (Page and Steele, 1999, Verscheure et al., 2000). A sports-bra should fit firmly enough to control breast motion, but not so tightly that it may interfere with breathing (Stamford, 1996). The inability to find a good supportive bra can lead to breast discomfort, poor adherence to exercise and the irreversible breakdown of breast tissue (Starr et al., 2005).

Early studies in breast motion during exercise only considered the vertical displacement of the breasts, in a small range of cup sizes and a single running speed (Gehlsen and Albohm, 1980, Lorentzen and Lawson, 1987). Vertical displacement of the breasts (cm) in a sports-bra condition was found to range between 1.37 and 3.66 in these studies (A-D cup size). Mason et al. (1999)

investigated vertical breast displacement and maximum downward deceleration force on the breast for three subjects during treadmill running, at four levels of intensity and in four breast support conditions. As expected, bare-breasted running caused the greatest vertical displacement and maximum downward deceleration force on the breast. The sports-bra reduced vertical breast movement by half compared to the no-bra condition.

Breasts do not simply move vertically, they also move simultaneously forwards and sideways (Griffin, 2004). Nevertheless, only vertical displacement of the breasts had been considered previously in breast motion research. Recently, a research group quantified three-dimensional (3D) breast displacement (Scurr *et al.,* 2007) and eliminated the movement of the body, i.e. normalised breast movement. Eight D cup subjects performed a ramped speed treadmill test. Mean total breast displacement was 0.12 m throughout one gait cycle for the bare-breasted activity. The authors proposed that mediolateral body motion present during gait resulted in considerable mediolateral breast movement; normalising only the vertical movement of the body would produce an overestimation of breast displacement during activity.

Ground reaction forces indicate the duration and intensity of stress that the body is subjected to whilst in contact with the ground (McClay *et al.,* 1994). Changes in vertical ground reaction forces due to varying levels of breast support during treadmill running were researched by Shivitz (2001). Seventeen subjects ran on an instrumented treadmill wearing low, medium and high support sports-bras; peak vertical force increased and breast motion and vertical stiffness decreased with increasing support, concluding insufficient breast support may lead to adaptations in a female's running mechanics (Shivitz, 2001). Boschma *et al.* (1994) investigated kinematic changes in gait during treadmill running with varying breast support. These authors found that average stride length, vertical centre of mass displacement and stride rate did not significantly differ between support conditions (including a bare-breasted condition) or cup size. This result was despite significant differences found in vertical breast displacement.

There is a shortage of literature that has successfully quantified breast motion. No previous researchers have investigated kinetic variables during overground running that may result from different types of breast support. It is important to understand the impact that differing modes of breast support may have on the female during exercise in order to assess implications for performance, injury and breast discomfort. Therefore, the aims of this investigation were to differentiate both kinetic variables and breast motion during overground running in a no-bra, everyday-bra and two sports-bra conditions. Valuable user opinions were also gained concerning breast comfort during testing. Firstly, it was hypothesised that vertical impact force increases as the level of breast support increased, coinciding with previous findings (Shivitz, 2001); secondly, mediolateral (m-l), anterioposterior (a-p) impact forces, loading rate and m-l impulse decrease as the level of breast support increased. Thirdly, it was hypothesised that vertical, m-l, a-p and resultant breast movement and breast discomfort increase as the level of breast support decreased, in-line with previous breast motion research (Gehlsen and Albohm, 1980; Lorentzen and Lawson, 1987; Mason *et al.*, 1999; Shivitz, 2001).

2. METHODS

2.1 Participants

Eight female participants with physical characteristics (mean ± SD): age 24.8 ± 6.4 years, height 1.66 ± 0.04 m, body mass 66.24 ± 7.13 kg, band size 34 ± 1.85, gave written informed consent to take part in the study. Exclusion criteria ensured participants were aged between 18-40 years, were currently active (exercised aerobically at least once a week), were not pregnant, had not given birth or breast-fed in the last year, were a D-cup breast size and had no previous surgery to the breasts. The study was approved by the Institutional Ethics Committee.

A trained bra fitter measured the participant's breast size; chest girth was determined underneath the breasts using a metal anthropometric tape, with 65-70 cm equating to a 32 band size, 70-75 cm equating to a 34 band size and so on. To begin with, cup size was assessed subjectively and a bra was fitted, then the following key parameters were used to ensure an acceptable bra fit i.e. shoulder strap tension, back strap tension and positioning of cup in relation to breast tissue.

2.2 Experimental Design

Participants completed a written consent form and pre-test health questionnaire; resting blood pressure was measured to ensure diastolic pressure was between 60 and 90 mmHg (Omron HEM705, Netherlands). Retro-reflective markers were placed on the clavicles (directly superior to the nipples), nipples (over the bra when applicable) and anterior superior iliac spines (ASIS). Following a familiarisation period, when participants were allowed sufficient practice to assure a consistent running speed and a normal foot placement on the platform, participants completed five successful running trials in each of the four randomised breast support conditions (no-bra, everyday-bra and two sports-bra conditions). The rest intervals between trials were dictated by the individuals.

The everyday-bra was the best selling bra from Marks & Spencer™ (seamfree plain underwired T-Shirt Bra, non-padded; made from 88% Polyamide and 22% elastane lycra). The two sports-bras were the leading UK sports bra manufacturers' top-selling compression (B515 racing back design; made from 57% polyester, 36% elastane, 4% polyamide and 3% xstatic silver nylon) and encapsulation (B109; made from 89% polyester, 6% nylon polyamide, 5% elastane) bras (Shock Absorber™, Gossard, UK).

Each trial consisted of the participant running at 3 $m\,s^{-1}$ (± 0.1 $m\,s^{-1}$) and striking the force platform with their right foot (Kistler 9281CA, Switzerland, 500 Hz). Running speed was set at 3 $m\,s^{-1}$ (10.8 $km\,h^{-1}$) (Sprint timer CM LSMEM, Brower, UK) to enable comparisons with previous breast motion studies (Gehlsen and Albohm, 1980; Lorentzen and Lawson, 1987; Mason *et al.*, 1999). Successful trials were those at the correct speed, where full contact was made with the force platform and when no targeting occurred. Five calibrated ProReflex infrared cameras (Qualisys, Sweden, 100 Hz) were positioned around the force platform;

force and visual data were synchronised using a 32ch USB Interface Synchronisation unit (Qualisys, Sweden).

Feedback was obtained from each participant after each breast support condition, to establish the comfort/perceived pain level. These data were collected quantitatively using a validated Likert scale of 1 to 10 (Mason *et al.,* 1999); with 1 being comfortable, 5 being uncomfortable, and 10 being painful.

2.3 Data Analysis

Maximum vertical, medial, lateral, anterior and posterior impact forces, mean loading rate and m-l impulse were normalised by body weight. All force data were averaged across five trials for each participant. For breast motion data analysis, markers were identified and three-dimensional data reconstructed in the Qualisys Track Manager Software (version 1.10.282, Qualisys, Sweden) using a grid calibration for internal camera parameters and a wand calibration for external camera parameters (bundle adjustment technique). Raw position-time data in mm (vertical, m-l, a-p) were exported into Excel (Microsoft Office, 2003). To establish independent resultant displacement of the right nipple and eliminate the six degrees of freedom movement of the body, a reference grid of left and right clavicles and ASIS' markers converted the global to a local coordinate system, with the origin at the right clavicle. Vertical, m-l, a-p and resultant displacement data for each nipple were averaged from two full gait cycles, converted in to m and averaged for the left and right nipple. Resultant, vertical, m-l and a-p breast displacement were averaged across the four individual phases of the gait cycle as suggested by Scurr *et al.* (2007). Comfort ratings for each breast support condition were averaged across subjects.

2.4 Statistics

Data were checked for normality and homogeneity of variance. Repeated-measures one-way analyses of variance (ANOVA) (SPSS, v12.0.1) was employed to examine differences in kinetic variables and breast displacement between the four support conditions. As the interaction between dependant variables was not of interest in this study, no multiple analyses of variance were considered. According to the Central Limit Theorem, sample means were normally distributed around the original distribution mean (Fallowfield *et al.,* 2005); therefore using an average of five trials was acceptable and suited the assumption of normal distribution for ANOVA. The significance level was set at 0.05. *Post-hoc* analysis took the form of multiple paired-samples t-tests with a Bonferroni adjustment (P = 0.008).

3. RESULTS

3.1 Force Data

All force data were averaged across five trials for each participant and then averaged for the eight participants (Table 1). Statistics indicated no condition effect on maximum vertical force ($F_{3, 21}$ = 1.00, P = 0.41), braking force ($F_{3, 2}$ = 2.09, P = 0.13), propulsion force ($F_{3, 21}$ = 1.08, P = 0.38) and lateral force ($F_{3, 21}$ = 0.81, P = 0.50). Mediolateral impulse was 47.4% larger in the no-bra condition (0.006 ± 0.002) compared with the everyday-bra (0.003 ± 0.002 bw s), however, there were no significant differences between support conditions (F $_{3, 21}$ = 1.70, P = 0.20). Loading rate was 26.5% higher for the no-bra condition (32.66 ± 16.82) compared with the compression bra (24.01 ± 4.51 bw s^{-1}), but there was no significant difference between support conditions (F $_{3, 21}$ = 2.18, P = 0.12). A significant difference was found between mean maximum medial force (F $_{3, 21}$ = 3.64, P = 0.03) and support condition; *post-hoc* paired-samples t-tests found this difference to lie between the compression and no-bra support conditions (t = -4.406, P = 0.003).

Table 1. Mean ± standard deviation (SD) values for all kinetic variables for each breast support condition (n = 8).

Variable		Encapsulation bra	Compression bra	Everyday bra	No bra
Maximum vertical force (bw)	Mean	2.09	2.11	2.12	2.05
	SD	0.17	0.18	0.14	0.16
Maximum braking force (bw)	Mean	-0.4	-0.4	-0.42	-0.43
	SD	0.04	0.05	0.05	0.06
Maximum propulsion force (bw)	Mean	0.23	0.24	0.22	0.23
	SD	0.03	0.03	0.03	0.03
Maximum medial force (bw)	Mean	0.13	0.12	0.12	0.15
	SD	0.03	0.03	0.05	0.04
Maximum lateral force (bw)	Mean	-0.04	-0.05	-0.05	-0.05
	SD	0.02	0.03	0.03	0.03
Loading rate (bw s^{-1})	Mean	26.32	24.01	25.3	32.66
	SD	7.83	4.51	6.48	16.82
Mediolateral impulse (bw s)	Mean	0.005	0.005	0.003	0.006
	SD	0.002	0.001	0.002	0.002

3.2 Breast Displacement Data

Resultant breast displacement data were averaged over five trials for each participant for two full gait cycles and subsequently averaged across the eight participants (Table 2). Repeated-measures one-way ANOVA results showed a significant effect of breast support condition on resultant displacement (F $_{1.14, 7.99}$ =

33.85, P < 0.01). Significant differences were also found between the a-p displacement ($F_{3,21} = 18.00$, P < 0.01), m-l displacement ($F_{1.14, 8.00} = 20.72$, P < 0.01) and vertical displacement ($F_{3,21} = 44.70$, P < 0.01) of the breasts between support conditions.

Table 2. Mean ± standard deviation (SD) values for breast displacement (m) across each breast support condition (n = 8).

Condition		Resultant displacement	A/p displacement	M/l displacement	Vertical displacement
Encapsulation	Mean	0.05	0.02	0.02	0.03
	SD	0.02	0.01	0	0.02
Compression	Mean	0.05	0.02	0.02	0.03
	SD	0.01	0	0	0.01
Everyday-bra	Mean	0.06	0.03	0.02	0.04
	SD	0.02	0.01	0.01	0.02
No-Bra	Mean	0.11	0.04	0.04	0.08
	SD	0.04	0.01	0.02	0.03

Post-hoc paired samples t-tests (Bonferroni adjustment, P = 0.008) revealed where the significant within-subject effects were found (Table 3).

Table 3. P values illustrating where significant (*) within-subject effects for breast displacement data were found.

Within-subject effect	Resultant displacement	A/p displacement	M/l displacement	Vertical displacement
Encapsulation - Compression	0.934	0.498	0.075	0.542
Encapsulation - Everyday	0.000*	0.014	0.036	0.000*
Encapsulation – No-bra	0.000*	0.002*	0.003*	0.000*
Compression - Everyday	0.001*	0.012	0.003*	0.001*
Compression – No-bra	0.001*	0.001*	0.001*	0.000*
Everyday – No-bra	0.001*	0.008*	0.004*	0.001*

*= p < 0.008

No significant differences were found for vertical, m-l, a-p or resultant breast displacement between the two sports-bra conditions; however, a substantial number of significant differences was found between the sports-bra conditions and the everyday-bra and no-bra conditions. Compared with the no-bra condition, the compression and encapsulation sports-bras significantly reduced breast movement by 56.6% and 56.4% respectively; the everyday-bra reduced breast movement by 41.3% (Figure 1).

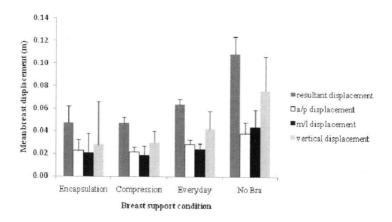

Figure 1. Mean breast displacement (m) in each breast support condition.

3.3 Comfort ratings

Mean comfort score ratings (Figure 2) illustrated that the compression sports-bra was considered to be the most comfortable breast support condition, whereas the no-bra condition was considered to be between uncomfortable and painful.

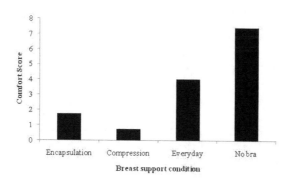

Figure 2. Mean comfort score ratings for each breast support condition.

4. DISCUSSION

In the present study, peak medial impact force in the no-bra condition (0.15 times body weight was significantly greater ($P = 0.003$) than in a compression sports-bra (0.12 times body weight). The increased medial force could be explained by significantly greater m-l breast displacement found (0.04 m) when wearing no-bra, compared with the compression sports-bra (0.02 m). The larger moment created about the foot-ground interface, as a consequence of more mediolateral breast displacement with no support, could have led to the significantly greater medial force measured during foot contact. A greater medial impact force indicates a higher intensity of stress for the participant, which could increase physiological demand, and over time could lead to the development of stress-related injuries (Keller *et al.*, 1996), confirming the need for effective breast support (Gehlsen and Albohm, 1980; Lorentzen and Lawson, 1987; Boschma *et al.*, 1994; Mason *et al.*, 1999).

Mean peak vertical impact force was lower when participants ran in no bra (2.05 times body weight) compared with the every-day (2.12 times body weight), encapsulation (2.09 times body weight) and compression (2.11 times body weight) bras. Results were non-significant; therefore the first hypothesis of this study is rejected. However, post priori power analysis (www.dssresearch.com/toolkit/ spcalc/power_a1.asp) demonstrated that results for vertical force had a low power (0.41). By increasing the sensitivity of the data, i.e. a larger sample size, a significant difference may be found in the future, verifying the findings of Shivitz (2001), which claimed peak vertical force was greater with increased support. A lower vertical impact force recorded in bare-breasted running could be attributable to the participant's attempt of greater force attenuation through flexion of the lower extremity joints; perhaps in an effort to reduce breast movement and discomfort, reflected in the comfort scores (Figure 2).

Greater impulses translate to a greater strain on the musculoskeletal system (McClay *et al.*, 1994). A low power for mediolateral impulse results (0.57) suggests that with greater participant numbers a significantly larger impulse may be seen during bare-breasted running, potentially increasing the participant's risk of injury. Lateral GRF did not differ significantly between breast support conditions; high variability within the results could be due to inconsistent foot placement between individuals (Hamill and Knutzen, 1995). Maximum a-p forces and loading rate did not differ significantly between breast support conditions. The force data collected suggest that insufficient breast support may prove to be detrimental to performance; therefore future research in this area is warranted. The second hypothesis is partially accepted as medial impact force was significantly lower with increased breast support.

The average cup size in the UK is a C (Lantin, 2003), therefore participants in this study may be considered large breasted. Vertical, m-l, a-p and resultant breast displacement were significantly greater in the no-bra condition ($P < 0.008$) than all other support conditions, confirming the third hypothesis of this study. Resultant and vertical breast displacement were significantly greater in the everyday-bra condition than the two sports-bra conditions ($P < 0.008$), reinforcing earlier research that found vertical breast motion decreased with increasing support (Gehlsen and Albohm, 1980; Lorentzen and Lawson, 1987; Mason *et al.,* 1999; Shivitz, 2001). *Post-hoc* analysis revealed no significant differences in breast displacement between the two sports-bras, challenging claims that encapsulation sports bras are better at reducing breast movement in larger breasted females (Page and Steele, 1999; Verscheure *et al.,* 2000).

The vertical breast motion data collected in this study compare well with previous breast motion studies (Gehlsen and Albohm, 1980; Lorentzen and Lawson, 1987). Mean resultant breast displacement of 0.12 m for D-cup subjects during bare-breasted running was reported by Scurr *et al.* (2007), which compares well with this study (0.11 m). This is the first study to quantify breast motion during overground as opposed to treadmill running; breast motion data suggest overground running did not have an influence on breast displacement, but further research is warranted.

Vertical breast displacement (0.08 m) was greater than m-l (0.04 m) and a-p (0.04 m) displacement in the no-bra condition and m-l and a-p breast displacement were significantly greater in the no-bra condition compared with the other support conditions. No previous studies have reported m-l or a-p breast displacement data; these data could have implications for sports bra design. Bras have been designed primarily to reduce vertical breast displacement, however, consideration should be given to reducing m-l breast displacement in particular, as in the no-bra condition m-l breast displacement contributes greatly to resultant breast displacement, which may be detrimental to performance due to the increases seen in medial impact force with no breast support. Mediolateral breast displacement did not significantly differ between the everyday-bra and the encapsulation sports-bra, suggesting the effectiveness of the sports-bra at reducing side-to-side movement is not dissimilar to an everyday-bra.

Mean comfort scores obtained from participants revealed that the compression sports-bra was considered the most comfortable; it also produced the least resultant breast displacement out of all support conditions. This is consistent with previous literature that proposed as breast support decreased, perceived pain increased (Mason *et al.,* 1999; Bowles and Steele, 2003). This finding provides validation for the use of a sports-bra during exercise, due to the lower perceived pain felt by participants.

This investigation has explored the effect that different breast support conditions have on kinetics and breast motion during overground running, and subjective feedback from participants was obtained. The findings suggest that there were differences in running kinetics and kinematics between the support conditions. Bra design needs to focus more on reducing mediolateral breast displacement to improve performance. Qualitative analysis provided justification for the use of sports-bras for females during exercise due to the increased comfort felt. Changes in kinetics and breast motion with insufficient breast support may have implications for females in terms of performance and discomfort during exercise.

References

Boschma, A.C., Smith, G.A., and Lawson, L., 1996, Breast support for the active woman: relationship to 3D kinematics of running. *Medicine and Science in Sports and Exercise,* **26**, S99.

Bowles, K., and Steele, J.R., 2003, Does inadequate breast support pose an injury risk? *Journal of Science and Medicine in Sport,* **6**, S67.

Fallowfield, J.L., Hale, B.J. and Wilkinson, D.M., 2005, *Using Statistics in Sport and Exercise Science Research,* (Chichester: Lotus Publishing).

Gehlsen, G. and Albohm, M., 1980, Evaluation of sports bras. *The Physician and Sportsmedicine,* **8**, pp. 89-96.

Griffin, M., 2004, *Beating the bounce.* Retrieved October 7, 2004, from http://www.theage.com.au/articles/2004/08/27/1093518087807.html?oneclick=true.

Hamill, J. and Knutzen, K.M., 1995, *Biomechanical Basis of Human Movement,* (UK: Williams & Wilkins).

Keller, T.S., Weisberger, A.M., Ray, J.L., Hasan, S.S., Shiavi, R.G. and Spengler, D.M., 1996, Relationship between vertical ground reaction force and speed during walking, slow jogging, and running. *Clinical Biomechanics,* **11**, pp. 253-259.

Lantin, B., 2003, *A weight off my shoulders.* Retrieved July 31, 2007, from http://www.telegraph.co.uk/core/Content/displayPrintable.jhtml;jses...

Lorentzen, D., and Lawson, L., 1987, Selected sports bras: A biomechanical analysis of breast motion while jogging. *The Physician and Sportsmedicine,* **15**, pp. 128-139.

Mason, B.R., Page, K., and Fallon, K., 1999, An analysis of movement and discomfort of the female breast during exercise and the effects of breast support in three case studies. *Journal of Science and Medicine in Sport,* **2**, pp. 134-144.

McClay, I.S., Robinson, J.R., Andriacchi, T.P., Frederick, E.C., Gross, T., Martin, P., Valiant, G., Williams, K.R., and Cavanagh, P.R., 1994, A profile of ground reaction forces in professional basketball. *Journal of Applied Biomechanics,* **10**, pp. 222-236.

Page, K., and Steele, J.R., 1999, Breast motion and sports brassiere design: implications for future research. *Sports Medicine,* **27**, pp. 205-211.

Scurr, J., Galbraith, H., and Hedger, W., 2007, Qualitative analysis of 3D breast displacement during treadmill activity. In *Proceedings of the 12th Annual Congress of the ECSS, 11-14 July 2007, Jyväskylä, Finland.*

Shivitz, N.L., 2001, *Adaptation of vertical ground reaction force due to changes in breast support in running,* (Eugene Oregon: Microform Publication).

Stamford, B., 1996, Sports bras and briefs: choosing good athletic support. *The Physician and Sportsmedicine,* **24**, Retrieved March 18, 2007, from http://www.physsportsmed.com/issues/1996/12_96/bra.htm

Starr, C., Branson, D., Shehab, R., Farr, C., Ownbey, S., and Swinney, J., 2005, Biomechanical analysis of a prototype sports bra. *Journal of Textile and Apparel, Technology and Management,* **4**, pp. 1-14.

Verscheure, S.K., Arata, A.W., and Hreljac, A., 2000, How effective are different sports bra designs at attenuating forces during jumping. *Medicine and Science in Sports and Exercise,* **32**, S267.

Part IV

Diet and fitness assessment

Prediction of maximal oxygen uptake from the Åstrand-Ryhming nomogram and ratings of perceived exertion

James Faulkner, Danielle Lambrick, Gaynor Parfitt,
Ann Rowlands and Roger Eston

School of Sport and Health Sciences, University of Exeter, Exeter, UK

1. INTRODUCTION

Maximal oxygen uptake ($\dot{V}O_2$max) is the most appropriate indicator of cardiovascular fitness and is often directly assessed during a graded exercise test (GXT). However, due to the practical concerns of conducting exhaustive exercise tests with non-athletic or patient populations a number of submaximal test protocols have been developed. These include the Åstrand-Ryhming nomogram (Åstrand and Ryhming, 1954), the Young Men's Christian Association (YMCA) cycle test (Golding et al., 1982), and the Rockport Fitness Walking Test (Kline et al., 1987), among others. Submaximal exercise tests provide a safe and practical means of assessing aerobic fitness for individuals who are likely to be limited by pain and fatigue, abnormal gait, or impaired balance (Noonan and Dean, 2000).

Developed over 50 years ago, the Åstrand-Ryhming nomogram is perhaps the most widely adopted method for predicting $\dot{V}O_2$max given the simplicity of its procedure and the minimal time requirements for its employment (Wagganer et al., 2003). Although the ACSM (American College of Sports Medicine, 2006) suggests that the Åstrand-Ryhming nomogram may be applied to both conditioned and unconditioned individuals, the nomogram was originally devised using only physically active participants (Åstrand, 1960). It is therefore unsurprising that much larger differences between measured and predicted $\dot{V}O_2$max have been reported when employing the Åstrand-Ryhming nomogram with untrained participants. Rowell et al. (1964) observed that $\dot{V}O_2$max was underestimated by 5.6 ± 4.2 % in highly-trained men, and 26.8 ± 7.2 % in untrained men. The Åstrand-Ryhming nomogram has also been shown to overestimate the $\dot{V}O_2$max of African-American men (Vehrs and Fellingham, 2006), and healthy physically active women (Zwiren et al., 1991).

Researchers have recently utilised Borg's (1998) 6-20 Ratings of Perceived Exertion (RPE) scale as a means of predicting $\dot{V}O_2$max from submaximal sensations of exertion. The Keele Lifestyle RPE nomogram (Buckley et al., 1998) was devised to be administered during an Åstrand-Ryhming cycle ergometer test to provide both perceptual and submaximal heart rate data, whilst simultaneously enabling a prediction of $\dot{V}O_2$max to be made from the ratings of perceived exertion. In a mixed group of sedentary, recreational and highly-trained men and women (n = 21, 18-43 years), Buckley *et al.* (1998) reported the RPE nomogram to underestimate measured $\dot{V}O_2$max (-0.308 ± 0.407 $\ell \cdot min^{-1}$), despite a high correlation between predicted and measured scores (r = 0.91). However, a derived regression equation ($\dot{V}O_2$max = 1.076 (RPE) + 0.085) provided suitable estimates of $\dot{V}O_2$max. To our knowledge, no further studies have assessed the validity of the RPE nomogram in estimating $\dot{V}O_2$max.

Accurate predictions of $\dot{V}O_2$max have been demonstrated when $\dot{V}O_2$ values have been extrapolated against submaximal RPE during passive estimation procedures (Faulkner and Eston, 2007). Recent research has also shown that submaximal, perceptually-regulated graded exercise tests (production protocol), using RPE 9, 11, 13, 15 and 17, provide accurate estimates of $\dot{V}O_2$max in high and low-fit men and women (Eston *et al.*, 2005, 2006, 2008; Faulkner *et al.*, 2007). Faulkner *et al.* (2007) observed that $\dot{V}O_2$max was predicted with acceptable error when $\dot{V}O_2$ values from the perceptually-regulated RPE range 9-13 were extrapolated to the theoretical maximal RPE 20.

Previous research by Eston and colleagues has assessed the efficacy of active production and passive estimation exercise tests which have been prescribed to a high intensity (i.e. RPE 17), and have followed a period of habituation. With sedentary and certain patient populations of low fitness, it would be of particular benefit if responses to low intensities of exercise (i.e. RPE 9-13) could be used as a suitable predictor of $\dot{V}O_2$max. A lower perceptual range would reduce the potential cardiovascular risk associated with the high intensity exercise (i.e. RPE 17) in sedentary and patient populations of low fitness. With regards to the duration, cost and participant motivation involved in exercise stress testing, it would also be of additional benefit to obtain accurate predictions of $\dot{V}O_2$max from a single submaximal exercise trial. In this regard, Faulkner *et al.* (2007) observed that the power outputs elicited from single perceptually-regulated bouts of exercise (production protocol), corresponding to RPE 13 and 15, were significantly correlated with $\dot{V}O_2$max and maximal power output.

The purpose of this study was to compare the prediction of $\dot{V}O_2$max from four submaximal exercise procedures to $\dot{V}O_2$max measured on a cycle ergometer. An estimation and production protocol (up to RPE 13), Keele Lifestyle RPE nomogram and Åstrand-Ryhming nomogram were employed with men and women of low-fitness. We hypothesised that the procedures involving the RPE provide predictions of $\dot{V}O_2$max that are at least as accurate, or more accurate than the Åstrand-Ryhming nomogram. Based on findings from previous research, we also hypothesised that the prediction of $\dot{V}O_2$max from the Åstrand-Ryhming nomogram significantly overestimates ($P < 0.05$) measured $\dot{V}O_2$max for women.

2. METHODS

2.1 Participants

Eleven men (Mean ± SD; age: 31.3 ± 9.0 years) and thirteen women (age: 40.8 ± 12.7 years) of low fitness volunteered to take part in the study (Table 1). All participants were healthy, asymptomatic of illness, and free from acute and chronic injuries and disease. Participants involved in the study were deemed to be of low fitness after self-reporting their activity status (no structured physical activity in the six months prior to the research study). Participants had no prior experience of perceptual scaling with the Borg 6-20 RPE scale. Each volunteer completed a medical history questionnaire and provided written informed consent prior to participation. Research was conducted in agreement with the guidelines and testing policies of the Ethics Committee of the School of Sport and Health Sciences at the University of Exeter.

Table 1. Participant descriptive statistics (Mean ± SD).

	N	Age (years)	Height (m)	Body mass (kg)	Body fat (%)	BMI (kg·m^2)	Blood pressure (mm Hg)
Men	N = 11	31.3 ± 9.0	1.80 ± .04	81.3 ± 12.1	19.9 ± 7.7	25.1 ± 3.9	125.3 / 79.9 ± 7.9 / 4.5
Women	N = 13	40.8 ± 12.7	1.64 ± .05	64.6± 10.2	27.5 ± 6.6	23.9 ± 3.4	124.5 / 77.7 ± 7.8 / 5.1

2.2 General procedures for cycle ergometry testing

Each participant performed three laboratory-based exercise tests on an electronically-braked cycle ergometer (Lode Excalibur Sport V2, Lode BV, Groningen, The Netherlands), separated by a 48 to 72 h recovery period. These included a maximal test (graded exercise test to establish $\dot{V}O_2$max), and two submaximal tests (RPE production protocol; Åstrand-Ryhming test) performed in a randomised order. Submaximal RPE data collected during the maximal exercise test provided a prediction of $\dot{V}O_2$max (estimation protocol). Heart rate and RPE data collected during the Åstrand-Ryhming test enabled two further predictions of $\dot{V}O_2$max (Åstrand-Ryhming nomogram and Keele Lifestyle nomogram, respectively). The final predictive procedure was the production protocol which involved cycling at perceptually-regulated intensities equating to RPE 9, 11 and 13.

Participants refrained from exercise (except for walking) during the recovery period. Resistance on the cycle ergometer was manipulated using the Lode workload programmer, accurate to ± 1 W, which was independent of pedal cadence. The display screen that specifies the resistance of the cycle ergometer was masked from the participants' view at all times. Pedal cadence was maintained at 60 rev·min⁻¹ throughout each exercise test. This frequency was in accordance with previous research that suggests that low pedal cadences (< 60 rev·min⁻¹) influence the magnitude of the local sensations of exertion (Jameson and Ring, 2000). Pedal frequency was displayed on a separate screen. Seat height and handle bar position were established and noted before the first exercise test and remained the same for each test thereafter. On-line respiratory gas analysis occurred every 10 s throughout each exercise test via a breath-by-breath automatic gas calibrator system (Cortex Metalyzer II, Biophysik, Leipzig, Germany). The system was calibrated prior to each test in accordance with manufacturer's guidelines using a 3-ℓ syringe for volume calibration and an ambient air measurement for gas calibration. Expired air was collected continuously using a facemask to allow participants to communicate verbally with the experimenters. A wireless chest strap telemetry system (Polar Electro T31, Kempele, Finland) continuously recorded heart rate during each exercise test. All physiological outputs were concealed from the participant.

Prior to the initial exercise test, the participants' height, mass, (SECA, Hamburg, Germany), body fat (Tanita, Body Composition Analyzer, BF-350, Tokyo, Japan) and blood pressure (Accoson, London, England) were measured. On the initial visit to the laboratory, participants were introduced to the Borg 6-20 RPE scale (Borg, 1998). Recent findings suggest that the overall sensation of exertion provides a more reliable and accurate prediction of $\dot{V}O_2$max than peripheral sensations of exertion (Faulkner and Eston, 2007). Thus, overall exertion was reported during each exercise test. Standardised instructions on how to implement the scale during the three exercise tests were given (Borg, 1998). There was no period of active familiarisation with the scale prior to participation. The scale was in full view of the participant for the duration of each test.

2.3 Exercise tests

2.3.1 Graded exercise test to establish $\dot{V}O_2$max (estimation)

The graded exercise test to establish $\dot{V}O_2$max was continuous and incremental in style, commencing at 40 W and increasing by 40 W every 3 minutes until $\dot{V}O_2$max. During the final 30 s of each increment of the graded exercise test, the participant reported the RPE. These submaximal RPE values, estimated throughout the exercise test (estimation protocol), were subsequently utilised to provide a prediction of $\dot{V}O_2$max. Termination of the exercise test resulted from the attainment of one or more of the following criteria for achieving $\dot{V}O_2$max: a plateau in oxygen consumption (± 150 ml·min⁻¹), a heart rate within ± 10 beats·min⁻¹ of the age-predicted maximum, a respiratory exchange ratio (RER) that was equal too or exceeded 1.15, a failure to maintain the required pedal cadence or if the participant reported volitional exhaustion (Cooke, 2001). After completing the exercise test, participants cycled against a light resistance to aid recovery. Due to the 'light' intensity employed during the initial stages of the graded exercise test, a warm-up was not implemented.

2.3.2 Perceptually-regulated graded exercise test (production)

The submaximal production protocol consisted of three perceptually-regulated exercise intensities, 9, 11 and 13 according to the Borg 6-20 RPE scale, being prescribed in that order. Participants were initially instructed to cycle at an intensity that was equivalent to an effort of RPE-9. Within a 2-min habituation period, participants instructed the experimenters to adjust the resistance until they were satisfied that the exercise intensity equated to an RPE-9. Participants then cycled at that intensity for 3 min. No further intensity adjustments were made unless requested by the participant. At the end of the 3-min period, the participant instructed the experimenters to adjust the exercise intensity so that an RPE-11 was attained. An initial increment of 25 W was used when increasing the power output between the RPE stages. Further increases in power output were applied in increments of 10 W and 5 W until the target RPE was attained. If the target RPE was exceeded, power output was adjusted in decrements of 10 W and 5 W. This process was repeated for an RPE-13. Participants had a 5-min cool- down period following completion of the exercise test. Due to reasons stipulated in the graded exercise test to establish $\dot{V}O_2$max, a warm-up period was not included in the test. The submaximal $\dot{V}O_2$ values elicited at each of the three RPE production levels (9, 11 and 13) were extrapolated to RPE 20 and RPE 19 to predict $\dot{V}O_2$max.

2.3.3 Modified Åstrand-Ryhming test (incorporating the Keele-Lifestyle RPE procedures)

There was no period of familiarisation (warm-up) prior to the Åstrand-Ryhming test (Åstrand and Ryhming, 1954). A pilot study involving six men and three women was undertaken to assess whether the work-rates suggested by the ACSM (2006) would be suitable for a sample of low-fit men and women. Although the upper threshold suggested by the ACSM was suitable for women of low-fitness (75 W), the suggested work rates for men of low-fitness (50 or 100 W) did not allow heart rate to exceed the specified threshold (> 125 beats·min^{-1}) that would permit an acceptable prediction of $\dot{V}O_2$max. Manipulation of the exercise intensity during the pilot study suggested that a work rate of 130 W was more appropriate for men of low-fitness, for the age range involved in this study. Consequently, during the Åstrand-Ryhming test, men and women cycled at a constant pedal cadence (60 rev·min^{-1}) for 6 minutes at 130 W and 75 W, respectively. According to the Åstrand-Ryhming criteria, heart rate was required to be between 125-170 beats·min^{-1} within the 5th and 6th minute of the test. An additional minute was administered if the heart rate did not exceed 125 beats·min^{-1} in the final minute of the exercise test. If the participant's heart rate was not within the desired range during the final two minutes of the exercise test (6th-7th minute), the test was terminated. Participants reported the RPE at the completion of each minute of the exercise test to enable $\dot{V}O_2max$ to be predicted from the Keele-Lifestyle RPE nomogram (Buckley *et al.*, 1998). Heart rate and oxygen uptake were recorded continuously. A cool-down period was allowed following completion of the test, at the discretion of each participant. The modified Åstrand-Ryhming test therefore allowed two submaximal predictive procedures to be utilised for the purpose of $\dot{V}O_2max$ estimation (Åstrand-Ryhming nomogram and Keele Lifestyle RPE nomogram).

2.4 Data analysis

2.4.1 Methodology to predict $\dot{V}O_2max$

The mean $\dot{V}O_2$ ($\ell{\cdot}min^{-1}$) values collated over the final 30 s of each stage of the graded exercise test to $\dot{V}O_2max$ (estimation protocol; EST) and the RPE production protocol (PROD) were used in the subsequent analyses. Firstly, linear regression analysis was used to determine the submaximal $\dot{V}O_2$ value which corresponded to an RPE of 13 ($\dot{V}O_2 = a + b\ (RPE_{13})$) during the estimation test. Individual linear regression analyses using the submaximal $\dot{V}O_2$ values elicited prior to and including an RPE_{13} for the estimation and perceptually-regulated exercise tests were extrapolated to an RPE of 19 (average peak RPE value reported at the termination of the GXT to $\dot{V}O_2max$; EST_{19}, $PROD_{19}$, respectively), and 20 (theoretical maximal RPE; EST_{20}, $PROD_{20}$, respectively), to enable a prediction of $\dot{V}O_2max$ ($\ell{\cdot}min^{-1}$).

2.4.2 Åstrand-Ryhming nomogram

The Åstrand-Ryhming nomogram (Åstrand and Ryhming, 1954; Åstrand, 1960) provided a prediction of $\dot{V}O_2max$ based on the average submaximal heart rate (125 -170 beats$\cdot$min^{-1}) recorded during the 5th and 6th minute of the Åstrand-Ryhming test and the corresponding exercise intensity (power output). Age correction factors were applied to account for the decrease in age-predicted maximal heart rate with increasing age, according to the original methods of Åstrand (1960) and as described by the ACSM (2006).

2.4.3 Keele-Lifestyle RPE nomogram

The RPE and the corresponding exercise intensity reported at the termination of the Åstrand-Ryhming test, combined with the associated correction factor reported by Buckley *et al.* (1998), enabled the RPE nomogram to provide an estimate of $\dot{V}O_2max$ (Table 2).

2.5 Statistical Analysis

Individual correlation coefficients obtained following linear regression analysis, performed on the $\dot{V}O_2$ and RPE data prior to and including an RPE_{13}, were converted to Fisher Zr values to approximate the normality of the sampling distribution during the estimation and production protocols.

A one sample t-test compared the average RPE reported at the termination of the Åstrand-Ryhming test with the average submaximal RPE intensity that was used to provide predictions of $\dot{V}O_2max$ from the estimation and production exercise tests (RPE_{13}).

2.5.1 Comparison of measured and predicted $\dot{V}O_2max$

A one-factor repeated measures ANOVA was used to compare measured $\dot{V}O_2max$ with predictions of $\dot{V}O_2max$ from the Åstrand-Ryhming nomogram, Keele Lifestyle RPE nomogram, estimation (EST_{19}, EST_{20}) and production exercise tests ($PROD_{19}$, $PROD_{20}$), for both men and women. Where assumptions of sphericity were violated, the critical value of F was adjusted by the Greenhouse-Geisser epsilon value from the Mauchley test of sphericity. Where significant differences were reported between measured and predicted $\dot{V}O_2max$, post-hoc analyses in the form of paired samples t-tests were performed using a Bonferroni adjustment to protect against type I error. Alpha was set at 0.05 and adjusted accordingly. Intra-class correlation coefficients (ICC) were calculated via the two-way mixed effects model to quantify the reproducibility between measured and predicted $\dot{V}O_2max$. A Bland and Altman (1986) 95 % limits of agreement (LoA) analysis quantified the agreement (bias and random error) between measured $\dot{V}O_2max$ and each predicted $\dot{V}O_2max$. In accordance with recommendations for performing LoA analysis (Bland and Altman, 1986; Nevill and Atkinson, 1997), the data were checked for heteroscedastic error by conducting correlation analysis on the measurement error and the mean of the measured and predicted $\dot{V}O_2max$ scores.

Table 2. The Keele Lifestyle (RPE) Nomogram for predicting $\dot{V}O_2$max from RPE during submaximal ergometer cycling. Published with permission from the Taylor & Francis Group, Routledge Journals. From Buckley et al., (1998). The development of a nomogram for predicting $\dot{V}O_2$max from ratings of perceived exertion during submaximal exercise on a cycle ergometer. *Journal of Sports Sciences*, 16, p. 15.

RPE	Pr $\dot{V}O_2$max (L·min⁻¹)	Cycle workrates (W)																			
		30	40	50	55	60	65	70	75	80	85	90	100	105	110	120	130	140	150	160	170	180
9	0.5	1.01	1.3	1.62	1.76	1.91	2.08	2.23	2.37	2.54	2.69	2.83	3.15	3.3	3.44	3.76	4.05	4.37	4.68	4.97	5.29	5.58
10	0.55	0.92	1.2	1.48	1.62	1.75	1.89	2.03	2.17	2.31	2.45	2.59	2.87	3.01	3.15	3.42	3.7	3.98	4.26	4.54	4.81	5.09
11	0.6	0.84	1.1	1.36	1.49	1.62	1.74	1.87	2	2.13	2.25	2.37	2.63	2.76	2.89	3.15	3.4	3.65	3.9	4.16	4.42	4.68
12	0.65	0.79	1.03	1.26	1.38	1.5	1.62	1.73	1.85	1.97	2.09	2.2	2.44	2.56	2.67	2.91	3.15	3.38	3.62	3.85	4.09	4.32
13	0.7	0.74	0.96	1.18	1.28	1.39	1.51	1.62	1.72	1.83	1.95	2.05	2.27	2.37	2.49	2.71	2.92	3.15	3.36	3.58	3.8	4.01
14	0.75	0.69	0.9	1.1	1.21	1.3	1.41	1.51	1.62	1.71	1.82	1.91	2.13	2.23	2.33	2.54	2.74	2.94	3.15	3.35	3.55	3.76
15	0.8	0.66	0.84	1.04	1.13	1.23	1.33	1.42	1.52	1.62	1.71	1.81	2.0	2.1	2.18	2.37	2.57	2.76	2.95	3.15	3.34	3.52
16	0.85	0.62	0.8	0.98	1.07	1.17	1.25	1.35	1.43	1.53	1.62	1.7	1.88	1.98	2.06	2.25	2.43	2.6	2.78	2.96	3.15	3.33
17	0.9	0.6	0.77	0.93	1.02	1.1	1.19	1.27	1.36	1.44	1.53	1.62	1.79	1.87	1.96	2.13	2.29	2.46	2.62	2.8	2.97	3.15
18	0.95	0.57	0.73	0.89	0.97	1.05	1.13	1.21	1.29	1.37	1.45	1.43	1.69	1.78	1.85	2.02	2.18	2.34	2.5	2.66	2.82	2.98
19	1.0	0.55	0.69	0.84	0.93	1.01	1.08	1.16	1.23	1.3	1.38	1.47	1.62	1.69	1.76	1.91	2.08	2.23	2.37	2.54	2.69	2.83

Instruction: Read down the left-hand column to the RPE given during the sixth minute of the exercise test and then across to the column of the cycle work rate. For example, an RPE of 12 at 90 W gives an estimated $\dot{V}O_2$max of 2.2 L·min⁻¹. *Note:* The nomogram is based on the following equation: Predicted $\dot{V}O_2$max = (WR / 70) / (pr $\dot{V}O_2$max: RPE). Where WR is the cycle work rate in watts, 70 is a constant oxygen cost ($\dot{V}O_2$) for ergometer cycling of 70 W L⁻¹ min⁻¹ (Astrand & Rodahl, 1986, *Textbook of Work Physiology*. New York: McGraw-Hill), and pr $\dot{V}O_2$max: RPE = the proportion of $\dot{V}O_2$max represented by the RPE after 6 min of the cycle test. The following regression equation is used to correct the difference between measured and predicted scores: Corrected Predicted $\dot{V}O_2$max = 1.076x + 0.085, where x is the predicted $\dot{V}O_2$max calculated from the earlier stated predictive equation.

2.5.2 *Comparison of absolute and relative exercise intensities between submaximal tests*

A series of one-factor ANOVAs were used to identify differences in the absolute ($\dot{V}O_2$, HR, power output) and relative (%$\dot{V}O_2$max, %HRmax) exercise intensities reported at the termination of the Åstrand-Ryhming test and at an RPE_{13} in the estimation and production tests. Paired sample t-tests were used to assess differences in %$\dot{V}O_2$max and %HRmax between men and women for each of the submaximal exercise tests.

All data were analyzed using the statistical package SPSS for Windows, PC software, version 13.

3. RESULTS

The physiological ($\dot{V}O_2$max, HRmax, RER), physical (power output), and perceptual (RPE) values reported at the termination of the graded exercise test to $\dot{V}O_2$max, are shown in Table 3.

Individual linear correlations from (typically) three submaximal $\dot{V}O_2$ data points elicited prior to and including an RPE_{13} produced average Fisher Zr values of 2.42 ± 0.61 and 2.33 ± 0.58 for the estimation and production exercise tests, respectively. When converted to correlation coefficients, a strong relationship between the RPE and $\dot{V}O_2$ values was demonstrated for both the estimation (0.97 ± 0.38) and production exercise tests (0.97 ± 0.38). A one sample t-test revealed no differences (t $_{(23)}$ = -0.42, $P > 0.05$) between the RPE reported at the termination of the Åstrand-Ryhming test (12.8 ± 1.95) and an RPE_{13}.

Table 3. The average (mean ± SD) $\dot{V}O_2$max, HRmax, RER, RPE and power output values reported during the graded-exercise test to $\dot{V}O_2$max, in men and women.

	n	$\dot{V}O_2$max (ℓ·min^{-1})	HRmax (beats·min^{-1})	RER	RPE	Power output (W)
Men	11	3.13 ± .27	185.6 ± 9.3	1.21 ± .06	19.1 ± .7	247 ± 24
Women	13	1.90 ± .29	179.1 ± 11.1	1.23 ± .09	19.3 ± 1.1	155 ± 30

3.1 Comparison of measured and predicted $\dot{V}O_2$max

A one-way ANOVA revealed no significant differences between measured and predicted $\dot{V}O_2$max (Åstrand-Ryhming nomogram, Keele-Lifestyle RPE nomogram, EST_{19}, EST_{20}, $PROD_{19}$, $PROD_{20}$) for men (F $_{1.8, 18.1}$ = 0.68, $P > 0.05$). The mean ($\pm$ SEE) measured (3.14 $\pm$ 0.08 $\ell \cdot min^{-1}$) and predicted $\dot{V}O_2$max values were 3.15 $\pm$ 0.21, 3.11 $\pm$ 0.14, 3.10 $\pm$ 0.27, 2.92 $\pm$ 0.25, 3.00 $\pm$ 0.12 and 2.83 $\pm$ 0.11 $\ell \cdot min^{-1}$, respectively (Figure 1A). However, a significant difference was observed between measured and predicted $\dot{V}O_2$max for women (F $_{(2.0, 23.7)}$ = 5.68, $P = 0.01$). The mean ($\pm$ SEE) measured (1.90 $\pm$ 0.08 $\ell \cdot min^{-1}$) and predicted $\dot{V}O_2$max was 2.35 $\pm$ 0.17, 1.86 $\pm$ 0.07, 2.05 $\pm$ 0.13, 1.95 $\pm$ 0.12, 1.86 $\pm$ 0.11 and 1.77 $\pm$ 0.10, respectively. As demonstrated in Figure 1b, post-hoc analyses ($P = 0.008$) revealed that the Åstrand-Ryhming nomogram significantly overestimated $\dot{V}O_2$max in women. The agreement between the measured and predicted $\dot{V}O_2$max (ICC and 95 % LoA) can be observed in Table 4.

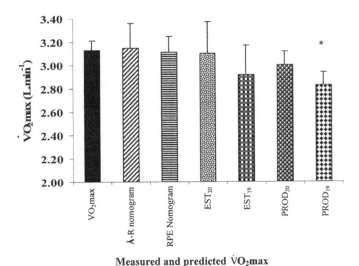

Figure 1a. Comparison of measured and predicted $\dot{V}O_2$max (mean $\pm$ SE) for men.
* Approaching significance ($P = 0.013$).

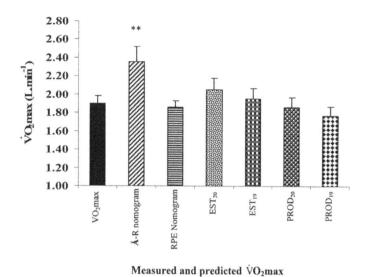

Figure 1b. Comparison of measured and predicted $\dot{V}O_2max$ (mean ± SE) for women.
** Significant difference between measured and predicted $\dot{V}O_2max$ ($P = 0.008$).

Table 4. The ICC and 95 % LoA analysis between measured $\dot{V}O_2max$ and predicted $\dot{V}O_2max$ ($\ell \cdot min^{-1}$) for men and women. The confidence intervals of the ICC are reported in parenthesis.

		Å-R Nomogram	RPE Nomogram	EST_{20}	EST_{19}	$PROD_{20}$	$PROD_{19}$
Men	ICC	.45 (-1.05 - .85)	.43 (-1.14 - .85)	.35 (-1.42 - .82)	.39 (-1.28 - .84)	.64 (-.34 - .90)	.65 (-.31 - .91)
	LoA ($L \cdot min^{-1}$)	-.03 ± 1.22	.02 ± .90	.03 ± 1.63	.21 ± 1.47	.12 ± .67	.30 ± .65
Women	ICC	.54 (-.51 - .86)	.72 (.07 - .91)	.78 (.28 - .93)	.79 (.33 - .96)	.83 (.45 - .95)	.84 (.48 - .95)
	LoA ($L \cdot min^{-1}$)	-.46 ± 1.04	.04 ± .52	-.15 ± .64	-.05 ± .60	.04 ± .52	.13 ± .48

3.2 Comparison of absolute and relative exercise intensities between submaximal tests

One-factor ANOVAs demonstrated no significant differences in the exercise intensity (power output) at the conclusion of the Åstrand-Ryhming test, and at an RPE_{13} in the estimation and production exercise tests, for both men ($F_{2,20} = 3.00$, $P > 0.05$) and women ($F_{2,24} = 2.90$, $P > 0.05$). As shown in Table 5, similar findings were observed between the three submaximal exercise tests when comparing the absolute ($\dot{V}O_2$, HRmax) and relative (%$\dot{V}O_2$max, %HRmax) exercise intensities for men and women ($P > 0.05$). However, the results of paired samples t-tests indicated higher %$\dot{V}O_2$max for women compared with men at an RPE_{13} during the estimation and production exercise tests ($P < 0.05$).

Table 5. The absolute and relative $\dot{V}O_2$, HR and power output values reported at an equivalent perceptual intensity (RPE_{13}) between the Åstrand-Ryhming test, estimation and production exercise tests, for men and women. Values are mean ± SD.

		$\dot{V}O_2$ (L·min^{-1})	% $\dot{V}O_2$max	HR (b·min^{-1})	% HRmax	Power Output (W)
Men	**Å-R test**	2.06 ± .07	70.8 ± 8.7	140.4 ± 4.5	75.7 ± 7.6	133 ± 2
	Estimation	1.88 ± .14	59.9 ± 12.2	136.0 ± 7.1	72.0 ± 12.0	124 ± 8
	Production	1.77 ± .06	57.0 ± 6.1	131.6 ± 4.3	69.6 ± 6.7	116 ± 5
Women	**Å-R test**	1.38 ± .08	66.2 ± 8.2	141.0 ± 3.9	79.4 ± 8.7	79 ± 2
	Estimation	1.31 ± .08	68.9 ± 9.6*	141.1 ± 3.2	78.8 ± 6.6	85 ± 4
	Production	1.22 ± .06	64.6 ± 8.5*	138.8 ± 3.8	77.7 ± 8.9	79 ± 4

4. DISCUSSION

The aim of this study was to assess whether the ratings of perceived exertion, corresponding to a low-moderate exercise intensity, could be used to provide accurate predictions of $\dot{V}O_2$max when compared to the Åstrand-Ryhming nomogram. As hypothesised, the estimates of $\dot{V}O_2$max provided by the Åstrand-Ryhming nomogram were substantially higher than measured values for women. The present study provided further evidence for the validity of predicting $\dot{V}O_2$max from the ratings of perceived exertion in men and women (Eston *et al.*, 2005, 2006, 2008; Faulkner and Eston, 2007; Faulkner *et al.*, 2007). The extrapolation of submaximal $\dot{V}O_2$ values (prior to and including RPE_{13}) from both the graded exercise (passive estimation) and active production protocols provided accurate predictions of $\dot{V}O_2$max from a single, low-moderate exercise session in men and women of low-fitness. The study also supports previous research that has shown the RPE to be highly correlated with $\dot{V}O_2$ at low-moderate RPE intensities (Faulkner and Eston., 2007; Faulkner *et al.*, 2007).

The extrapolation of $\dot{V}O_2$ values from the estimation test (RPE_{13}) to EST_{19} and EST_{20} provided accurate estimates (within 0.01 - 0.03 $\ell \cdot min^{-1}$, $P > 0.05$) of $\dot{V}O_2$max for both men and women. Using these procedures, measured $\dot{V}O_2$max was underestimated in men by 1 % (EST_{20}) and 7 % (EST_{19}), but overestimated in women by 8 % and 3 %, respectively. As predictions of $\dot{V}O_2$max were derived from individual linear regression equations ($\dot{V}O_2$max = a + b (RPE)), estimates of $\dot{V}O_2$max were expected to be higher when extrapolated to the theoretical maximal RPE on the Borg 6-20 scale (EST_{20}) for both men (0.18 $\ell \cdot min^{-1}$) and women (0.10 $\ell \cdot min^{-1}$). Similar findings were demonstrated with the perceptually-regulated exercise test for both men and women (0.17 and 0.09 $\ell \cdot min^{-1}$, respectively). As can be seen from Table 4, the perceptually-regulated exercise test revealed the greatest consistency between predicted and measured $\dot{V}O_2$max, demonstrated by high ICCs and small LoA for both men and women. Similar ICCs ($\sim$ 0.72-0.79) and LoA were revealed for the estimation test for women. However, much lower ICCs ($\sim$ 0.35-0.39) and larger LoA were revealed for men. Although predictions of $\dot{V}O_2$max from the production exercise tests were similar to measured values, it should be noted that $PROD_{19}$ underestimated measured $\dot{V}O_2$max by approximately 10 %, in the men.

The study findings demonstrated that extrapolation of submaximal $\dot{V}O_2$ values from an RPE_{13} (estimation and production exercise tests) may provide estimates of $\dot{V}O_2$max that are similar to, or more accurate than the Åstrand-Ryhming nomogram. In this study, the Åstrand-Ryhming nomogram was observed to be a valid predictor of $\dot{V}O_2$max in men. Predictions of $\dot{V}O_2$max from the Åstrand-Ryhming nomogram were accurate to within 1 % ($\sim$ 0.01 $\ell \cdot min^{-1}$) of measured $\dot{V}O_2$max in men. However, LoA between the predictions of $\dot{V}O_2$max from the Åstrand-Ryhming nomogram and measured $\dot{V}O_2$max (-0.03 ± 1.22 $\ell \cdot min^{-1}$) were much larger than those demonstrated from the perceptually-regulated exercise tests ($PROD_{19}$ and $PROD_{20}$) and the RPE nomogram (0.30 ± .65, 0.12 ± 0.67, 0.02 ± 0.90 $\ell \cdot min^{-1}$, respectively). For the women in this study, the Åstrand-Ryhming nomogram overestimated $\dot{V}O_2$max by 19 % ($\sim$ 0.45 $L \cdot min^{-1}$). This is in accordance with previous research that has demonstrated the $\dot{V}O_2$max of women to be overestimated by approximately 20 % during cycle ergometry (Siconolfi *et al.*, 1985; Zwiren *et al.*, 1991). Additionally, with the women of this study, the Åstrand-Ryhming nomogram revealed the lowest ICC and largest LoA than any of the submaximal exercise tests that utilised the ratings of perceived exertion. However, the margin of error between measured and predicted $\dot{V}O_2$max from each submaximal protocol was relatively wide. This may be a consequence of the limited sample size that was utilised in this study. Such predictive protocols may be more appropriate when assessing larger samples of a similar population.

This is the first study to provide independent support for the use of the Keele-Lifestyle RPE nomogram (Buckley *et al.* 1998) as an acceptable means of predicting $\dot{V}O_2$max. Such a nomogram may be of value to a clinician, given that accurate predictions of $\dot{V}O_2$max may be obtained following a 6-min bout of exercise. The results offer further evidence that the RPE provides predictions of $\dot{V}O_2$max that are as good as, or better than heart rate (Morgan and Borg, 1976; Faulkner *et al.*, 2007).

Although the predictions of $\dot{V}O_2$max from the Keele-Lifestyle RPE nomogram were within acceptable limits, it is clear that the nomogram contains a fundamental error. The estimation of the fractions of $\dot{V}O_2$max elicited at each successive RPE in the left hand column of the nomogram (pr$\dot{V}O_2$max: RPE), particularly at the lower and moderate RPE values, are too high. These fractions are overestimated by approximately 10 %. The overestimation of pr$\dot{V}O_2$max: RPE during cycle ergometry is confirmed by previous and current research on healthy individuals (Eston and Williams, 1988; Eston *et al.*, 2005, 2008; Faulkner and Eston, 2007; Faulkner *et al.*, 2007). A further concern with the Keele-Lifestyle RPE nomogram is that a correction factor is necessary to compensate for the significant differences between predicted and actual $\dot{V}O_2$max. The development of a nomogram which takes these factors into account and is based on a larger sample (the nomogram used in this study was developed on only 21 subjects) is therefore recommended.

The RPE has been shown to be more strongly correlated with physiological markers of exertion ($\dot{V}O_2$ and HR) at higher exercise intensities (Eston *et al.*, 2005, 2006, 2008; Faulkner *et al.*, 2007). As such, slightly less accurate estimates of $\dot{V}O_2$max are generated when predicting $\dot{V}O_2$max from a lower perceptual range (i.e. RPE 9-13; Faulkner *et al.*, 2007). As expected, the ICCs between measured and predicted $\dot{V}O_2$max revealed in this study were lower than previously reported findings. A greater degree of agreement between measured and predicted $\dot{V}O_2$max is obtained following a period of familiarisation, and when using higher perceptual ranges (i.e. 9-17; Eston *et al.*, 2005, 2006, 2008; Faulkner *et al.*, 2007). The ICCs derived from the production exercise test are comparable to those reported by Faulkner *et al.* (2007) from an initial bout of exercise (~ 0.84), particularly for the women in this study. Exercise tests that terminate at such submaximal intensities, for example RPE$_{13}$, may be considered beneficial for individuals of low-fitness, especially as low-fit individuals are more likely to be negatively affected by the demand of high intensity exercise (Hardy and Rejeski, 1989; Parfitt *et al.*, 1996). Men and women in this study chose to exercise between 57-69 % of $\dot{V}O_2$max when requested to perform a single exercise session up to an RPE$_{13}$. These intensities are similar to those reported by Eston *et al.* (2005) and Faulkner *et al.* (2007).

In this study, the Åstrand-Ryhming test was modified for men to ensure that heart rate achieved the specified range (125 – 170 beats·min^{-1}) to facilitate a prediction of $\dot{V}O_2$max. The consequence of this was to increase power output to 130 W, which was above the recommended level of 100 W (ACSM, 2006). The average heart rate corresponding to 130 W was 140 beats·min^{-1}. Davies (1968) reported that predictions of $\dot{V}O_2$max are more accurate from heart rates of 140-180 beats·min^{-1}.

The duration of the estimation and production protocols were longer than the Åstrand-Ryhming test (6-7 min) due to (typically) being a three stage exercise test. On average, the power output corresponding to a production intensity equivalent to an RPE$_{13}$ was attained after approximately 11 min, which accounts for the adjustment period given between each 3-min RPE stage. During the graded exercise test (estimation), the power output that corresponded to a derived RPE$_{13}$ was approximately 6 min for women and 9 min for men after the onset of exercise. Despite a slightly longer duration, the estimation and production exercise tests may provide a more appropriate submaximal measure of performance than the Åstrand-Ryhming test, given the more accurate estimate of $\dot{V}O_2$max at a comparable physiological and perceptual workload.

The present study concurs with previous research that observed RPE values which were less than the theoretical maximal RPE of 20 at maximal exertion (St Clair Gibson *et al.*, 1999; Noakes *et al.*, 2004; Eston *et al.*, 2007; Faulkner and Eston, 2007; Faulkner *et al.*, 2007). However, in accordance with Faulkner and Eston (2007), extrapolation of submaximal $\dot{V}O_2$ from RPE values prior to and including an RPE$_{13}$ to theoretical maximal RPE (20) provided a more accurate prediction of $\dot{V}O_2$max in comparison to extrapolation to a terminal RPE of 18 or 19. Conversely, when performing exercise at a higher perceptual intensity (i.e. RPE 17), research has revealed that extrapolation to an RPE of 19 provides more accurate predictions of $\dot{V}O_2$max (Eston *et al.*, 2006; Faulkner *et al.*, 2007). It is therefore feasible that during low-intensity exercise, participants gauge their perception of exertion in relation to a perceptual framework using an observed maximal value of 20 on the Borg 6-20 RPE scale. However, during high-intensity exercise (i.e. RPE 17) the perceptual frame of reference is modified slightly, which may transpire as a reduction in the terminal RPE (i.e. RPE 19). This observation is to some extent in conflict with previous interpretations of the teleoanticipation hypothesis. On the basis of their observations during the times to exhaustion at a given submaximal work rate, Eston *et al.* (2007) proposed that the maximal RPE is anticipated at the onset of exercise, and that the brain increases the RPE as a proportion of the amount of exercise completed, or the amount that remains. The teleoanticipatory process is considered to underpin the length of the exercise bout, and incorporate prior exercise experience and substrate utilisation to avoid catastrophic failure (Noakes *et al.*, 2004; St Clair Gibson *et al.*, 2006). It would appear that further research is necessary to examine the rate of change in RPE at low, moderate and severe exercise intensities, and the individual setting of the maximal RPE of these intensities.

In conclusion, the results of the present study support the use of the ratings of perceived exertion for providing accurate predictions of $\dot{V}O_2$max. The Keele Lifestyle RPE nomogram, estimation and production exercise tests provided estimates of $\dot{V}O_2$max that were similar to the measured $\dot{V}O_2$max during a graded exercise test to exhaustion. Although somewhat less accurate than a procedure involving habituation trials, this study has shown that a single submaximal exercise session that incorporates the RPE (estimation or production) has the potential to provide an appropriate means of estimating maximal functional capacity in healthy men and women of low-fitness. Furthermore, as this study has shown that the Åstrand-Ryhming nomogram significantly overestimates $\dot{V}O_2$max in women of low fitness, a submaximal exercise test encompassing the ratings of perceived exertion may be more appropriate. Further research is required to identify whether the consistency and repeatability of predictions of $\dot{V}O_2$max provided by the ratings of perceived exertion may be improved.

References

American College of Sports Medicine, 2006, *ACSM's Guidelines for Exercise Testing and Prescription*, 7th ed., (Philadelphia: Lippincott, Williams and Wilkins), pp. 70-75.

Åstrand, I., 1960, Aerobic work capacity in men and women with special reference to age. *Acta Physiologica Scandinavica,* **49** (Suppl. 169), pp. 1-92.

Åstrand, P.-O. and Rodahl, K., 1986, *Textbook of Work Physiology*, (New York: McGraw-Hill).

Åstrand, P.-O. and Ryhming, I., 1954, A nomogram for calculation of aerobic capacity (physical fitness) from pulse rate during submaximal work. *Journal of Applied Physiology,* **7**, pp. 218-221.

Bland, J. M. and Altman, D.G., 1986, Statistical methods for assessing agreement between two methods of clinical measurement. *Lancet, i,* pp. 307-310.

Borg, G., 1998, *Borg's Perceived Exertion and Pain Scales*, (Champaign, IL: Human Kinetics).

Buckley, J., Cannon, E. and Mapp, G., 1998, The development of a nomogram for predicting VO_2max from ratings of perceived exertion during submaximal exercise on a cycle ergometer. *Journal of Sports Sciences,* **16**, pp. 14-15.

Cooke, C. B., 2001, Maximal oxygen uptake, economy and efficiency. *In:* Eston, R.G. and Reilly, T., eds. *Kinanthropometry and Exercise Physiology Laboratory Manual: Tests, Procedures and Data. Volume 2: Exercise Physiology,* (London: Routledge), pp. 161-191.

Davies, C. T. M., 1968, Limitations to the prediction of maximum oxygen intake from cardiac frequency measurements. *Journal of Applied Physiology,* **24**, pp. 700-706.

Eston, R.G. and Williams, J.G., 1988, Reliability of ratings of perceived effort regulation of exercise intensity. *British Journal of Sports Medicine,* **22**, pp. 153-155.

Eston, R., Lambrick, D., Sheppard, K. and Parfitt, G., 2008, Prediction of maximal oxygen uptake in sedentary males from a perceptually-regulated, sub-maximal graded exercise test. *Journal of Sports Sciences,* **26**, pp. 131-139

Eston, R. G., Faulkner, J. A., Mason, E. A. and Parfitt, G., 2006, The validity of predicting maximal oxygen uptake from perceptually-regulated graded exercise tests of different durations. *European Journal of Applied Physiology,* **97**, pp. 535-541.

Eston, R. G., Faulkner, J. A., St Clair Gibson, A., Noakes, T. and Parfitt, G., 2007, The effect of antecedent fatiguing activity on the relationship between perceived exertion and physiological activity during a constant load exercise task. *Psychophysiology*, **44**, pp. 779-86.

Eston, R. G., Lamb, K. L., Parfitt, G. and King, N., 2005, The validity of predicting maximal oxygen uptake from a perceptually-regulated graded exercise test. *European Journal of Applied Physiology,* **94**, pp. 221-227.

Faulkner, J. A. and Eston, R. G., 2007, Overall and peripheral ratings of perceived exertion during a graded exercise test to volitional exhaustion in individuals of high and low fitness. *European Journal of Applied Physiology*, **101**, pp. 613-20.

Faulkner, J. A., Parfitt, G. and Eston, R. G., 2007, Prediction of maximal oxygen uptake from the ratings of perceived exertion and heart rate during a perceptually-regulated sub-maximal exercise test in active and sedentary participants. *European Journal of Applied Physiology,* **101**, pp. 397-407.

Golding, L. A., Myers, C. R. and Sinning, W. E., 1982, *The Y's Way to Physical Fitness: A Guidebook for Instructors*, (Rosemont, IL: YMCA of the USA).

Hardy, C. J. and Rejeski, W. J., 1989, Not what, but how one feels: The measurement of affect during exercise. *Journal of Sport and Exercise Psychology,* **11**, pp. 304-317.

Jameson, C. and Ring, C., 2000, Contributions of local and central sensations to the perception of exertion during cycling: Effects of work rate and cadence. *Journal of Sports Sciences,* **18**, pp. 291-298.

Kline, G. M., Porcari, J. P., Hintermeister, R., Freedson, P. S., Ward, A., McCarron, R. F., Ross, J. and Rippe, J. M., 1987, Estimation of VO_2max from a one-mile track walk, gender, age, and body weight. *Medicine and Science in Sports and Exercise,* **19**, pp. 253-259.

Morgan, W. P. and Borg, G., 1976, Perception of effort in the prescription of physical activity. *In*: T. Craig, ed. *The Humanistic Aspects of Sports, Exercise and Recreation*, (Chicago: American Medical Association), pp. 126-129.

Nevill, A. M. and Atkinson, G., 1997, Assessing agreement between measurements recorded on a ratio scale in sports medicine and sports science. *British Journal of Sports Medicine,* **31**, pp. 314- 318.

Noakes, T., St Clair Gibson, A. and Lambert, E. V., 2004, From catastrophe to complexity: a novel model of integrative central neural regulation of effort and fatigue during exercise in humans. *British Journal of Sports Medicine,* **38**, pp. 511-514.

Noonan, V. and Dean, E., 2000, Submaximal exercise testing: clinical application and interpretation. *Physical Therapy,* **80**, pp. 782-807.

Parfitt, G., Eston, R. and Connolly, D., 1996, Psychological affect at different ratings of perceived exertion in high- and low-active women: A study using a production protocol. *Perceptual and Motor Skills,* **82**, pp. 1035-1042.

Rowell, L. B., Taylor, H. L. and Wang, Y., 1964, Limitations to prediction of maximal oxygen intake. *Journal of Applied Physiology,* **19**, pp. 919-927.

Siconolfi, S. F., Garber, C. E., Lasater, T. M. and Carleton, R. A., 1985, A simple, valid step test for estimating maximal oxygen uptake in epidemiology studies. *American Journal of Epidemiology,* **121**, pp. 382-390.

St Clair Gibson, A., Lambert, M. I., Hawley, J. A., Broomhead, S. A. and Noakes, T. D., 1999, Measurement of maximal oxygen uptake from two different laboratory protocols in runners and squash players. *Medicine and Science in Sports and Exercise,* **31**, pp. 1226-1229.

St Clair Gibson, A., Lambert, E. V., Rauch, L. H. G., Tucker, R., Baden, D. A., Foster, C. and Noakes, T. D., 2006, The role of information processing between the brain and peripheral physiological systems in pacing and perception of effort. *Sports Medicine,* **36**, pp. 705-722.

Vehrs, P. R. and Fellingham, G. W., 2006, Heart rate and VO_2 responses to cycle ergometry in white and African American men. *Measurement in Physical Education and Exercise Science,* **10**, pp. 109-118.

Wagganer, J. D., Pujol, T.J., Langenfield, M. E., Sinclair, A. J., Tucker, J. E. and Elder, C. L., 2003. A workload selection procedure for the Astrand-Ryhming test. *Medicine and Science in Sports and Exercise,* **35**, pp. 257.

Zwiren, L. D., Freedson, P. S., Ward, A., Wilke, S. and Rippe, J.M., 1991, Estimation of VO_2max: a comparative analysis of five exercise tests. *Research Quarterly for Exercise and Sport,* **62**, pp. 73-78.

CHAPTER SEVENTEEN

Reproducibility of ratings of perceived exertion soon after myocardial infarction: responses in the stress-testing clinic and the rehabilitation gymnasium

J.P. Buckley[1], J. Sim[2] and R. Eston[3]

[1]Centre for Exercise and Nutrition Science, University of Chester, Cheshire, UK
[2]School of Health and Rehabilitation, Keele University, Staffordshire UK
[3]School of Sport and Health Sciences, University of Exeter, Devon UK

1. INTRODUCTION

Ergonomics focuses on evaluating human performance while interfacing with specific devices or machines within a specified environment. The standardized exercise ECG treadmill (walking/running ergometer) test (exECG) that takes place daily in a typical hospital in developed countries, is an excellent example of such an interface between human, machine and a specified environment. In light of such an evaluation, the exECG stress test focuses on two outcomes: i) it is a routine diagnostic and prognostic assessment procedure for individuals with coronary artery disease (CHD), myocardial infarction (MI) and heart failure; and ii) it is an assessment of a patient's ability to cope (or not) with or without symptoms or clinical complications in his/her required activities of daily living (British Association for Cardiac Rehabilitation, 1995; American Association of Cardiovascular and Pulmonary Rehabilitation, 2004; American College of Sports Medicine, ACSM, 2005). Both these outcomes are used to determine the level of medical, surgical, or therapeutic intervention that a patient may require immediately or in the future.

The use of Borg's rating of perceived exertion (RPE) scales (Borg, 1998) is recommended as an integral part of these exECG testing protocols (American College of Sports Medicine, 2005; British Association of Sport and Exercise Sciences, 2006). The results of the exECG, including cardiac conductivity changes, heart rate (HR), RPE, estimated metabolic equivalents (METs) and rate-pressure product, not only provide a marker of the upper limits of exercise tolerance, diagnosis and prognosis related to myocardial dysfunction, but can also provide a means of setting subsequent exercise prescriptions as part of patient rehabilitation (Froelicher and Myers, 2000). The RPE scale, relative to a given percentage of maximum exercise tolerance and/or a specific point below critical ECG changes, has been demonstrated to be an effective means of representing an appropriate training intensity for patients undergoing rehabilitation after myocardial infarction (Robertson and Noble, 1997; Buckley *et al.*, 1999).

In a study to compare RPE responses between exECG and exercise training in cardiac patients, Gutmann *et al.* (1981) reported that for a given HR, RPE was the same during a modified Bruce exECG and during a standard rehabilitation exercise training session. They concluded that a given RPE was reliably related to a given HR over an eight-week training period. However, the MET values elicited during the gymnasium sessions were significantly lower compared to the exECG ($P<0.01$). This mismatch in METs to HR and RPE was due to the fact that during the exECG, the RPE was recorded after 6 minutes of exercise, whereas it was recorded after at least 30 minutes of exercise in the gymnasium sessions. It is known that RPE is not solely a function of external work output. It is also a function of activity duration when the workload is held constant (Horstman *et al.*, 1979; Eston *et al.*, 2007). In a later study, where patients self-regulated their exercise intensity by a pre-determined RPE during the gymnasium training sessions, the HR during these sessions was lower for a given RPE than when compared to the exECG (Brubaker *et al.*, 1994).

Current standards within the British National Service Framework for Coronary Heart Disease (Department of Health, 2000) recommend that rehabilitation exercise should be commenced four weeks after myocardial infarction or coronary artery bypass surgery (CABG). A cardiac event is often associated with subsequent state and trait psychological morbidity (Todd *et al.*, 1992) and individuals with heightened anxiety and depression tend to inflate estimation mode RPE (Rejeski, 1981). Borg and Linderholm (1970) observed that during exercise testing, cardiac patients reported higher RPE values for a given HR compared to age-matched control subjects. Thus a potential recommendation, when testing cardiac patients during standard exECG procedures, is that the measure of RPE relative to a given proportion of maximal exercise capacity is likely to be inflated compared to expected responses of healthy or younger populations.

The purpose of this study was therefore to compare RPE responses to the same exercise intensity between a standard exercise-ECG treadmill stress test (exECG) performed in a clinical setting and two subsequent bouts of treadmill exercise in a cardiac rehabilitation gymnasium.

2. METHODS

2.1 Participants and general testing protocol

Following local ethical approval and individual patient written informed consent, a convenience sample of 11 patients (9 males, 2 females) with recent MI participated in the study (Table 1). These participants met previously reported (Wilmer *et al.*, 1999) criteria for entering the "fast track" rehabilitation programme at the hospital where this study took place. The general aim of the "fast track" programme is to commence exercise-based rehabilitation within 14–28 days following MI.

Important components of these criteria include an uncomplicated recovery during the in-patient period (*Phase I* of rehabilitation) and achieving ≥ 7 METs during a standard 12-lead electrocardiogram exECG 7–14 days post-MI. Depending upon the *Phase I* clinical evaluation of the patient by the cardiologist, either a Bruce or Modified Bruce (BACR, 1995) treadmill protocol was chosen. Within three to six days of completing the initial exECG, patients began their exercise program at the hospital out-patient cardiac rehabilitation gymnasium. The initial aerobic exercise program included 10–20 min of activity, divided between treadmill, cycle and rowing ergometry, at an intensity equivalent to 65–80% of maximum METs achieved in the exECG. This intensity also had to be below the ischaemic threshold as determined by 2.0 mm or more of S-T segment depression on the ECG. The gymnasium testing aimed to reflect normal practice by including a standardized low intensity warm-up (BACR, 1995; AACVPR, 2004) prior to commencing the treadmill exercise. The initial period of the treadmill exercise replicated the stages of the initial exECG, but only up to the penultimate stage of what was achieved in the exECG. This initial gymnasium-based exercise protocol (gym-1) was repeated within two to five days (gym-2). During both of these gymnasium sessions, RPE and HR were recorded, as in the exECG, at the end of each completed three-minute stage. All tests, including the exECG, were performed between 10:00 and 14:00 hours and under the same pharmaceutical conditions. It is important to note that although adrenergic beta-blockade lowers HR for a given submaximal exercise intensity, during exercise lasting less than one hour the RPE at a given intensity is unaffected (Davies and Sargeant, 1979; Squires *et al.*, 1982; Eston and Connolly, 1996; Head *et al.*, 1997; Eston and Thompson, 1997). Wilcox *et al.* (1984) noted that RPE was increased when beta-blockade is first administered but with chronic treatment such an affect was abated.

2.2 Equipment

The exECG treadmill was a *Quinton* 12-lead system (Bothell, WA, USA) and the gym-1 and gym-2 treadmill was a *PowerJog GX100* (Sport Engineering Ltd., Birmingham, UK). Comparative calibrations were made to check for equal belt speeds and gradients between the two treadmills. During gym-1 and gym-2, HR was measured using both a finger-probe pulse oximeter and wireless chest strap telemetry and watch receiver system (Polar, Kempele, Finland).

2.3 Ratings of perceived exertion

Prior to each of the three testing sessions (exECG, gym-1 and gym-2), patients were instructed in the use of Borg's 6–20 RPE scale, with the following procedure in keeping with standard recommendations (Noble and Robertson, 1996; Borg, 1998):

- clarifying that participants understood the definition of RPE,
- "anchoring" the top and bottom ratings to previously experienced sensations of *no exertion at all* and *extremely hard/maximal exertion,*
- being made aware of giving an "all-over" integrated rating which incorporated both peripheral muscular and central cardio-respiratory sensations,
- focusing attention on the verbal descriptors of the scale as much as the numerical values,
- understanding that there was no right or wrong rating, and that it represented how hard the subject felt he/she was working at the time of giving the rating,
- having the scale in full view at all times.

2.4 Data collection points

As per standard clinical practice, HR and RPE were recorded in the last 15 seconds of each three-minute stage.

2.5 Ecological validity

Unlike many previous studies on RPE, ecological validity was assumed by the fact that none of the testing situations in this study were contrived; all measures were taken under normal hospital conditions and as part of standard patient assessment and therapy. Previous work has reported RPE across a range of intensities (Brubaker *et al.*, 1994). However, our study also aimed to achieve ecological validity by evaluating the results at typically recommended "target" intensities; those which represent a safe (below ischaemia) but effective (60%–80% peak exECG work-rate) level.

2.6 Statistical analyses

A repeated measures one-factor analysis of variance (ANOVA) was performed to determine inter-trial differences between exECG, gym-1 and gym-2 for both RPE and HR responses. As an effect size statistics for these analyses we used partial eta-squared (η^2), which represents the proportion of variance in the outcome measure accounted for by the within-subjects factor (Cohen, 1988). *Post-hoc* analyses to determine the location of any inter-trial differences in RPE or HR were performed using a paired *t*-test. To restrict the type I error rate, alpha was reduced for these pairwise analyses using a Bonferroni correction; in this case $P \leq 0.017$ $(0.05 \div 3)$. In the event of there being no significant inter-trial difference in RPE (e.g. between exECG and gym-1 or between gym-1 and gym-2), the *reproducibility* of RPE across exECG, gym-1 and gym-2 was then assessed using an analysis based on an intraclass correlation coefficient $(ICC_{2,1})$ (Shrout and Fleiss, 1979) and the bias $\pm$ 95% limits of agreement (95%LoA) (Bland and Altman, 1986). The 95% LoA indicate the range within which 95% of normally distributed differences in paired ratings are expected to lie, such that a smaller range indicates closer agreement in ratings. These same statistics were used to assess the *reliability* of the measurement of RPE across the two gymnasium sessions, gym-1 and gym-2.

3. RESULTS

Table 1 summarizes the patient demographics ($n = 11$), the time intervals between trials and the treadmill exercise intensity data. The RPE and HR results that have been analysed were those that related to a mean intensity of 6 METs, which is ~67% of peak METs (9 METs) that were achieved during the exECG.

Table 1. Patient demographic variables, time intervals between treadmill tests and treadmill work-rates.

Variable	Value: mean $\pm s$
Age (years)	
Male ($n = 9$)	62.0 ± 4.5
Female ($n = 2$)	55.5 ± 12.0
All	60.8 ± 6.1
Body mass index	
Male	$26.1 + 3.9$
Female	22.0 ± 2.8
All	25.4 ± 4.0
Days between MI and exECG	12.5 ± 7.1
Days between exECG and gym-1	5.0 ± 1.3
Days between gym-1 and gym-2	4.2 ± 1.3
Maximal treadmill work-rate (METs) value of exECG	9.2 ± 2.1
Treadmill work-rate (METs) at end of penultimate exECG stage from which inter-trial RPE was assessed	6.0 ± 1.0

MI = myocardial infarction; exECG = 12-lead exercise ECG treadmill stress test; gym-1 and gym-2 = gymnasium sessions 1 and 2, respectively; METs = estimated metabolic equivalents (1 MET = VO_2 of 3.5 ml.kg^{-1}.min^{-1}).

Figure 1 illustrates the RPE responses at the same work-rate (~67% peak METs) during each of the three trials (exECG, gym-1, and gym-2). RPE (mean $\pm s$) for the exECG, gym-1 and gym-2 were 15.8 ± 2.7, 13.3 ± 3.6 and 13.0 ± 3.4, respectively, which were significantly different ($F_{2,20} = 9.8$; $P = 0.001$; $\eta^2 = 0.50$). The *post hoc* t-tests revealed that RPE was significantly lower at gym-1 and gym-2 compared to the exECG ($P \leq 0.008$) but there was no difference between gym-1 and gym-2. The intra-participant agreement in RPE between gym-1 and gym-2 was substantial; ICC = 0.85 (95% confidence interval = 0.70 to 0.98; $P < 0.001$). The bias $\pm$ 95% LoA for this same pairwise comparison were 0.27 $\pm$ 3.8 RPE scale points.

Figure 2 illustrates the HR responses, at the same work-rate, during each of the three trials (exECG, gym-1, and gym-2). The HR for the exECG, gym-1 and gym-2 was 112 $\pm$ 22, 105 $\pm$ 16 and 99 $\pm$ 16 beats.min^{-1}, respectively, which were not significantly different ($F_{2, 18} = 2.98$; $P = 0.076$; $\eta^2 = 0.25$). Due to equipment error when measuring one of the patients, complete HR data for the three trials are based on 10 patients.

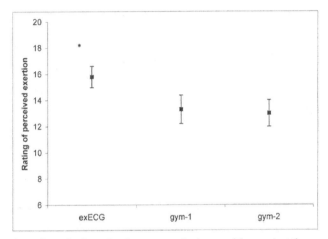

Figure 1. Ratings of perceived exertion (mean $\pm$ *standard error of the mean*), at the same work-rate (METs) during treadmill exercise performed in a post-MI ECG (exECG) clinic and during two subsequent cardiac rehabilitation exercise sessions (gym-1 and gym-2).
*Significantly greater than gym-1 and gym-2 ($P = 0.008$).

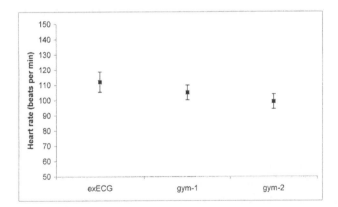

Figure 2. Heart rate (mean $\pm$ *standard error or the mean*), at the same work-rate (METs) during treadmill exercise performed in a post-MI ECG clinic (exECG) and during two subsequent cardiac rehabilitation exercise sessions (gym-1 and gym-2).

4. DISCUSSION

This study highlighted changes in standard response measurements (RPE and HR) that occur when patients perform exercise initially in a typical diagnostic clinic and subsequently in a rehabilitation gymnasium. For the same external work-rate (treadmill speed and gradient) RPE decreased significantly from the first testing point (exECG) to the second test (gym-1) but did not subsequently differ between gym-1and gym-2. The HR showed similar decrements, but these were of smaller magnitude and were not statistically significant ($P = 0.076$). However, the number of participants in the study limited the statistical power of the HR results. If HR truly did decrease across these three time points as a result of improved patient movement economy and/or reduced negative psychological stimuli, then RPE has proven to be valuable in reflecting elements of these factors. It is clear that the patient's perceptions of effort for the same amount of external work involved did change from the stress test to the gymnasium setting. Potential factors influencing such change – for RPE and possibly also for HR – include familiarization with the skill of performing exercise on a treadmill, the psychosocial milieu of a testing clinic while performing a maximal exercise test less than two-weeks after a heart attack, familiarization with perceiving one's exercise effort during a test, and familiarization with using Borg's RPE scale to rate such effort. As a result of these findings, it is recommended that cardiac rehabilitation practitioners consider such familiarization factors when using exECG results to guide subsequent physical activity and prescribe structured exercise.

At first, it would appear our findings contrast with those of previous reports, such as Gutmann *et al.* (1981), who reported RPE values to be unchanged between testing and subsequent exercise in a group of cardiac patients. However, these authors measured RPE at very different time points; after 6 min in the exECG and after at least 30 min in the gymnasium. Upward "drifts" in RPE have been shown to be a function of exercise duration, especially beyond 30 min, with a more pronounced effect in unfit individuals (Green *et al.*, 2007).

The most important observation from this study was that RPE was significantly greater ("inflated") in the initial exECG stress test compared with two subsequent cardiac rehabilitation gymnasium exercise sessions at the same externally-measured exercise intensity. There are two important clinical/practical implications of these results. First, with an inflated RPE, had patients been told it was safe to exert themselves to the RPE measured in their initial exECG – which represented a level of ~67% of peak work-rate and/or a level below the ischaemic threshold – subsequent bouts of activity eliciting this same RPE could have resulted in higher work-rates. Second, in the initial recovery stages from MI, at least two subsequent exercise sessions are required to establish a reliable relationship between RPE and work-rate. Once patients were familiarized with RPE from their first exposure in the exECG and second exposure in the gym setting, the exercise intensity-RPE relationship then remained constant between the two gymnasium sessions. These results concur with previous studies which have indicated that the RPE needs to be used at least once to ensure that it represents a given and appropriate exercise intensity (Eston and Williams, 1988; Lamb *et al.*,

1999; Buckley *et al.*, 2000). Furthermore, the mean RPE of 13 in the two gym sessions reflected 67% of peak work-rate, which accords with the recommended training intensity and physiological relationship in clinical populations (Borg, 1998; ACSM, 2005).

An obvious area for future study is to explore the potential psychological mechanisms for the inflated RPE during the exECG, as suggested by numerous reports (Morgan, 1973; Rejeski, 1981; Kohl and Shea, 1988; Morgan, 1994; Biddle and Mutrie, 2001). Anxiety and depression are typically observed in patients soon after MI (Todd *et al.*, 1992). As the exECG is often the patients' first experience of moderate to higher intensity physical exertion, it is likely that they may inflate RPE as a means of gaining some control over the testing situation. Kohl and Shea (1988) reported that those with an external locus of control give higher RPEs for a given work-rate than those with an internal locus of control.

Although Brubaker *et al.* (1994) reported that RPE was a reliable indicator of exercise intensity during cardiac rehabilitation, they did not use an analysis of intra-participant inter-trial agreement to support their conclusion of "good reproducibility". In the present study, the reliability of RPE between gym-1 and gym-2 is illustrated by the substantial ICC ($r = 0.85$; $P < 0.001$). The bias $\pm$ 95% LoA analysis of RPE between gym-1 and gym-2 (0.27 ± 3.8) helped to further illustrate the practical/clinical inter-trial reliability by providing information on an intra-participant basis.

It is notable that the 95%LoA was $\pm$ 3.8 RPEs. This magnitude of variation was only true for one of the patients. In the other 10 patients the difference in RPE between gym-1 and gym-2 was no greater than two scale points for two of the patients, one scale point for four of the patients and no difference in the remaining four patients. The poor intra-participant reliability of one patient out of 11 agrees with Borg and Borg (2001), according to whom it can be expected that RPE will be useable in up to 90% of a group of individuals. In practical settings, especially with clinical populations, it is therefore important to evaluate RPE reliability on an individual basis.

5. CONCLUSION

In conclusion, although the reproducibility of RPE between an initial exECG and its use within a typical cardiac rehabilitation setting is questionable, it became reliable and concurrently valid with a relative physiological strain after a second exposure in the cardiac rehabilitation gymnasium. The use of RPE in the exECG test is an important part of familiarizing the patient, along with learning to perceive exertion and then being able to rate this exertion on an RPE scale, so as to ensure the scale's subsequent reliable and valid use in the cardiac rehabilitation gymnasium. Future work should consider whether or not the differences found in the RPE-work-rate relationship between exECG and gymnasium settings are a function of familiarization with a given mode of exercise (e.g. treadmill walking), the differing psychosocial milieu, familiarization with perceiving effort and using RPE, or some combined effect of these three factors.

Acknowledgements

We are grateful to Navroz Amlani (Keele University physiotherapy student) for his assistance with the data collection and to Jon Creamer, Kath Herity and Terry Simpson from the Cardiology and Cardiac Rehabilitation Departments of the University Hospital of North Staffordshire, for their valuable involvement in patient recruitment, screening and testing.

References

American Association of Cardiovascular and Pulmonary Rehabilitation (AACVPR), 2004, *AACVPR Guidelines for Cardiac Rehabilitation and Secondary Prevention Programs*, (Champaign, IL: Human Kinetics).

American College of Sports Medicine (ACSM), 2005, *ACSM guidelines for Exercise Testing and Prescription* (7th Ed), (Baltimore: Lippincott, Williams and Wilkins).

Biddle, S.J.H. and Mutrie, N., 2001, *Psychology of Physical Activity; Determinants, Well-Being and Interventions*, (London: Routledge).

Bland, J.M. and Altman, D.G., 1986, Statistical methods for assessing agreement between two methods of clinical measurement. *Lancet,* **i**, pp. 307–310.

Borg, G. and Linderholm, H., 1970, Exercise performance and perceived exertion in patients with coronary insufficiency, arterial hypertension and vasoregulatory asthenia. *Acta Medica Scandinavica*, **187**, pp. 17–26.

Borg, G.A.V., 1998, *Borg's Perceived Exertion and Pain Scales*, (Champaign, IL: Human Kinetics).

Borg, G. and Borg, E., 2001, A new generation of scaling methods: level-anchored ratio scaling. *Psychologica*, **28**, pp. 15–45.

British Association for Cardiac Rehabilitation (BACR), 1995, *BACR Guidelines for Cardiac Rehabilitation*, (Oxford: Blackwell Science).

British Association of Sport and Exercise Sciences (BASES), 2006, *Sport and Exercise Physiology Testing Guidelines: Exercise and Clinical Testing vol 2.*, edited by Winter, E.M., Mercer, T., Bromley, P.D., Davison, R.C. and Jones, A.M., (London: Routledge).

Brubaker, P.H., et al., 1994, Cardiac patients' perception of work intensity during graded exercise testing. *Journal of Cardiopulmonary Rehabilitation,* **14**, pp. 127–133.

Buckley, J.P., Davis, J. and Simpson, T., 1999, Cardio-respiratory responses to rowing ergometry and treadmill exercise soon after myocardial infarction. *Medicine and Science in Sports and Exercise,* **31**, pp. 1721–1726.

Buckley, J.P., Eston, R.G. and Sim, J., 2000, Ratings of perceived exertion in Braille: validity and reliability in production mode. *British Journal of Sports Medicine,* **34**, pp. 297–302.

Cohen, J., 1988. *Statistical Power Analysis for the Behavioral Sciences*, 2nd edn, (Hillsdale, NJ: Lawrence Erlbaum).

Davies, C.T. and Sargeant, A.J., 1979. The effects of atropine and practolol on the perception of exertion during treadmill exercise. *Ergonomics*, **22**, pp. 1141–1146.

Department of Health, 2000. National Service Framework for coronary heart disease – modern standards and service models, (London: Department of Health (England) http://www.dh.gov.uk/PublicationsAndStatistics).

Eston, R.G. and Williams, J.G., 1988, Reliability of ratings of perceived exertion for regulation of exercise intensity. *British Journal of Sports Medicine*, **22**, pp. 153–155.

Eston, R.G. and Connolly, D., 1996, The use of ratings of perceived exertion for exercise prescription in patients receiving b-blocker therapy. *Sports Medicine*, **21**, pp. 176–90.

Eston, R.G. and Thompson, M., 1997, Use of ratings of perceived exertion for predicting maximal work rate and prescribing exercise intensity in patients taking atenolol. *British Journal of Sports Medicine*, **31**, pp. 114–119.

Eston, R.G., et al., 2007, The effect of antecedent fatiguing activity on the relationship between perceived exertion and physiological activity during a constant load exercise task. *Psychophysiology*, **44**, pp. 779–786.

Froelicher, V.F. and Myers, J.N., 2000, *Exercise and the Heart*, (Philadelphia: WB Saunders).

Green, J.M., et al., 2007, Influence of aerobic fitness on ratings of perceived exertion during graded and extended duration cycling. *Journal of Sports Medicine and Physical Fitness*, **47**, pp. 33–39.

Gutmann, M.C., et al., 1981, Perceived exertion-heart rate relationship during exercise testing and training in cardiac patients. *Journal of Cardiac Rehabilitation*, **1**, pp. 52–59.

Head, A., Maxwell, S. and Kendall, M.J., 1997, Exercise metabolism in healthy volunteers taking celiprolol, atenolol, and placebo. *British Journal of Sports Medicine*, **31**, pp. 120–125.

Horstman, D.H., et al., 1979, Perception of effort during constant work to self-imposed exhaustion. *Perceptual and Motor Skills*, **48**, pp. 1111–1126.

Kohl, R.M. and Shea, C.H., 1988, Perceived exertion: influences of locus of control and expected work intensity and duration. *Journal of Human Movement Studies*, **15**, pp. 225–272.

Lamb, K.L., Eston, R.G. and Corns, D., 1999, The reliability of ratings of perceived exertion during progressive treadmill exercise. *British Journal of Sports Medicine*, **33**, pp. 336–339.

Morgan, W.P., 1973, Psychological factors influencing perceived exertion. *Medicine and Science in Sports and Exercise*, **5**, pp. 97– 103.

Morgan, W.P., 1994, Psychological components of effort sense. *Medicine and Science in Sports*, **26**, pp. 1071–1077.

Noble, B.J. and Robertson, R.J., 1996, *Perceived Exertion*, (Champaign, IL: Human Kinetics).

Rejeski, W.J., 1981, The perception of exertion: a social psychophysiological integration. *Journal of Sport Psychology*, **4**, pp. 305–320.

Robertson, R.J. and Noble, B.J., 1997, Perception of physical exertion: methods, mediators, and applications. In *Exercise and Sport Sciences Reviews*, edited by

Holloszy, J.O., (Williams & Wilkins), pp. 407–452.

Shrout, P.E. and Fleiss, J.L., 1979, Intraclass correlations: uses in assessing rater reliability, *Psychological Bulletin*, **86**, pp. 420–428.

Squires, R.W., et al., 1982, Effects of propanolol on perceived exertion soon after myocardial revascularization surgery. *Medicine and Science in Sports and Exercise,* **14**, pp. 276–280.

Todd, I.C., et al., 1992, Cardiac rehabilitation following myocardial infarction; a practical approach. *Sports Medicine,* **14**, pp. 243–259.

Wilcox, R.G., et al., 1984, The effects of acute or chronic ingestion of propanolol or metoprolol on the physiological responses to prolonged, submaximal exercise in hypertensive men. *British Journal Clinical Pharmacol*ogy, **17**, pp. 273–281.

Williams, J.G. and Eston, R.G., 1989, Determination of the intensity dimension in vigorous exercise programmes with particular reference to the use of the rating of perceived exertion. *Sports Medicine,* **8**, pp. 177–189.

CHAPTER EIGHTEEN

Effects of the Zone-diet on training parameters in recreational master athletes

M.F. Piacentini[1,3], G. Salvatori[2], C. Di Cesare[2], F. Pizzuto[2], M. Olivi[1], C. Minganti[1], R. Meeusen[3] and L. Capranica[1]

[1]Department of Human Movement and Sport Sciences, University Institute for Movement Science, Rome, Italy
[2]Department of Health Science University of Molise, Campobasso, Italy
[3] Department of Human Physiology and Sportsmedicine, Vrije Universiteit Brussel, Belgium

1. INTRODUCTION

Runners competing in the 40-59 years age categories of the major marathons and half-marathons in Italy represent the largest proportion (50%) of the participants. Although these athletes regularly train and compete with the precise goal to improve their personal best in running events, they do not have comparable technical and scientific support as elite athletes and often adopt "word-of-mouth" training and nutritional programmes to optimize their performance. Unfortunately, incorrect or generalized training programmes often lead to overtraining (Meeusen et al., 2004) while nutritional "word of mouth" programmes could cause incorrect practices and insufficient quantities and qualities of food intake (Rosenbloom and Dunaway, 2007).

In Italy, the 'Zone diet' receives increasing attention from master and recreational athletes of all disciplines, despite the lack of direct evidence of better top level performances and severe criticism for its low carbohydrate content (Cheuvront, 1999). The Zone diet recommends athletes to consume 5 daily meals (3 main meals and 2 snacks), precisely organized in specific blocks (i.e., 1 block: carbohydrates = 9 g; protein = 7 g; and fat = 1.5 g) with a 1:1:1 ratio in order not to alter the 0.6-0.75 protein/carbohydrate ratio. Calculations based primarily on protein intake of 1.8-2.2 $g \cdot kg_{LBM}^{-1}$ (Sears, 1995, 2000) drastically increase the proportion of protein and reduce the proportion of carbohydrate in the athlete's diet. These calculations limit the caloric content of the diet, which becomes hypocaloric with respect to previous regimens. On the other hand, the Zone diet promises lower insulin concentrations, thus promoting free fatty acid utilisation and sparing muscle glycogen (Sears, 1995). A primary limiting factor in endurance events such as marathon running, triathlon, and cycling is the depletion of carbohydrate in the blood, muscle and liver (Erlenbusch et al., 2005), as confirmed by numerous studies (for reviews see Burke *et al.*, 2001; Hargreaves *et al.*, 2004). Moreover, as the exercise intensity increases the reliance on muscle and blood carbohydrate also increases. Therefore, reducing carbohydrate ingestion will greatly reduce glycogen content and impair not only training but also performance.

Studying the acute (1-week) effects of the Zone diet on performance, blood lipid levels and body composition in young (26±2years) recreational athletes, Jarvis and colleagues (2002) reported a 3- minute decrease in time to exhaustion on a treadmill run at 80% $\dot{V}_{O_2 \, max}$. The authors claimed that the Zone diet calculation led to a significant reduction of caloric intake with respect to the habitual athletes' nutritional regimen that consequently caused a reduction in their body weight. In preliminary studies from our group on the effects of an isocaloric Zone diet regimen in young (27±2years) and master (48±1.5years) endurance athletes (Piacentini *et al.*, 2004, 2007) no effect on exercise performance emerged and runners reported an altered mood state and a difficulty in maintaining their usual training regimen. Unfortunately, no data on training subjective ratings and on mood state were included to substantiate these reports. Therefore, the purpose of the present study was to evaluate the effects of a 2-week Zone diet regimen on training status, hormonal and mood parameters of master endurance athletes. We hypothesize that the new dietary regimen affects training, hormonal, and mood parameters of the master athletes.

2. METHODS

2.1 Participants

Six master athletes (52±4years, 74±6 kg, $\dot{V}_{O_2 \, max}$ 52±3 ml·kg·min^{-1}) provided their written consent to participate in the study. In addition, three co-aged athletes (53±1years, 87±2 kg, $\dot{V}_{O_2 \, max}$ 44±3 ml·kg·min^{-1}) were included as controls for comparisons of the collected weekly measures (cortisol, testosterone, and mood state).

2.2 Study Design

The study was performed after its approval by the University Ethical Committee. It was anticipated that examining the training parameters and the mood state of athletes would provide meaningful data for interpreting differences, if any, in performance. Since performance is affected by the previous training regimen, training variables were monitored daily by means of a validated on-line training diary (i.e., Blits®), which allowed the maintenance of a high ecological validity though controlling for confounding factors. If the training diary variables were found to be stable prior to the experimental period, differences emerging in the dependent variables would be attributed to the diet intervention.

During the four weeks prior to the study and the 2-week experimental period, the runners were instructed to maintain their normal lifestyle and to refrain from high-intensity physical activities the day before the experimental sessions. Furthermore, since different levels of training and changes in nutritional regimens could affect results, the criteria for inclusion were: 1) runners were regularly training (at least 5 days·week^{-1}) and competing in long distance races (i.e., 10-42 km) for the past 10 years; 2) athletes did not adopt diets that altered their regular nutritional regimen in the past three years.

2.3 Maximal Exercise Test

The athletes were requested to perform a graded incremental exercise test to exhaustion on a treadmill. During the test heart rate (HR) was monitored (Sport Tester, Polar Electro, Kempele, Finland) and oxygen consumption ($\dot{V}O_2$), carbon dioxide production ($\dot{V}CO_2$) and ventilation were recorded on a breath by breath basis (PFT ERGO COSMED, Italy). The flow meter was calibrated with a 3-l syringe (Hans Rudolph Inc, Dallas), and the gas analyzer was calibrated with known gas mixtures (O_2: 16% and 20.9%; CO_2: 5% and 0.03%). Participants started running at the speed of 5.4 km/h and every 3 min the speed was increased by 1.8 km.h^{-1} until voluntary exhaustion. Borg's scale (Borg, 1998) of perceived exertion for the whole body (RPE) between 6 (no exertion at all) and 20 (maximal exertion) was administered at the third minute of each step. A 10-µl capillary sample was taken from the ear lobe to measure blood lactate concentration [La] at rest, at every stage of the test, and after 3, 6 and 9 minutes of recovery (EBIO-Plus, Eppendorf, Hamburg, Germany).

2.4 Exercise test

Since previous studies on endurance athletes undergoing a Zone diet regimen (Piacentini et al., 2004; 2007) reported no difference in time to exhaustion during 30-minute and 60-minute running tests and participants in the present study were not available for laboratory tests exceeding one hour, to evaluate the training status and the recovery capacity of the master athletes a validated two-bout maximal exercise to exhaustion protocol (Meeusen et al., 2004) was selected. Athletes suffering from fatigue are expected to show decrements in performance and/or higher RPE values during their second running test.

Before (Pre) and after (Post) 15 days on the zone-diet, the athletes performed two exercise tests to exhaustion as described in the previous paragraph with a 3-hour interval (Meeusen *et al.*, 2004). The day before the experimental sessions athletes were asked to refrain from hard physical training, caffeine and alcohol intake. On the experimental days, the athletes reported in the laboratory between 08:30 and 10:30 hours and each participant performed his test at the same time of the day. After collection of their baseline data, body mass was determined with an accuracy of 100 g and percentage body fat was calculated by means of the skinfold method (Jackson and Pollock, 1985).

Peak velocity was calculated as follows:
Peak velocity $(km \cdot h^{-1})$ = 1.8/180*s of last stage + $km \cdot h^{-1}$ of previous stage
Where 1.8 $(km \cdot h^{-1})$ is the increase between one step and the other, 180 is the time (s) of each step.

2.5 Measurement of Energy Expenditure

Using standard forms, the participants were asked to record the weight of all food items and beverages consumed (other than water) for 7 consecutive days and to keep a detailed training log to measure their caloric intake, diet composition, and energy expenditure due to exercise. Before recording, the participants were given standardized instructions by a dietician. Brand names, methods of preparation and recipes for consumed mixed dishes were requested. Moreover, athletes were requested to wear a SenseWear® System Armband (SWA, version 5.0) for one week in order to assess energy expenditure (Jakicic *et al.*, 2004). As recommended by the manufacturer, the SWA was placed on the right arm over the triceps muscle at the midpoint between the acromion and olecranon processes. The SWA was placed on the athlete's arm, while seated, and kept on for 15 minutes before collecting the data, thus allowing adaptation to the skin temperature. Runners wore the SWA twenty-four hours a day for a week. The SWA incorporates a variety of measures (acelerometry, heat flux, galvanic skin response, skin temperature, near-body temperature) and demographic characteristics (gender, age, height, weight) into proprietary algorithms to estimate energy expenditure (Fruin and Rankin, 2004; Jakic *et al.*, 2004). Acelerometry is measured using a two-axis micro-electronic- mechanical sensor. Heat flux is measured using a proprietary sensor that incorporates low thermal resistant materials and thermocouple arrays. Galvanic skin response is used as an indicator of evaporative heat loss and is measured using two hypoallergenic stainless steel electrodes. Skin temperature is used to reflect the body's core temperature activity and is measured using a thermistor-based sensor. The near-body temperature sensor measures the temperature of the cover on the side of the armband. Energy expenditure was estimated using proprietary equations developed by the manufacturer. Results were analyzed with Innerview Research Software.

2.6 Dietary Control

The individual Zone diet to be elaborated should meet the recommended (Sears, 1995; 1997) percentages between macronutrients (i.e., 40% carbohydrates, 30% protein, 30% fat), starting from the protein requirements of 1.8 to 2.2 $g.kg_{LBM}^{-1}$ (Lemon, 2000). The administered diet was isocaloric compared to the energy expenditure measured with the armband. To maintain protein intake within the recommended range, according to the literature (Bosse *et al.*, 2004) the lipid intake was increased. Therefore, different percentages of macronutrients emerged even though during each meal (i.e., breakfast, morning and afternoon snacks, lunch and dinner) a protein to carbohydrate ratio between 0.6-0.75 was maintained according to the correct Zone diet guidelines (Sears, 1995).

2.7 Training Monitoring

During the four-weeks prior and the 2-week Zone-diet period, the runners filled a daily online training diary (BLITSâ) and a POMS questionnaire for objective and subjective evaluations of training and mood state, respectively. The training diary consists of three separate pages (Figure 1) to be filled in every day. The morning page (Figure 1a) includes objective (i.e., weight, hours of sleep, morning HR) and subjective data (i.e., muscle soreness) to be reported on a 10 cm visual analogue scale (VAS). The training page (Figure 1b) includes objective training data (i.e., volume, intensity, number of repetitions and HR) and two VAS scales for subjective ratings of training (i.e., intensity and attractiveness). The evening page (Figure 1c) includes five VAS (i.e., the overall subjective ratings of training intensity and attractiveness, physical and mental well-being, and muscle soreness). To prevent feedback for the athletes that would confound results, during the experimental period they had no access to their personal data normally provided by Blits® (www.blits.org). In fact, the program provides individual graphs representing the single training parameters. Furthermore, parameters that quantify and summarize training and adaptation to training, such as training load (Foster *et al.*, 1995), monotony of training and the overall strain (Foster, 1998), are provided. In particular, the parameter Load represents the magnitude of the training for both a single session and the weekly training, calculated by multiplying the training intensity (based on the VAS) by the duration of training. The parameter Monotony represents the day by day variability in training, calculated by dividing the average weekly training load by the standard deviation of the weekly load. Finally, the parameter Strain reflects adaptation or maladaptation to training and is calculated by multiplying the weekly Load x Monotony. Although the interpretation of the training diary data is strictly personal and individual, in the present study group trends are provided.

2.8 Profile of mood state

To evaluate the profile of mood state (POMS) of the runners a validated 32-item version questionnaire reflecting the individual's mood on five primary dimensions (i.e., depression, fatigue, vigor, tension and anger) was administered. On a weekly basis, the runner was required to describe "how have you felt during the past week" using a 5-point scale (i.e., not at all = 0; somewhat =1; moderately so =2; very much so =3; very very much so =4). The POMS data were analyzed separately for each specific dimension. Since sound analysis and interpretation of the specific training parameters need several months of data collection while confounding factors might arise interpreting data gathered over a shorter period, no comparison between experimental and control groups was considered appropriate. Thus, only data relative to pre-diet and post-Zone diet were analyzed, using participants as their own controls. Furthermore, percentages of changes were calculated between data collected during the two weeks on the Zone diet and the last week on the habitual nutritional regimen.

a

b

c

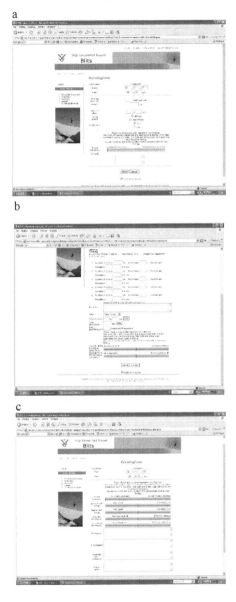

Figure 1. On-line Blits® training diary for morning (a), training (b), and evening (c) data.

2.9 Measurement of salivary cortisol and testosterone

Every week, samples of saliva were collected (Starstedt, Germany) in the morning after awakening, for cortisol (sC) and testosterone (sT) measurements. Samples were centrifuged at 3000 rev.min^{-1} for 15 min at 4 °C and stored at –80 °C until they were assayed using an enzyme immunoassay kit (Salimetrics, LLC, US), with an intra-essay coefficient of variation of 3.5±0.5% and an inter-essay reproducibility of 5.08±1.33% were accepted. A standard plate reader (Power Wave XS, Bio-Tech Instruments, US) was used.

2.10 Statistical analysis

Data are presented as mean ± standard deviations and/or as percentage of variation. Throughout the study the level of significance was set at P< 0.05. Normal distribution of the data was tested using the Kolmogorov-Smirnov test. Due to violation of normality assumption, the non-parametric Friedman test was used to detect differences across multiple measure tests and the Wilcoxon signed-rank test for single comparisons. Furthermore, to provide meaningful analysis for comparisons from small groups, Cohen's effect sizes (ES) with respect to pre-diet values were also calculated (Cohen 1988). An ES ≤ 0.2 was considered trivial, from 0.3 to 0.6 small, <1.2 moderate and >1.2 large.

3. RESULTS

3.1 Diet and energy expenditure

Table 1 represents the macronutrient and micronutrient composition of the diet before and during the Zone diet. Prior to the experimental period the runners reported a macronutrient composition of 54% carbohydrates, 18 % proteins and 28% fat, while the relative picture during the Zone diet was 32%, 21%, and 47% for carbohydrates, proteins and fat, respectively. Although the actual proportion of the macronutrient composition differed from that recommended by Sears (1995) for sedentary individuals (i.e., 40% carbohydrates, 30 % proteins and 30% fat), it was deemed necessary to maintain the athlete's diet isocaloric with respect to his energy expenditure (3173±119 kcal·day^{-1}) while respecting the protein intake (range: 1.8-2.2 g·kg$_{LBM}$$^{-1}$) and the carbohydrate to protein ratio (range: 0.6-0.75) of the Zone diet guidelines. Therefore, the total carbohydrate intake decreased significantly from 328±83 g.day^{-1} to 252±10 g.day^{-1}.

After two weeks on the Zone diet significant decreases in body weight (Pre: 72.5±5.0 kg, Post: 69.3±4.7 kg, P=0.0003, ES=0.970), body fat (Pre: 9.9±2.4 kg, Post: 9.0±2.3 kg, P=0.0008, ES=1.0), and fat free mass (Pre: 62.7±6.0 kg, Post: 60.3±5.7 kg, P<0.0014, ES=1.0) were observed for the experimental group. The two daily graded incremental tests showed no significant difference for time to exhaustion, RPE, peak velocity, peak HR and peak lactate concentrations before and after the zone diet period (Table 2).

Table 1. Comparison of food intake reported by the athletes and the zone diet expressed as mean±sd.

	Pre-experimental diet	*Zone Diet*
Protein, %	18±3	21±1
Carbohydrate, %	54±5	32±5
Lipids, %	28±4	48±4
Kcal	2418±554	3169±240
Na, mg	2659±902	3717±594
K, mg	2955±246	4314±358
Fe, mg	12±2	17±2
Ca, mg	767±162	1832±175
P, mg	1326±219	2512±235
Cu, mg	2±1	3±1
Mg, mg	217±30	368±23
Cl, mg	794±323	1325±264

Table 2. Means and standard deviations of peak values measured during the 1st and the 2nd exercise test performed pre- and post-Zone diet.

	Pre zone diet		Post zone diet	
	1st test	*2nd test*	*1st test*	*2nd test*
RPE	1919±1	19±1	18±1	18±1
HR (beats.min^{-1})	167±10	171±10	171±8	169±13
Time to exhaustion (s)	1127.4±186	1148.3±187	1173.4±152	1220.0±166
Peak Velocity (km·h^{-1})	16.6±2.1	16.9±2.1	17.1±1.5	17.6±1.7
Peal lactate (mmol·l^{-1})	6.0±2	6.3±1	6.1±1	7.1±1

3.2 Salivary cortisol and testosterone

No difference emerged between groups for sC (experimental group: 14 ± 6 nmol·l^{-1}, controls: 7 ± 5 nmol·l^{-1}) and sT (experimental group: 1.0 ± 0.2 nmol·l^{-1}, controls: 1.1 ± 0.3 nmol·l^{-1}) values collected during the 4 weeks prior to beginning on the Zone diet. For comparative purposes, sC and sT samples collected during the last week of the normal diet and the second week on the Zone diet were considered. A significant (P=0.048, ES=0.6) decrease in sC (6 ± 5 nmol·l^{-1}, 56%) was observed only in the experimental group, while the control group showed a slight increase (9 ± 6 nmol·l^{-1}). No difference emerged for sT for both groups (experimental: 0.8 ± 0.3 nmol·l^{-1}, control: 0.9 ± 0.2 nmol·l^{-1}).

3.3 Training parameters and profile of mood state

During the four-week period prior to the Zone diet, no difference emerged between POMS and daily training parameters in the experimental group. Despite no significant change in training volume between the week before and the end of the experimental period, taking into consideration the runner's perception of training intensity an overall 30% increase in Load was observed (Figure 2), while the relative picture for Monotony of training was 108%. As a result, Strain showed a 73% increase. Changes in the POMS scores (Figure 3) were opposite for the experimental and the control data. With respect to individual basal values, controls showed increases in Vigor (14%) and decreases in the other variables (Tension: 12%, Depression: 16%, Fatigue: 22%, and Anger: 25%), while runners on the Zone diet showed decreases in Vigor (15%) and increases in the other dimensions (Tension: 2%, Depression: 21%, Fatigue: 12%, and Anger: 12%).

a   b

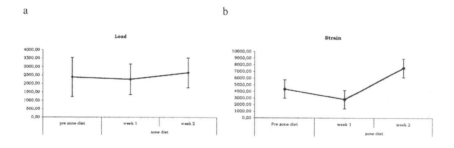

Figure 2. Means and standard deviations of Load (a) and Strain (b) training parameters before (Pre Zone diet) during (week 1) and after (week 2) the Zone diet.

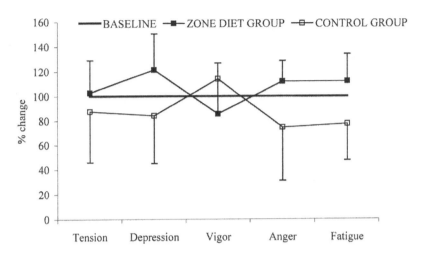

Figure 3. Means and standard deviations of the percentages of changes in the profile of mood state dimensions for controls and experimental group during the experimental period.

4. DISCUSSION

It is well recognised that nutrition has a relevant impact on endurance performance. Therefore, master athletes tend to adopt dietary regimens suggested by friends or specific magazines without consulting sport nutritionists. This places master runners at risk of using incorrect training schedules or nutritional patterns. At present there is no information regarding this important issue also in relation to their performances. The research paradigm used in this study aimed at investigating the effects of a dietary regimen on physiological, psychological and the performance aspects of master runners. The strict selection criteria and the fact that senior athletes are reluctant to take part in experimental settings that include stringent measurement schedules for several weeks limited the sample size, thus affecting the statistical power and urging some caution in making generalizations for the present results. Nonetheless, this study provides a useful model for examining interactive effects of various training and nutritional stressors.

In the present study the runners were provided accurate details for preparing their meals. Since they were not fed under controlled conditions and no prepared meals were provided, the compliance with the diet might be questioned. However, the athletes were highly motivated by expecting improvements in their performance and were allowed to drop out of the study if they felt that their training was compromised or if they had a hard time adhering to the diet.

Although the Zone diet is publicized to improve athletic performance, reduce body fat, increase muscle mass and improve mental state (Sears, 1995), in accordance to previous reports (Jarvis *et al.*, 2002; Bosse *et al.*, 2004; Piacentini *et al.*, 2004; 2007) the present results do not fully substantiate this claim. In fact, no difference in performance was observed, while the mood and Blits® parameters of runners worsened. Furthermore, the reduction in body weight was also due to a reduction in fat free mass.

The literature clearly shows that a relevant factor limiting endurance events is the depletion of carbohydrate in the blood, muscle and liver (Burke *et al.*, 2001; Hargreaves *et al.*, 2004; Erlenbusch *et al.*, 2005). To study the effect of the Zone diet on endurance performance, athletes should be administered tests long enough to tax their carbohydrate stores (Piacentini *et al.*, 2004, 2007). The master athletes were not available for overstressing tests that would have a negative impact on their training plans. Therefore, to evaluate their training status and the recovery capacity in this study a protocol including two bouts of maximal exercise to exhaustion was selected (Meeusen *et al.*, 2004).

Although the Zone diet is not merely based on reductions in carbohydrate intake and a relevant aspect is a 0.6-0.75 carbohydrate/protein ratio, it is possible that the drop-out of one runner and the worsened POMS and training parameters of the experimental group observed in the present study strongly depend on the reduction of the carbohydrate intake allowed during the Zone diet. Although we hypothesized a decrease in the second daily test performance as evidence of the reduced recovery capacity of the athletes (Meeusen *et al.*, 2004), no difference emerged as a result of the Zone diet regimen. These results might be due to the fact that this test is not sensitive enough to detect changes over such a short experimental period or that changes in performance might be more resilient than subjective feelings of training intensity and mood.

The most interesting and novel results of the present study emerged from analysis of the training Blits® diaries and POMS values recorded before and during the period on the Zone diet. The day to day variability in training parameters has been seen to be a good indication of monitoring training to avoid overreaching or overtraining (Foster, 1998). Training load has been originally calculated by multiplying session RPE by the duration of training (Foster, 1998), while the Load parameter of Blits® takes into consideration the VAS scale that might better reflect the runner's perception of training intensity. Both high Load and high Monotony scores are related to a negative adaptation to training, and the product of the two (i.e., Strain) has been related to early stages of overreaching, being a good indicator of how the athlete is adapting to that training regimen. Although each athlete's Strain value is calculated on the basis of his own threshold resulting from data collection over a long period of time, in the present study for simplicity, we presented the average trend in the group. In fact, normal Strain values were reported during the weeks prior to the Zone diet while a significant overall increase in Strain units between the first and the second week of the experimental period emerged (169%). Strain represents the product of Monotony and Load, the latter depending on training volume and subjective perception of training intensity. Given the lack of significant difference in training volume, the 30% increase in Load is mainly due to the amplified subjective perception of training intensity. Thus, the large raise in Strain is attributable to the athlete's feelings rather than training parameters per se. Although these are very subjective data and interpretation is sometimes hard because traditional statistics cannot help in the individualisation of a training maladaptation in a short period of two weeks, it seems to be a valid ecological tool and may be the only way to quantify training using a single term that is immediately visible to the trainer's eye.

The results of training diaries are in line with those observed from the percentages of changes in the athletes' mood state between baseline and the experimental period. This data reduction was deemed appropriate to avoid confounding factors that might emerge due to a short period of data collection and a high inter-individual variability, especially evident in the control group that did not show a typical baseline "iceberg" profile. Despite the lack of significant results, the experimental data showed changes not in the desired direction as seen in the control condition. In the runners who were administered the Zone diet, all the scores increased except for vigor which dropped by 15% during the second week on the zone diet. The control data instead showed the typical changes due to a normal training: increase in the positive scores and decrease in fatigue. However, Morgan *et al.* (1988) already demonstrated that the POMS reflects training distress caused by an increase in training and therefore it is considered a valid tool for monitoring of training. The effects of diet on mood has been widely explored; consumption of a diet containing 8.5 $g\;kg\;day^{-1}$ of carbohydrates (65% total energy) compared with a diet of 5.5 $g\;kg\;day^{-1}$ of carbohydrates (40% of total energy) not only resulted in a better maintenance of physical performance but also in a better mood state over the course of a period of intensified training (Achten *et al.*, 2004). Keith *et al.* (1991) demonstrated that after changing the carbohydrate content of a meal from 54% to 72% or 13% of total caloric intake, the low carbohydrate diet adversely affected depression, tension anger and vigor in well trained cyclists measured with the POMS.

The reason why alterations of macronutrient intake might have an effect on mood is not well known. Wurtman *et al.* (2003) hypothesized that a diet rich in carbohydrates relieves depression by increasing serotonin synthesis. High carbohydrate meals increase the ratio of tryptophan to large neutral amino acids that compete with each other in order to pass through the blood-brain barrier.

No difference emerged in either group for testosterone values measured weekly prior and during the Zone diet. However, only the experimental group showed a significant decrease in the cortisol concentrations during the period on the Zone diet. Despite the experimental group showing rather high baseline cortisol concentrations, the 56% reduction observed after two weeks on the Zone diet is relevant, with post-diet values lower than those registered in the control group. Volek *et al.* (2002) administered a restricted carbohydrate diet and showed a trend towards a decrease in cortisol concentrations. However, this diet allowed < 50 g of carbohydrates per day, not comparable to the present nutritional regimen. Provision of a high carbohydrate diet before an experimental stress has been shown to inhibit cortisol and feelings of depression (Markus *et al.*, 2000), and carbohydrate loading prior to prolonged exercise has been shown to increase performance and reduce the normal rise in cortisol (Deuster *et al.*, 1992). Moreover, due to the modifications of the macronutrient intake to match energy expenditure, the diet proposed to our runners was with a high fat content (47% of total energy intake). Venkatraman *et al.* (2001) observed that increasing fat content in the diet caused an increase in cortisol at baseline and post-exercise concentrations. Apparently the observed decrease in cortisol cannot be explained by the modification in the macronutrient intake of the athletes and is more related to the high pre-diet values, the increase in Strain observed during the training, and the mood disturbances observed during the Zone diet. Furthermore in a preliminary study (Piacentini *et al.*, 2004), we found that after 4 weeks on a Zone diet, ACTH concentrations were suppressed during exercise and this could also explain the decreased cortisol concentrations we found in the experimental group compared to the control group.

In conclusion the results of the present study do not support the hypothesis of an enhancement in performance or an increase in free fat mass due to the Zone diet, confirming previous work. Furthermore, due to its low carbohydrate content this nutritional regimen cannot be recommended for master endurance runners because it leads to a misperception of their training intensity and an increase in the negative scales of their mood states.

References

Achten, J., Halson, S. L., Moseley, L., Rayson, M. P., Casey, A. and Jeukendrup A.E., 2004, Higher dietary carbohydrate content during intensified running training results in better maintenance of performance and mood state. *Journal of Applied Physiology*, **96**, pp. 1331-1340.

Borg, G., 1998, *Borg's Perceived Exertion and Pain Scales*, (Champaign, IL: Human Kinetics).

Bosse ,M .C., Davis, S. C., Puhl, S. M., Pedersen, M., Low, V., Reiner, L., Dominguez, T. and Seals, N., 2004, Effects of Zone diet macronutrient proportions on blood lipids, blood glucose, body composition, and treadmill exercise performance. *Nutrition Research,* **24**, pp. 521-530.

Burke, L.M., Cox, G.R., Culmmings, N. K. and Desbrow, B., 2001, Guidelines for daily carbohydrates intake. Do athletes achieve them? *Sports Medicine,* **31**, pp. 267-299.

Cheuvront, S.N., 1999, The Zone Diet and athletic performance, *Sports Medicine,* **27**, pp. 213-228.

Cohen, J., 1988, *Statistical Power Analysis for the Behavioral Sciences*, (Hillsdale, NJ: Lawrence Eribaum Associates), 2nd ed.

Deuster, P., Singh, A., Hotman, A., Moses, F. M. and Chrousos, G.C., 1992, Hormonal responses to ingesting water or carbohydrate beverage during 24 run. *Medicine and Science in Sports and Exercise*, **24**, pp. 72-79.

Erlenbusch, M., Haub, M., Munoz, K., Macconnie, S. and Stillwell, B., 2005, Effect of high-fat carbohydrate diets on endurance exercise: A meta-analysis. *International Journal of Sport Nutrution and Exercise Metabolism*, **15**, pp. 1-14.

Foster, C., 1998, Monitoring training in athletes with reference to overtraining syndrome, **30**, pp. 1164-1168.

Foster, C., Hector, L., Welsh, R., Scharager, M., Green, M. A. and Snyder, A.C., 1995, Effects of specific versus cross training on running performance.

European Journal of Applied Physiology, **70**, pp. 367-372.

Fruin, M. L. and Rankin, J.W., 2004, Validity of a multi sensor armband in extimating rest and energy expenditure. *Medicine and Science in Sports and Exercise*, **36**, pp. 1063-1069.

Hargreaves, M., Hawley, J.A., and Jeukendrup, A., 2004, Pre-exercise carbohydrate and fat ingestion: effects on metabolism and performance. *Journal of Sports Sciences,* **22**, pp. 31-38.

Jackson, A.S. and Pollock, M.L., 1985, Practical assessment of body composition. *The Physician and Sports Medicine,* **13**, pp. 76-80.

Jakicic, J. M., Marcus, M., Gallagher, K. I., Randall., C., Thomas, E., and Goss. F.L., 2004, Evaluation of the SenseWear pro Armband to assess energy expenditure during exercise. *Medicine and Science in Sports and Exercise*, **36**, pp. 897-904.

Jarvis, M., McNaughton, L., Seddon, A. and Thompson, D., 2002, The acute 1-week effects of the Zone diet on body composition, blood lipid levels and performance in recreational endurance athletes. *Journal of Strength and Conditioning Research,* **16**, pp. 50-57.

Keith, R. E., O'Keeffe, K.A.., Blessing, D. L. and Wilson, G.D., 1991, Alterations in dietary carbohydrate, protein, and fat intake and mood state in trained female cyclist, *Medicine and Science in Sports and Exercise*, **23**, pp. 212-216.

Lemon, P.W.R., 2000, Beyond the zone: Protein needs of active individuals. *Journal of the American College of Nutrition,* **19**, pp. 513-521.

Markus, R., Panhuysen, A., Tuiten, G. and Koppeschaar, H., 2000, Effects of food and cortisol and mood in vulnerable subjects under controllable and uncontrollable stress. *Physiology and Behavior,* **70**, pp. 333-342.

Meeusen, R., Piacentini, M. F., Busschaert, B., Buyse, L., De Schutter, G. and Stray-Gundersen J., 2004, Hormonal responses in athletes: the use of a two bout exercise protocol to detect subtle differences in (over)training status. *European Journal of Applied Physiology,* **91**, pp. 140-146.

Morgan, W., Costill, D., Flynn, M., Raglin, J. and O'Connor, P., 1988, Mood disturbance following increased training in swimmers. *Medicine and Science in Sports and Exercise*, **20**, pp. 408-414.

Piacentini, M.F., Parisi, A., Bonanni, E. and Capranica, L., 2004, Hormonal response to a 30 minute time trial after 4 weeks on the zone diet. *Proceedings of the 9th Annual Congress of the ECSS, Clermont Ferrand, France*, 234.

Piacentini, M.F., Parisi, A., Bonanni, E. and Capranica, L., 2007, Hormonal response to a 60 minute time trial after 4 weeks on low carbohydrate diet. *Proceedings of the 12th Annual Congress of the ECSS, 11-14 July, Jyvaskyla Finland*, 123.

Rosenbloom, C.A. and Dunaway, A.., 2007, Nutrition recommendations of masters athlete. *Clinical Sports Medicine,* **26**, pp. 91-100.

Sears, B., 1995, *Mastering the Zone*, (New York, NY: Harper Collins).

Sears, B., 2000, The Zone diet and athletic performance correspondence. *Sports Medicine*, **29**, pp. 289-291,

Venkatraman, J. T., Feng, X. and Pendergast, D., 2001, Effects of dietary fat and endurance exercise on plasma cortisol, prostaglandin e_2, interferon-γ and lipid peroxides in runners. *Journal of the American College of Nutrition,* **20**, pp. 529-536.

Volek, J. S., Sharman, M. J., Dawn, M.L., Neva, G. A., Gòmez, A. L., Scheet, T. P. and Kraemer, W.J., 2002, Body composition and hormonal responses to a carbohydrate-restricted diet. *Metabolism,* **51**, pp. 864-870.

Wurtman, R. J., Wurtman, J. J., Regan, M.M., McDermott, J. M., Tsay, R. H. and Breu, J.J., 2003, Effects of normal meals rich in carbohydrates or proteins on plasma tryptophan and tyrosine ratios, *American Journal of Clinical Nutrition,* **77**, pp. 128-132.

CHAPTER NINETEEN

Does completion of an exercise referral scheme affect anthropometric measures and physical activity levels?

Karen Williams[1], Suzanne Taylor[2] and Katherine Bond[3]

[1]Research Institute for Sport and Exercise Sciences, Liverpool John Moores University, Liverpool, UK
[2]Sport and Exercise Sciences, Glyndwr University, Wrexham, UK
[3]University of Chichester, Chichester, UK

1. INTRODUCTION

The prevalence of being overweight or obese has recently been recognised as a serious health problem in most developed nations (WHO, 2000a). There are implications for the transition from being classed as overweight to being considered obese; these conditions are linked to increases in health problems such an increased incidence of cardiovascular disease, type 2 diabetes mellitus, hypertension, stroke, dyslipidaemia, osteoarthritis, and some cancers (Burton et al., 2005). The prevalence of obesity is approximately 15-20% of the population in Europe, and estimated to be as high as 22% in children, 26% in men and 31% in women (Seidell and Flegal, 1997). Obesity and its related diseases have become a major health concern in the United States (Wilborn et al., 2005) and in the Asia-Pacific region (Cockram, 2000) with evidence suggesting that obesity-related mortality may differ among racial and ethnic groups (Savage et al., 1991; Deurenberg and Deurenberg-Yap, 2003). Data from the England Health Survey indicate that the proportion of overweight or obese men between the ages of 16 and 64 increased from 45% in 1986–1987 to 64% in 2002-2003. The same statistics calculated for women indicate a rise from 36 to 58% during the same period (Leicester and Windmeijer, 2004). It has been suggested that 70% of men and 80% of women in the UK are insufficiently active to benefit their health (Sports Council, Health Education Council, 1992).

The most current definitions of overweight and obesity for international use in adults are those of the World Health Organization (WHO), with overweight defined as a BMI of 25 or greater and obesity defined as a BMI of 30 or greater (WHO, 1998). Body Mass Index (BMI), whilst not without limitations (Deurenberg and Deuerenberg-Yap, 2003; DeLorenzo et al., 2003), is a simple index of body mass-for-height that is commonly used to classify underweight, overweight and obesity in adults (WHO, 1995). Although BMI is the most widely used and simplest index of body size and is frequently used to estimate the prevalence of obesity within a population (WHO, 1995; Colditz *et al.*, 1995), measurements of waist circumference and waist–hip ratio (WHR) have been viewed as alternatives to BMI as excess intra-abdominal fat is associated with a greater risk of obesity-related morbidity than is overall adiposity (Ho *et al.*, 2001). In particular, waist circumference has been shown to be the best and most simple index of both intra-abdominal fat mass and total fat (Han *et al.*, 1997). Widely used in both clinical and research settings, waist measurement is useful as an indicator of responses to exercise and dietary interventions in GP referral schemes. With regards to distribution of fat, both higher waist circumference and waist-to-hip ratio (WHR) are shown to be associated with poor mobility and poor physical performance (Bannerman *et al.*, 2002; Guo *et al.*, 2002). This is pertinent when obesity is considered to be one of the strongest predictors of onset of mobility difficulty in middle-aged adults (Clark *et al.*, 1998).

The global estimate of physical inactivity levels has been reported as 17%, whereas the estimate for insufficient levels of physical activity (< 150 minutes moderate or < 60 minutes of vigorous activity per week) is 40% (US Department of Health and Human Resources, 1996). Several observational studies have determined a relationship between higher baseline levels of physical activity or improvements in physical activity and either reduced incidence of weight gain (French *et al.*, 1994; DiPietro *et al.*, 1998) or lower odds of significant weight gain (Williamson *et al.*, 1993; DiPietro *et al.*, 1998). Increasing physical activity in the population has the potential to modify risk factors such as obesity amongst other health problems (Taylor *et al.*, 1998). Evidence suggests that declining amounts of physical activity have contributed importantly to a rising prevalence of overweight and obesity (Grundy *et al.*, 1999).

In response to the concerns about the high population levels of physical inactivity and the recent evidence on the benefits of moderate physical activity (Anderson *et al.*, 2000), policy recommendations aimed at the promotion of active lifestyles in all members of the population have been published (Pate *et al.*, 1995; Health Education Authority, 1996). Currently, there is no single national strategy to promote physical activity within the United Kingdom (Dugdill *et al.*, 2005). More commonly, targets for physical activity promotion are set within other government health policies including the Department of Health's National Service Frameworks for coronary heart disease (2000) and diabetes (2002). There is an emerging interest in the promotion of physical activity through primary health care (Biddle *et al.*, 1994; Hillsdon *et al.*, 1995) such as 'exercise on prescription schemes' (EoP) or exercise referral schemes (Biddle *et al.*, 1994; Lord and Green, 1995; Riddoch *et al.*, 1998; Stevens *et al.*, 1998; Harland *et al.*, 1999; Lawlor and Hanratty, 2001). Typically, exercise referral schemes offer the opportunity for individuals who have medical conditions to undertake a structured programme of physical activity, usually within the confines of a leisure facility and extends over a period of 6 – 14 weeks. Inclusion in the scheme is via referral from a health professional (DoH, 2001). At present, research to support their effectiveness remains inconclusive (Stevens *et al.*, 1998; Taylor *et al.*, 1998; Harland *et al.*,

1999; Hillsdon *et al.*, 2002).

Physical activity is recognised as a major contributor to good health and is an important focus for health promotion (Department of Health and Human Services, 1996) and as such the promotion of physical activity is an issue of increased importance within public health. Therefore the aim of this study was to investigate the effectiveness of attendance of an exercise referral scheme on anthropometric and physical activity measures.

2. METHODS

2.1 Participants

Thirty one patients (males $n = 14$, females $n = 17$) adhered to the scheme and were recruited from General Practices in the Wrexham area of Wales. All patients had been referred by their general practitioner (GP) to the Exercise Referral Scheme as they demonstrated cardiovascular disease risk factors. Subjects provided informed consent prior to the study. Ethics approval was granted through the local NHS ethics board.

Table. 1 Descriptives statistics of subjects (Mean ± SD).

	Males	**Females**
Subjects (n)	14	17
Age (years ± SD)	57.3 ± 6.3	50.7 ± 19.1
Height (m ± SD)	1.73 ± 0.06	1.65 ± 0.12

2.2 Data collection

Data were collected in three phases: baseline assessment (i.e. just before they started the GP referral scheme); post-intervention follow up (i.e. when they had completed 6 sessions over 3 weeks) and then a 4-months post-intervention repeat assessment (i.e. 4 months after the initial 6 sessions). All assessments consisted of completion of the short form International Physical Activity Questionnaires (IPAQ) and physiological measurements which included height, body mass, waist circumference, hip circumference and waist-hip ratio. A randomised controlled design could not be implemented as subjects were recruited following GP referral; therefore ethics would not allow the opportunity to exercise to be removed for the purposes of the study.

2.3 Outcome measures

Measurement of physical activity is a pre-requisite for monitoring population health and evaluating effective interventions, with the preferred and most practical method in epidemiological studies being self-report questionnaires. Self-reported physical activity was assessed by using the short version of the International Physical Activity Questionnaire (IPAQ). Use of the IPAQ assesses physical activity undertaken across a comprehensive set of domains including leisure time physical activity, domestic and gardening (yard) activities, work-related physical activity and transport-related physical activity. The IPAQ asks about three specific types of activity undertaken in the four domains. The specific types of activity that were assessed are walking, moderate-intensity activities and vigorous-intensity activities. The items in the IPAQ form were structured to provide separate scores on walking, moderate-intensity and vigorous-intensity activity. Computation of the total score for the short form requires summation of the duration (in minutes) and frequency (days) of walking, moderate-intensity and vigorous-intensity activities. The reliability and validity of the IPAQ have been demonstrated (Craig *et al.,* 2003). Anthropometric and physiological measures included height and body mass, calculation of BMI. Waist and hip circumference were measured to calculate waist-hip ratio.

2.4 Intervention

Participants were requested to attend six supervised one-to-one exercise sessions with a GP referral trained exercise instructor at a reduced cost of £6.00 for six sessions. The exercise sessions incorporated cardiovascular exercise using equipment including treadmills, cycle ergometers, recumbent cycle ergometers, rowing ergometers, cross trainers and a number of weighted machines for resistance exercise. Following the final supervised session, participants repeated the anthropometric and physiological measures along with the IPAQ questionnaires. Participants were then requested to continue attending supervised sessions at the leisure centre for a reduced cost (when compared to standard public rates) of £1.50. Four months following baseline testing, participants were contacted and were requested to return for the final follow up anthropometric, physiological and questionnaire measurements/assessments. The results of the IPAQ were not made available to the subjects during the course of the study.

2.5 Statistical analysis

Normality of the data was analysed using the Shapiro-Wilk test. The test revealed that the data for the waist, hip and physical activity measures (METS) were not normally distributed. These data were analysed using the Friedman test. Post hoc analysis was performed using Bonferroni corrected Wilcoxon tests. The normally distributed data were analysed using a one-way repeated measures analysis of variance. Violations to the sphericity were corrected using the Greenhouse Geiser correction as the Greenhouse Geiser epsilon was below 0.75 in each case (see Atkinson, 2001). The dependent variables were the IPAQ measurement of physical activity and the anthropometric measurements (height, body mass, waist and hip circumference and BMI. The independent variables were gender and adherence to the activity. The null hypothesis was that changes in self- reported physical activity and physiological measures at follow up is the same as baseline.

3. RESULTS

Physical activity levels, expressed as total met minutes per week, revealed significant changes from baseline (χ^2_2 = 22.13, P = 0.01). Wilcoxon tests revealed that mean ± SD physical activity was significantly different from baseline following six sessions, increasing from 650 ± 994.5 MET-minutes per week to 981 ± 1049.8 MET-minutes per week (P = 0.01). Physical activity then decreased from 981 ± 1049.8 MET-minutes per week at the mid-point to 725 ± 836.3 MET-minutes per week at the four-month follow up assessment. The results revealed no significant difference between levels at baseline and four moths post-assessment.

Mean ± SD body mass changed significantly over time from baseline to the four month follow-up assessment ($F_{1.4,41.2}$ = 17.02, P < 0.0005). Post hoc tests demonstrated that body mass continued to decrease significantly at each time point from 82.4 ± 14.2 kg at baseline, to 80.7 ± 13.1 kg following six sessions (P < 0.0005), and 78.9 ± 12.3 kg four months post-assessment (P < 0.0005). Mean ± SD waist and hip measurements also decreased significantly over time (χ^2_2 = 28.86, P < 0.0005) and (χ^2_2 = 24.8, P < 0.0005) respectively. Wilcoxon tests revealed that mean ± SD waist and hip measurements significantly decreased from 97.6 ± 13.3 cm and 109.7 ± 12.1 cm respectively at baseline to 96.0 ± 13.4 cm and 108.7 ± 11.3 cm respectively following six sessions (P < 0.0005). Waist and hip measurements continued to decrease significantly from the six session point to the four month post assessment point (P < 0.0005) to 94.6 ± 13.0 cm and 107.6 ± 10.9 cm respectively. The results at baseline for both hip and waist measurements were also significantly different from the values measured at four months post-assessment (P < 0.0005). The mean ± SD waist-hip ratio changed significantly over time from baseline to the four month follow-up assessment ($F_{1.5,44}$ = 3.9, P = 0.03). Post hoc tests demonstrated that waist-hip ratio decreased significantly from 0.89 ± 0.07 at baseline to 0.88 ± 0.07 following six sessions (P = 0.03), it was also significantly different from baseline at the four month post-assessment measuring 0.87 ± 0.07 (P < 0.0005). There was no significant difference between the measures of waist-hip ratio at the six-session time point and the four-month follow up assessment.

Mean ± SD body mass index (BMI) changed significantly over time from baseline to the four-month follow up assessment ($F_{1.5,44}$ 14.1, P < 0.0005). Pairwise comparisons displayed that BMI decreased from 29 ± 4.8 at baseline to 28 ± 4.4 following six sessions (P = 0.005) and continued to decrease significantly to 27 ± 4.2 at the four month follow up assessment (P = 0.001). The BMI value at the sixth session was also significantly different from the value measured at the four-month assessment point (P = 0.019).

Table 2. Mean and Standard Deviations of changes in BMI, waist, hip, waist-hip ratio, and body mass over time 1 (baseline), time 2 (post 6 sessions) and time 3 (4 months post), * = significant difference from baseline (P < 0.05)

	Pre	Post	4 months post
Body mass (kg ± SD)	82.4 ± 14.2	80.7 ± 13.1*	78.9 ± 12.3*
Waist (cm ± SD)	97.6 ± 13.3	96.0 ± 13.4*	94.6 ± 13.0*
Hip (cm ± SD)	109.7 ± 12.1	108.7 ± 11.3*	107.6 ± 10.9*
Waist-hip ratio (± SD)	0.89 ± 0.07	0.88 ± 0.07*	0.87 ± 0.07*
BMI (kg.m^2 ± SD)	29 ± 4.8	28 ± 4.4*	27 ± 4.2*
PA levels (MET-min.week^{-1} ± SD)	650 ± 994.5	981 ± 1049.8*	725 ± 836.3

4. DISCUSSION

The aim was to investigate the effectiveness of attendance of an exercise referral scheme on anthropometric and physical activity measures. Mean ± SD physical activity, expressed as MET-minutes/week increased from a baseline value of 650 ± 994.5 MET-minutes/week to 981 ± 1049.8 MET-minutes per week after the initial six sessions but thereafter decreased to 725 ± 836.3 MET-minutes per week at the four-month follow-up assessment (P = 0.01). These values, when expressed as scores on IPAQ activity scale ranges demonstrated that physical activity scores increased from 'low' at the baseline score to 'moderate' following the six sessions, returning to 'low' at the four-month follow-up assessment point. The drop in activity levels has been attributed to a high level of mechanisation occurring both at home and in the workplace coupled with an increase in the number of individuals using alternative means of transport other than walking or cycling (Rosengren and Lissner, 2008). The UK exercise guidelines (DoH, 2004) for maximising health benefit are that every adult should 'accumulate 30 minutes, or more, of moderate intensity physical activity on most (at least 5 days), and preferably all, days of the week', i.e. a minimum of 150 min of moderate physical activity a week. This level was achieved following the initial six sessions where the average activity score was 'moderate'but was not maintained. The UK exercise guidelines (DoH, 2004) also recommend that, in order to sustain cardiovascular fitness, individuals should aim to participate in at least three 20-min bouts of continuous vigorous activity per week, in addition to the accumulated moderate activity in order to maintain cardiovascular fitness. There has been a shift from the promotion of vigorous only activity to induce cardiovascular benefits, as prescribed by the American College of Sports Medicine (ACSM) in 1978, to suggestions of more moderate physical activity levels. The lack of maintenance of the physical activity levels following the six sessions may be due to the mode of the exercise performed. A concept of health-enhancing physical activity (HEPA) has been discussed (WHO, 2002) which includes activities that reflect lifestyle such as aspects of leisure, work, transport and home activities. Moderate activities are also considered to be more attractive and sustainable to individuals as a form of physical activity. Physical activity as performed in a gym-based setting may not be a reflection of lifestyle activities and therefore adherence is reduced. This view is consistent with the results of this study as physical activity levels decreased following the initial six sessions from 'moderate' to 'low' at the four-month post-assessment point.

The adoption of an active lifestyle later in life is related to prolonged life (Paffenbarger *et al.,* 1993) and a reduced risk of cardiovascular disease (CVD) with regular exercise over a five-year period (Blair, 1995). In order to monitor the effect of physical activity on indices of cardiovascular risk, accurate measurements of the level of physical activity must be undertaken along with anthropometric measurements to ascertain the level of change. Physical activity was assessed using the International Physcial Activity Questionnaire (IPAQ), the results of which indicated that activity increased until the six session point and declined thereafter. It may be plausible to consider potential errors in the reporting of the data and more useful, when possible, to collect quantitative data on physical activity levels. There are various methods of measuring body composition but there are limitations associated with each (Wagner and Heyward, 1999). At population level, questionnaires are the most commonly used with the majority of existing questionnaires focused on physical activity during leisure time or within the workplace, thus limiting the use of these subjective tools (Hagstomer *et al.,* 2005). There are limited questionnaires that analyse the physical activity that occurs during daily situations, such as transportation, occupation, household, family care, and leisure time (Hagstromer *et al.,* 2005). One such tool is the International Physical Activity Questionnaire (IPAQ) which is thought to be a reliable and valid method for the subjective assessment of physical activity (Craig *et al.,* 2003); however, there is also research to suggest that some elements of the IPAQ show levels of unreliability and validity (Macfarlene *et al.,* 2007; Maddison *et al.,* 2007).

The anthropometric measures all decreased from baseline through to the four-month follow-up assessment. This decrease is typical of any individual that raises his or her level of activity from sedentary to moderate, with such changes most likely to occur within the first six months of a change in activity behaviour (Wing, 1999). A close relationship exists between abdominal adiposity and risk of CVD; however, the current waist circumference cut-off points suggested by the World Health Organization (WHO) are not based on associations with CVD risk factors, but rather on their correlation with corresponding values of BMI (WHO, 2000b). Although the anthropometric values decreased during the course of the study, when evaluated against the WHO (2000) classification scores, the subjects failed to change their classification, remaining at the 'pre-obese' category throughout. The values for waist and hip circumference were also above the classification for increased CVD risk based upon the calculation of waist-hip ratio (Rosenbaum *et al.,* 1997) throughout the study. However, the results demonstrated a trend towards a decrease in values, suggesting that the positive changes required for the benefit of health have been instigated, and thus, a possible reduction in the incidence of CVD risk factors.

Limitations included issues such as a four-month data collection point. Without further study it is not possible to determine whether the positive changes would continue subsequent to the four-month assessment point. As measures were based upon attendance of an exercise referral scheme, it is pertinent to discuss the role of the scheme that was used. A control group was not utilised which is in line with the majority of exercise referral research since very few studies use assessment-only control groups (Morgan, 2005). The measurement of physical activity is also a factor that varies considerably amongst studies. The likelihood of bias from physical activity measures is considered to be one of the greatest limitations in research (Paffenberger *et al.*, 1993), with very few studies attempting to validate the measured recall data using physiological methods (Morgan, 2005). This was limiting as only anthropometric data was collected in conjunction with the physical activity questionnaires. However, what must be considered is that the aim was to monitor the effectiveness of attendance to an exercise referral scheme on physical activity and anthropometric measures. Adherence to exercise referral schemes is relatively low (Lord and Green, 1995; Stevens *et al.*, 1998; Taylor *et al.*, 1998) and the evaluation of subjects who do adhere has not been studied to a great extent, with a high level of studies looking at the effectiveness of schemes as a whole (See Dugdill *et al.*, 2005 for a review). With respect to the inclusion criteria, based upon the exercise referral guidelines in place on the scheme that was evaluated and based upon the aims, similar to many other studies, risk factors related to CVD were the inclusion criteria. Exclusion criteria for the purposes of this study included non-attendance to the scheme. The medication use of subjects was not considered and is a further limitation since it is not known whether the changes that occurred were affected by medication use. A randomised controlled design would have increased the strength of the research but ethical considerations did not allow for this. Further studies should aim to monitor medication and use a randomised design.

5. CONCLUSIONS

The study was performed as an evaluation of the service efficacy of attendance within one specifically identified Exercise on Referral Scheme. Although data related to both the physical activity and anthropometric data were statistically significant, the implications of these changes were negligible, hence no change in WHO classification for risk factors related to CVD emerged. The trends observed were in a positive direction, with evidence that adherence to an exercise on referral scheme has significant effects on both anthropometric and physical activity variables. The results of this study provide a foundation of evidence upon which future studies can be based, and statistical validity better substantiated.

References

American College of Sports Medicine, 1978, Position statement on the recommended quantity and quality of exercise for developing and maintaining fitness in healthy adults. *Medicine and Science in Sports and Exercise*, **10**, pp. 7-10.

Anderson, L.B., Schnorl, P., Schroll, M. and Hein, H.O., 2000, All-cause mortality associated with physical activity during leisure time, work, sports and cycling to work. *Archive of International Medicine*, **160**, pp. 1621-1628.

Atkinson, G., 2001, Analysis of repeated measurements in physical therapy research. *Physical Therapy in Sport*, **2**, pp. 194-208.

Bannerman, E., Miller, M.D., Daniels, L.A., Cobiac, L., Giles, L.C., Whitehead, C., Andrews, G.R. and Crotty, M., 2002, Anthropometric indices predicts physical function and mobility in older Australians: the Australian Longitudinal Study of Ageing. *Public Health Nutrition*, **5**, pp. 655– 62.

Biddle, S., Fox, K. and Edmund, L., 1994, Physical Activity in Primary Health Care in England, (London: Health Education Authority).

Blair, S.N., 1995, Exercise prescription for health. *Quest*, **47**, pp. 338-353.

Burton, B.T., Foster, W.R., Hirsch, J. and VanItallie, T.B., 1995, Health implications of obesity: NIH consensus development conference. *International Journal of Obesity and Related Metabolic Disorders*, **9**, pp. 155-169.

Clark, D.O., Stump, T.E. and Wolinsky, F.D., 1998, Predictors of onset of and recovery from mobility difficulty among adults aged 51–61 years. *American Journal of Epidemiology*, **148**, pp. 63–71.

Cockram, C.S., 2000, The epidemiology of diabetes mellitus in the Asia-Pacific region. *Hong Kong Medicine Jounral*, **6**, pp. 43–52.

Colditz, G., Willett, W., Rotnitzky, A. and Manson, J., 1995, Weight gain as a risk factor for clinical diabetes mellitus in women. *Annals of Internal Medicine*, **122**, pp. 481–6.

Craig, C.L., Marshall, A.L., Sjostrom, M., Bauman, A.E., Booth, M.L., Ainsworth, B.E., Pratt, M., Ekelund, U., Yngve, A., Sallis, J.F. and Oja, P., 2003, International Physical Activity Questionnaire:12-Country Reliability and Validity. *Medicine and Science in Sports and Exercise*, **35**, pp. 1381-95.

De Lorenzo, A,, Deurenberg, P., Pietrantuono, M., Di Daniele, N., Cervelli, V. and Andreoli, A., 2003, How fat is obese. *Acta Diabetologica*, **40**, pp. 254-7.

Department of Health, 2000, National Service Framework for Coronary Heart Disease, (London: Department of Health).

Department of Health, 2001, National Service Framework for Older People, (London: Department of Health).

Department of Health, 2002, National Service Framework for Diabetes, (London: Department of Health).

Department of Health, 2004, At least five a week: Evidence on the impact of physical activity and its relationship to health, (London: Department of Health).

Deurenberg, P. and Deurenberg-Yap, M., 2003, Validity of body composition methods across ethnic population groups. *Nutrition Forum*, **56**, pp. 299-301.

DiPietro, L., Kohl, III, H.C., Barlow, C.E. and Blair, S.N., 1998, Improvements in cardiorespiratory fitness attenuate age-related weight gain in healthy men and women: the Aerobics Center Longitudinal Study. *International Journal Obesity*, **22**, pp. 55-62.

Dugdill, L., Graham, R.C. and McNair, F., 2005, Exercise Referral: the public health panacea for physical activity promotion? A critical perspective of exercise referral schemes; their development and evaluation. *Ergonomics,* **48**, pp. 1390-1410.

French, S. A., Jeffrey, R.W., Forster, J.L., McGovern, P.G., Kelder, S.H. and Baxter, J.E., 1994, Predictors of weight change over two years among a population of working adults: The Healthy Worker Project. *International Journal of Obesity*, **18**, pp. 145-154.

Grundy, S.M., Blackburn, G., Higgins, M., Lauer, R., Perri, M.G. and Ryan, D., 1999, Physical activity in the prevention and treatment of obesity and its comorbidities: evidence report of independent panel to assess the role of physical activity in the treatment of obesity and its comorbidities. *Medicine and Science in Sports and Exercise*, **31**, pp. 1493.

Guo, X., Matousek, M., Sundh, V. and Bertil S., 2002, Motor performance in relation to age, anthropometric characteristics, and serum lipids in women. *Journal of Gerontolology Series A, Biological Sciences and Medical Sciences*, **57**, pp. 37–44.

Hagstromer, M., Pekka, O. and Sjostrom, M., 2005, The International Physical Activity Questionnaire (IPAQ): a study of concurrent and construct validity. *Public Health Nutrition*, **9**, pp. 755-762.

Han, T.S., McNeill, G., Seidell, J.C. and Lean, M.E., 1997, Predicting intra-abdominal fatness from anthropometric measures: the influence of stature. *International Journal of Obesity and Related Metabolic Disorders*, **21**, pp. 587–93.

Harland, J., White, M., Drinkwater, C., Chinn, D., Farr, L. and Howel, D., 1999, The Newcastle exercise project: a randomised controlled trial of methods to promote physical activity in primary care. *British Medical Journal*, **319**, pp. 828-832.

Health Education Authority, 1996, Promoting physical activity in primary care-guidance for the primary health care team, (London: Health Education Authority).

Hillsdon, M., Thorogood, M., Anstiss, T. and Morris, J., 1995, Randomised controlled trials of physical activity promotion in free living populations: a review. *Journal of Epidemiology and Community Health*, **49**, pp. 448–53.

Hillsdon, M., Thorogood, M., White, I. and Foster, C., 2002, Advising people to take more exercise is ineffective: a randomized controlled trial of physical activity promotion in primary care. *International Journal of Epidemiology*, **31**, pp. 808-815.

Ho, S.C., Chen, Y.M., Woo, J.L., Leung, S.S., Lam, T.H. and Janus, E.D., 2001, Association between simple anthropometric indices and cardiovascular risk factors. *International Journal of Obesity and Related Metabolic Disorders*, **25**, pp. 1689–97.

Lawlor, D.A. and Hanratty, B., 2001, The effect of physical activity advice given in routine primary care consultations: a systematic review. *Journal of Public Health Medicine*, **23**, pp. 219-226.

Leicester, A. and Windmeijer, F., 2004, The 'fat tax': economic incentives to reduce obesity. IFS. Briefing Notes: BN49. Available at: http://www.ifs.org.uk/bns/bn49.pdf.

Lord. J.C. and Green, F., 1995, Exercise on prescription: does it work? *Health Education Journal*, **54**, pp. 453-64.

Macfarlene, D.J., Lee, C.C., Ho, E.Y., Chan, K.L. and Chan, D.T., 2007, Reliability and validity of the Chinese version of IPAQ (short, last 7 days). *Journal of Science and Medicine in Sport*, **10**, pp. 45-51.

Maddison, R., Mhurchu, C.L., Jiang, Yannan., Vander Hoorn, S., Rodgers, A. and Lawes, Mm, C., 2007, International Physical Activity Questionnaire (IPAQ) and New Zealand Physical Activity Questionnaire (NZPAQ): A doubly labelled water validation. *International Journal of Behavioural Nutrition and Physical Activity*, **4**, pp. 62.

Morgan, O., 2005, Approaches to increase physical activity: reviewing the evidence for exercise referral schemes. *Public Health*, **119**, pp. 361-370.

Paffenberger, R.S., Blair, S.N., Lee, I.M. and Hydr, R.T., 1993, Measurement of physical activity to assess health effects in free-living populations. *Medicine and Science in Sports and Exercise*, **25**, pp. 60-70.

Pate, R.R., Pratt, M., Blair, S.N, *et al.*, 1995, Physical activity and public health. A recommendation from the Centres for Disease Control and Prevention and the American College of Sports Medicine. *Journal of the American Medical Association*, **273**, pp. 402-407.

Riddoch, C., Puig-Ribera, A. and Cooper, A., 1998, Effectiveness of Physical Activity Promotion schemes in Primary Care: A review, (London: Health Education Authority).

Rosenbaum, M., Leibel, R.L. and Hirsch, J., 1997, Obesity. *New England Journal of Medicine*, **337**, pp. 396-407.

Rosengren, A. and Lissner, L., 2008, The sociology of obesity. *Frontiers of Hormone Research*, **36**, pp. 260-270.

Savage, P.J. and Harlan, H.R., 1991, Racial and ethnic diversity in obesity and other risk factors for cardiovascular disease: implications for studies and treatment. *Ethnicity and Disease*, **1**, pp. 200-211.

Seidell, J.C. and Flegal, K.M., 1997, Assessing obesity: classification and epidemiology. *British Medical Bulletin*, **53**, pp. 238–52.

Sports and Health Education Council, 1992, Allied Dunbar national fitness survey: main findings. Northampton: Sports Council, Health Education Authority.

Stevens, W., Hillsdon, M. and Thorogood, M., 1998, The cost effectiveness of a primary care-based physical intervention in 45-74 year old men and women: a randomised controlled trial. *British Journal of Sports Medicine*, **32**, pp. 236-41.

Taylor, A.H., Doust, J. and Webborn, N., 1998, Randomised controlled trial to examine the effects of a GP exercise referral programme in Hailsham, East Sussex, on modifiable coronary heart disease risk factors. *Journal of*

Epidemiology and Community Health, **52**, pp. 595-601.

US Department of Health and Human Resources, 1996, A report for the surgeon general. Physical activity and health, (Atlanta: National Centre for Chronic Disease Prevention and Health Promotion).

Wagner, D.R. and Heyward, V.H., 1999, Techniques of body composition assessment: a review of laboratory and field methods. *Research Quarterly for Exercise and Sport*, **70**, pp. 135-49.

Wilborn, C., Beckham, J., Campbell, B., Harvey, T., Galbreath, M., La Bounty, P., Nassar, E., Wismann, J. and Kreider, R., 2005, Obesity: prevalence, theories, medical consequences, management, and research directions. *Journal of the International Society of Sports Nutrition*, **2**, pp. 4–31.

Williamson, D. F., Madans, J., Anda, R.F., Kleinman, J.C., Kahn, H.S. and Byers, T., 1993, Recreational physical activity and 10-year weight change in a U.S. national cohort. *International Journal of Obesity*, **17**, pp. 279-286.

Wing, R.R., 1999, Physical activity in the treatment of the adulthood overweight and obesity: current evidence and research issues. *Medicine and Science in Sports and Exercise*, **31**, pp. 547-552.

World Health Organisation, 1995, Physical status: the use and interpretation of anthropometry. Report of a WHO Expert Committee. WHO Technical Report Series 854, (Geneva: World Health Organisation).

World Health Organisation, 2000a, Obesity: Preventing and Managing the Global Epidemic. Report of a WHO Consultation. WHO Technical Report Series.

World Health Organisation, 2000b, Health systems: improving performance. Available at: http://www.who.int/whr/2000/en/index.html

World Health Organisation, 2002, Reducing risks, promoting healthy life. http://www.who.int/whr/2002/en/

CHAPTER TWENTY

Correlation between inter-limb coordination, strength and power performances in active and sedentary individuals across the life span

Cristina Cortis[1,2], Antonio Tessitore[2,3], Caterina Pesce[4], Corrado Lupo[2],
Fabrizio Perroni[2] and Laura Capranica[2]

[1]Department of Health Sciences, University of Molise, Campobasso, Italy
[2]Department of Human Movement and Sport Science, IUSM, Rome, Italy
[3]Vrije Universiteit, Brussel, Belgium
[4]Department of Education in Sport and Human Motion, IUSM, Rome, Italy

1. INTRODUCTION

Western societies are challenged with an increasing incidence of dependency in everyday life contexts of older individuals. The deterioration of neuromuscular control in older individuals has been attributed to the reduction of strength, the slowing of behaviour and the modification in the quality of coordination (Spirduso et al., 2005). Conventionally, neuromuscular control and function are evaluated by means of handgrip strength and functional muscular power of the lower limbs (i.e., counter-movement jump, CMJ), also in relation to chronological age. Age-related changes show an inverted-U trend (Laforest et al., 1990), with a development throughout childhood, peak values at 20-30 years of age and a decrement with advancing age, more evident in the lower limbs with respect to the upper limbs (Larsson et al., 1979; Lafargue et al., 2003). However, a wide variation between individuals of a similar age is often observed, limiting the interpretation of the role played by chronological age also related to physical activity. In fact, training proved to be effective in increasing strength and power performances of young (Rowland, 2005) and old (Spirduso et al., 2005) individuals.

Although strength and power decrements in older individuals have been studied quite frequently, much less in known about age-related changes in motor control during motor tasks (Spirduso et al. 2005). Inter-limb coordination is of primary concern to ageing adults being intrinsic to many activities of daily living, which could represent a critical factor for independence and safety. The stability of inter-limb coordinated movements is governed by a variety of constraints at different levels of the motor system (Carson and Kelso, 2004), in relation to frequency of execution (Kelso, 1984; Jeka and Kelso, 1995; Serrien and Swinnen, 1998; Capranica et al., 2004, 2005), amplitude (Peper and Beek, 1998), loading (Baldissera et al., 1991), direction (Capranica et al., 2004, 2005, Meesen et al.,

2006), mechanical perturbations (Kay *et al.*, 1991), and sensory feedback (Kelso *et al.*, 2001; Salesse and Temprado, 2005; Salesse *et al.*, 2005). In particular, cyclic flexion-extensions of the hand and its ipsilateral foot are easily associated when the two segments oscillate with the same direction (i.e. in-phase), while greater control and attentional efforts are required when they move in opposite directions (i.e. anti-phase; Baldissera *et al.*, 1982, 1991, 2000; Swinnen *et al.*, 1997; Serrien and Swinnen, 1998; Serrien *et al.*, 2001), especially with increased frequency of movement (Kelso, 1984; Kelso *et al.*, 1986; Carson *et al.*, 1995; Capranica *et al.*, 2004, 2005). Although laboratory analysis of hand and foot coordination provides accurate and complete information on the potentially relevant factors affecting foot control (Baldissera *et al.*, 1982, 1991, 1994, 2000; Carson *et al.*, 1995; Salesse *et al.*, 2005; Meesen *et al.*, 2006), the cost and/or complexity of instrumental monitoring do not allow large-scale evaluations. Capranica *et al.* (2004) validated a field inter-limb coordination test in children, adults, and older individuals using the time of correct execution as the main dependent variable. The authors reported better values in adults, intermediate values in ten-year old children, and worst scores in individuals older than 60 years of age. This test also discriminated performances of sedentary and trained older individuals (Capranica *et al.*, 2005), with the inverted-U trend attenuated following chronic practice of rhythmic gymnastics that significantly reduced decrements with advancing age. Although the relationship between inter-limb coordination and strength has been hypothesized (Baldissera *et al.*, 1994), there is a lack of information on this issue.

Strength is often recognized as a relevant factor in the performance of physical skills and becomes even more important as individuals age (Spirduso *et al.*, 2005). Usually, static or dynamic strength assessments are carried out to evaluate the capacity of the active muscles. Large-scale longitudinal strength evaluations make use of isometric grip strength as an important measure of general health in young (Hager-Ross and Rosblad, 2002; Visnapuu and Jurimae, 2007) and older individuals (Sayer *et al.*, 2006), and the assessment of the interaction between force of contraction and speed of movement (i.e., power performances) is highly related to physical performance (Bean *et al.*, 2003) and functional decline (Kuo *et al.*, 2006). From the early study of Bosco and Komi (1980) counter-movement jump performances has become popular in evaluating functional power in children, adults and older individuals, being a complex movement that greatly depends on proper timing (Bobbert and Van Ingen Schenau, 1988; Bobbert and Van Soest, 2001). Therefore, it is conceivable that inter-limb coordination could share common demands of adequacy of temporal parameters for motor coordination with the counter-movement jump rather than isometric strength.

To investigate the relationship of inter-limb coordination, strength and power performances across the life-span and to provide functionally-relevant assessments of neuromuscular performance along with a profile of factors likely to be associated with them, in this study field tests were favoured. Understanding these factors may provide clues to the health of future older populations. Thus, the present study aimed at verifying whether (1) inter-limb coordination, strength, and power performances differ between active and sedentary individuals as a function of age and expertise, and (2) inter-limb coordination is predicted by strength and power performances and this relation is moderated by age and expertise.

2. METHODS

2.1 Participants

The local Institutional Research Committee approved this study designed to investigate the relationship between age, physical activity, inter-limb coordination, strength and power in children, adults and older individuals. Prior to the evaluation, 187 male participants (65 adolescents: 13.2 ± 1.5 years, range 13-17 years; 99 adults: 34.4 ± 14.4 years, range 18-63 years; 23 old individuals: 75.3 ± 7.0 years, range 65-89 years) provided their informed consent and answered the AAHPERD exercise/medical history questionnaire (Osness *et al.*, 1996) to ascertain their activity level, educational background, dietary habits, tobacco smoking and alcohol consumption, medication use and history of physical activity. For participants under 18 years of age, consent was obtained from their parents.

Since many of the ascertained factors seem to influence the relationship of interest between physical activity and motor coordination and cognitive functioning (Schuit *et al.*, 2001), individuals with any of the following conditions were excluded from the study: evidence, or a known history, of neuromuscular disorders, stroke, cervical myelopathy, Parkinson's disease, cognitive impairment, wrist and/or ankle arthritis, or use of medications that would affect the test performances. Older participants lived a fully independent and non-institutionalized lifestyle. None of the participants had prior experience with the test setting. Prior to the experiment, the participants were informed about the goal of the evaluation.

To be included in the active or sedentary groups the following criteria were established: sedentary young participants were engaged only in 2 h.week^{-1} school physical education, while active subjects had to ensure regular participation in three 1.5 h.week^{-1} training sessions besides school physical education (Strong *et al.*, 2005); sedentary adult and older volunteers were engaged in < 2 h.week^{-1} physical activity, while active subjects had to ensure regular participation in ≥ 2 h.week^{-1} physical activity programmes.

2.2 Procedures

To prevent testing order effects, in the present study inter-limb coordination of the individual's preferred body side, handgrip, and counter-movement jump tests were administered in a randomized order. Since strength of the preferred hand tends to decrease less with advancing age (Spirduso *et al.*, 2005) and no difference has been reported in inter-limb coordination of the preferred and non-preferred limb (Capranica *et al.*, 2004), these tests were administered only on the preferred body side. In determining the preferred body side in older individuals, two caveats were considered: 1) hand preference might have shifted from left-hand to right hand due to constrictions imposed during their youth (Spirduso *et al.*, 2005); 2) a greater discrepancy between the use of the preferred and non-preferred hand than that of the preferred and non-preferred leg as a consequence of frequent use. According to the literature (Capranica *et al.*, 2004) the preferred leg of the participants was considered more appropriate to mirror the individual's body side preference. Therefore, leg preference was determined by observing which foot the individual placed first while attempting to climb a flight of stairs, beginning from a standing position with feet together.

2.3 Inter-limb coordination

To allow independent motion of the hands and lower limbs in the sagittal plane, participants were seated on a table with elbows and knees flexed at 90°. They were instructed to make the cyclic homolateral hand and foot movements in a continuous fashion for the total duration of a trial (60 s) preserving spatial and temporal requirements of the movement patterns (i.e., in-phase mode: associations of hand extension with foot dorsal flexion and hand flexion with foot plantar-flexion; and anti-phase mode: association of hand flexion with foot dorsal flexion and hand extension with foot plantar-flexion) with a 1:1 ratio (Figure 1a,b).

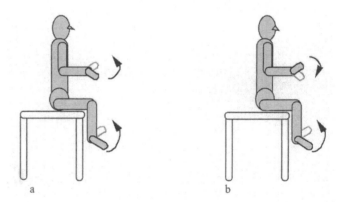

a   b

Figure 1. Schema of the in-phase (a) coordination mode (associations of wrist extension with the ankle dorsal flexion and wrist flexion with the ankle plantar flexion) and anti-phase (b) coordination mode (association of wrist flexionwith the ankle dorsal flexion and wrist extension with the ankle plantar flexion).

Each test condition was performed at three frequencies (80, 120, and 180 times per minute), dictated by a metronome. During the 2-min rest given between test conditions, participants were allowed to stand. Following 15 s at the required metronome pace, a "ready-go" command led to the start of a trial with a single observer measuring the time (s) of correct execution (i.e., the time from the beginning of the movement up to when the individual failed to meet either the spatial and/or the temporal task requirements). Although this parameter could be criticised because it relies strongly on subjective evaluations, Capranica *et al.* (2004) showed significantly high test-retest stability coefficients for both in-phase (range: 0.95-0.96) and anti-phase (range: 0.72-0.98) data and this measurement of gross time and phase transitions proved to be effective in discriminating for age (Capranica *et al.*, 2004) and expertise (Capranica *et al.*, 2005). Finally before data collection, the observer showed high intra-individual reliability coefficients (range: 0.89-0.95) for test-retest evaluations from video-recordings.

2.4 Handgrip

Grip strength was measured using a mechanical handgrip dynamometer (Lode, Groningen, The Netherlands). The Lode handgrip dynamometer consists of an ergonomically formed measuring beam (range 0-1000 N) connected to an amplifier. The force exerted on the strain gauges is displayed on-line. Before the test, the grip width was individually adjusted to achieve a 90-degree angle with the proximal-middle phalanges. Each participant stood upright with his arm vertical and the measuring beam close to the body. The maximal peak pressure (expressed in N) was recorded for a set of two contractions. The participants were asked to make the strongest grip possible and verbal encouragement was given each time to obtain their maximal score. Two trials, with a 1-min pause in between were allowed for each handgrip assessment, and the highest value was used for statistical analysis.

2.5 Counter-movement jump

Counter-movement jump (CMJ) was evaluated by means of an optical acquisition system (Optojump, Microgate, Udine, Italy), developed to measure with 10^{-3} s precision all flying and ground contact times. The Optojump photocells are placed at 6 mm from the ground and are triggered by the feet of the participant at the instant of take-off and are stopped at the instant of contact upon landing. Then, calculations of the height of the jump are made. For the counter-movement jump verbal encouragement was provided. From the standing position the participants were required to bend their knees to a freely chosen angle, which was followed by a maximal vertical thrust. The effect of the arm swings was minimized by requesting the athletes to keep their hands on their hips. Since it was assumed that participants maintained the same position at take-off and landing, they were instructed to keep their body vertical throughout the jump, and to land with knees fully extended. Any jump that was perceived to deviate from the required instructions was repeated. For each test participants were allowed two trials with a 5-min recovery period between trials. Their best performance was used for statistical analysis.

2.6 Statistical analysis

Means and standard deviations were calculated for each studied parameter and an alpha level of 0.05 was selected throughout the study. A 2 (Activity Levels: active and sedentary) x 3 (Ages: adolescent, adult, and older individuals) Analysis of Variance (ANOVA) was applied to jump (cm), and handgrip (N) performances, while a 2 (Activity Level: active and sedentary) x 3 (Age: adolescent, adult, and older individuals) x 2 (Coordination Modes: in-phase, anti-phase) x 3 (Execution Frequencies: 80, 120, and 180 cycles.min^{-1}) ANOVA with repeated measures was applied to the inter-limb coordination test (s). If the overall F test was significant, post hoc Fisher protected least significant difference comparisons with Bonferroni corrections were used.

In addition, we examined whether CMJ and handgrip performances are predictors of inter-limb coordination and whether this relation is moderated by age and expertise. With this aim, we conducted two hierarchical regression analyses with in-phase and anti-phase performances as dependent variables, respectively. To obtain a single value for each coordination mode, we computed the number of movements to be performed per second (mov.s^{-1}) for each frequency of execution (1.3 mov.s^{-1} at 80 cycles.min^{-1}, 2 mov.s^{-1} at 120 cycles.min^{-1}, and 3 mov.s^{-1} at 180 cycles.min^{-1}). Then, we calculated the individual number of correct movements executed at 80, 120, and 180 mov.s^{-1} and summed them to pool data across execution frequencies separately for in-phase (IP) and anti-phase (AP) coordination.

We chose to use hierarchical regression in order to evaluate the main effects of the hypothesized predictors (i.e., CMJ and Hangrip performances) before estimating any effect a predictor could exert jointly with the potential moderating factors (i.e., age and activity). Thus, the estimation was performed via a stepwise-forward method involving three hierarchical regression blocks. First, coordination (i.e., IP or AP) was independently predicted by the main effects accrued by Handgrip and CMJ (Block 1). Then, the interactive effects created by multiplying the factors Handgrip and CMJ with Age (Block 2) and Activity (Block 3) were estimated. The order of entry of the interaction factors in Blocks 2 and 3 was selected to evaluate first the potential role of Age as moderator of the relation between CMJ, Handgrip and inter-limb coordination performances and, thereafter, the moderating role played by chronic Activity. Since the activity level in the senior population typically co-varies with age according to an inverse relationship (Spirduso *et al.*, 2005), the selected hierarchical sequence of blocks allowed isolating any effect of Activity from the effect of Age on the relation between CMJ, Handgrip and inter-limb coordination.

Overall, the presence of statistically significant regression effects for any of the four interaction terms included in Blocks 2-3 would prove that the relations between IP-AP and CMJ or Handgrip were differently moderated by Age and Activity. A Pearson test was applied to evaluate correlations between variables in the case of significant effects of the predictors in the regression model.

3. RESULTS

The AAHPRED Activity questionnaire distinguished 95 active (adolescent: n=35; adult: n=54, old: n=6) and 92 sedentary (adolescent: n=30; adult: n=45, old: n=17) individuals.

3.1 Effects of age and activity on grip strength, counter-movement jump, and inter-limb coordination performances

Figure 2 shows the handgrip performances of the active and sedentary group for the three age classes. A main effect was found for Age ($F_{2, 177}$=90.79, P<0.0001), Activity Level ($F_{1,177}$=7.72, P=0.006) and their interaction ($F_{2, 177}$=4.53, P=0.01). Performances were better in active (387±152 N), and worst in sedentary (339±128 N) individuals. Post-hoc analysis showed a difference between all age groups, with better performances in adults (456±115 N), intermediate in old (316±92 N) and worst in adolescent (243±82 N) individuals. Post-hoc analysis for the Age x Activity interaction showed differences only in adults and older individuals.

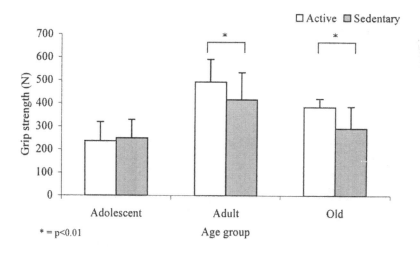

Figure 2. Means and standard deviations of the handgrip performances of active and sedentary adolescent, adult, and old individuals.

Counter-movement performances showed main effects for Age ($F_{2, 177}$=47.91, P<0.0001) and Activity level ($F_{1, 177}$=13.09, P=0.0004). Performances were best in active (27.4±8.6 cm) and worst in sedentary (21.5±10.2 cm) individuals. Post-hoc analysis of the Age effect showed significantly worst performances of older individuals (8.9±5.5 cm), intermediate in adolescent (23.4±5.4 cm), and best in adults (28.8±9.1 cm) individuals.

Inter-limb coordination showed main effects for Age ($F_{2, 180}$=12.45, P<0.0001), Activity Level ($F_{1, 180}$ = 12.58, P=0.0005), Coordination Mode ($F_{1, 180}$=317.05, P<0.0001), Execution Frequency ($F_{2, 360}$ = 280.05, P<0.0001), and the interactions Activity Level x Age ($F_{2, 180}$ = 3.50, P=0.03), Execution Frequency x Age ($F_{4, 360}$ = 5.29, P=0.0004), Coordination Mode x Execution Frequency ($F_{2, 360}$ = 8.39, P=0.0003), Coordination Mode x Execution Frequency x Activity Level ($F_{2, 360}$ = 5.39, P=0.005), and Coordination Mode x Execution Frequency x Age ($F_{4,360}$ = 5.41, P=0.0003). As expected, correct executions were better during the in-phase condition (49.9±18.1 s) than in anti-phase (25.9±25.0 s), with a significant decrement between frequencies (80 cycles.min^{-1}: 49.6±20.0 s; 120 cycles.min^{-1}: 42.2±23.4 s; 180 cycles.min^{-1}: 22.0±22.5 s). Active individuals showed better performances (41.3±24.0 s) than their sedentary counterpart (34.4±25.4 s). Performances were worst in older individuals (22.8±25.3 s), intermediate in adolescents (38.2±23.9 s), and best in adults (41.2±24.2 s), all the age groups being significantly different. Only the three-way interaction was analyzed further, because it includes the former two-way interactions. Regarding activity level (Figure 3), post-hoc analysis of the in-phase performances showed differences only for 120 cycles.min^{-1} ($F_{1, 183}$=9.97, P=0.002) and 180 cycles.min^{-1} ($F_{1, 183}$=12.37, P=0.0006) frequencies of execution, while anti-phase performances differed only at 80 cycles.min^{-1} ($F_{1, 183}$=9.69, P=0.002). Regarding age (Figure 4), post-hoc analysis of in-phase performances always showed differences ($F_{2, 184}$=20.15, P<0.0001) between older individuals and the younger age groups, while adolescents differed from adults only at 180 cycles.min^{-1} ($F_{2, 183}$=9.75, P<0.0001). Older individuals always differed ($F_{2, 183}$=16.46, P<0.0001) from adults for the anti-phase performances. At 80 beats.min^{-1} a difference ($F_{2, 183}$=21.87, P<0.0001) emerged also between adolescents and the older individuals, while at 120 cycles.min^{-1} differences ($F_{2, 183}$=10.53, P<0.0001) were found between all age groups.

3.2 Correlation between inter-limb coordination, grip strength and jump performances

To provide meaningful information on the role of activity level in determining handgrip, jump and inter-limb coordination performances, Table 1 shows the percentages of decrements calculated for each age group. Differences between age and activity groups were low in adolescents (range 0-9%), high in older individuals (17-85%) and intermediate in adults (range 2-37%). Generally, differences between co-aged groups were more evident in CMJ (range 9-55%) and in anti-phase inter-limb coordination performances with increasing frequency of execution.

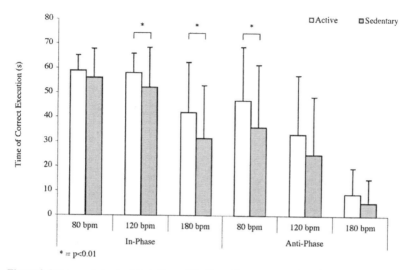

Figure 3. Means and standard deviations of the time of correct executions for in-phase and anti-phase inter-limb coordination performances in active and sedentary individuals.

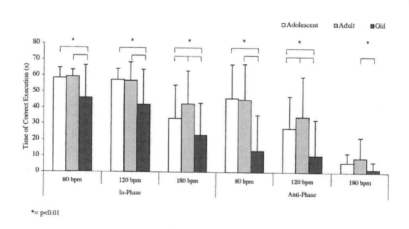

Figure 4. Means and standard deviations of the time of correct executions for in-phase and anti-phase inter-limb coordination performances in adolescents, adults and older individuals.

Table 1. Percentages of the differences between active and sedentary group means.

	Differences (%)		
	Adolescent	Adult	Old
Handgrip	0	15	24
CMJ	9	15	55
In-phase coordination (80 cycles.min^{-1})	0	2	17
In-phase coordination (120 cycles.min^{-1})	3	5	32
In-phase coordination (180 cycles.min^{-1})	8	25	51
Anti-phase coordination (80 cycles.min^{-1})	1	18	76
Anti-phase coordination (120 cycles.min^{-1})	0	16	85
Anti-phase coordination (180 cycles.min^{-1})	5	37	73

To estimate the main and joint effects of CMJ and handgrip on coordination performances, two 3-block hierarchical regressions were initially performed for in-phase coordination and anti-phase coordination, respectively. Significant effects at each step were found for both in-phase (Block 1, $F_{1, 181}$=88.63, P<0.0001, Block 2, $F_{3, 178}$=22.89, P<0.0001, Block 3, $F_{3, 175}$=13.61, P<0.0001) and anti-phase (Block 1, $F_{1, 181}$=71.66, P<0.0001, Block 2, $F_{3, 178}$=19.31, P<0.0001, Block 3, $F_{3, 175}$=11.27, P<0.0001) coordination.

Tables 2 and 3 show the predictors' significant regression coefficients and the amount of variance explained (R^2 and adjusted R^2) associated with each regression block. The hierarchical regression yielded a statistically significant model of CMJ effect at Block 1 for in-phase (r=0.6, P<0.0001) and anti-phase (P<0.0001) coordination. The interaction between CMJ and age was significant (P=0.04) at Block 2 for adults (r=0.5, P<0.0001) and old individuals (r=0.6, P=0.003) only. Handgrip did not provide significant increments in the explained variance. Overall, the final model significantly explained 33% and 28% of the variance for IP and AP, respectively.

4. DISCUSSION

The aim of the present study was to investigate the relationship between inter-limb coordination, strength and power performances as a function of age and expertise. However, several caveats have to be considered when comparing performance differences of young, adults, and older individuals. First, sedentary adolescents undergo 2-hour compulsory physical education classes weekly. Second, although activity level questionnaires are generally used to describe populations at large, they might be insufficient to describe the typology, frequency and intensity of exercise regimens that might determine differences in motor capabilities of the individual.

Table 2. Hierarchical Multiple Regression Analysis Model of In-phase (IP) inter-limb coordination as a function of Counter-movement jump (CMJ), Handgrip (HG) performances, Age and Activity factors.

		Block 1		*Block 2*			*Block 3*		
Factors	Beta	t	p	Beta	t	p	Beta	t	p
CMJ	0.57	9.42	<0.0001						
HG	0.05	0.78	ns						
CMJ x Age				0.38	1.27	ns			
HG x Age				0.08	0.66	ns			
CMJ x Activity							-0.06	-0.78	ns
HG x Activity							-0.01	-0.05	ns
R^2 change		0.33		0.01			0.01		
F ratio for R^2 change		88.63		0.98			1.16		
P		<0.0001		0.40			0.33		
Total R^2	0.35								
Adjusted R^2	0.33								

Table 3. Hierarchical Multiple Regression Analysis Model of Anti-phase (AP) inter-limb coordination as a function of Counter-movement jump (CMJ), Handgrip (HG) performances, Age and Activity factors.

		Block 1		*Block 2*			*Block 3*		
Factors	Beta	t	p	Beta	t	p	Beta	t	p
CMJ	0.53	8.47	<0.0001						
HG	-0.06	-0.79	ns						
CMJ x Age				0.63	2.06	0.04			
HG x Age				-0.09	-0.72	ns			
CMJ x Activity							-0.09	-0.26	ns
HG x Activity							0.14	0.58	ns
R^2 change		0.28		0.02			0.02		
F ratio for R^2 change		71.66		1.62			0.69		
P		<0.0001		0.19			0.56		
Total R^2		0.31							
Adjusted R^2		0.28							

As expected, a strong association between chronic practice of physical activity and actual performances was found, confirming the importance of active lifestyles for enhancing strength, power and coordination performances and protecting from age-related decrements in neuromuscular function in older years. However within the general pattern, marked differences in the relations of the three performance measures with age were found. Although adults always showed best values, the youngest group showed lower grip strength than older individuals while the reverse picture emerged for CMJ and coordination. This is not surprising. Handgrip is a simple isometric test of upper-body muscle strength, whereas the dynamic measures involving lower limbs require higher levels of neuromuscular speed and control. Our findings support and extend a previous study by Kuo *et al.* (2006) showing that decrements in complex dynamic tasks may be more effective in representing neuromuscular changes than just strength or basic skills. Therefore, the results indicate that these measures should not be used as surrogates for each other in assessments of physical performance. In fact, Syddall and colleagues (2003) claimed that in older individuals fewer factors impinge directly on muscle strength, while many have subtle deleterious effects on central coordination and control of dynamic movements.

The only measure that did not show better values for the active individuals was grip strength in the youngest group who showed performances comparable to those reported for co-aged boys weekly involved in a 2-hour physical education programme (Koutedakis and Bouziotas, 2003). Since training proved to elicit increments in handgrip in co-aged athletes (Hansen *et al.*, 1999; Oxyzoglou *et al.*, 2007; Visnapuu and Jurimae, 2007), it is possible to argue that at this age strength development might be highly influenced by the specificity of training. Unfortunately, the questionnaire used to assess the activity level of participants does not provide sufficient information regarding the type and intensity of the additional training performed by the active group, thus leaving this issue unresolved.

In agreement with the literature on old individuals (Spirduso *et al.*, 2005), the present results showed age-related decrements less pronounced for the upper limb. Although grip strength is regarded as an important measure of general health and a powerful predictor of disability, morbidity and mortality with advancing age (Sayer *et al.*, 2006), the maintenance of muscle power has been related to life independency. In fact, power training has been found to be more effective than strength training for improving physical function and intervention programs that include high-velocity exercises have been considered essential in older ages (Spirduso *et al.*, 2005). In the present study, the CMJ performances of the older group were lower that those reported for senior team sport athletes (Tessitore *et al.*, 2005, 2006). Since the AAHPERD activity questionnaire provides no information regarding the qualitative and quantitative aspects of training, it is possible to argue that the quality of training is the most relevant factor for the maintenance of this performance from the consideration that the best CMJ values were reported for basketball players (Tessitore *et al.,* 2006) who trained with a frequency (i.e., 1.5 hour·week^{-1}) lower than that reported (i.e., 3.0 hour.week^{-1}) for senior soccer players (Tessitore *et al.*, 2005). Basketball players need good jumping capacity, especially to get rebounds and to block shots. During a basketball match, the repeated all-out jumping performances have beneficial training effects on muscle mass, neural activation, time available to develop force, storage and re-utilization of elastic energy, and contribution of reflexes.

Generally, the observed reductions in homolateral hand and foot coordination with advancing age (Serrien *et al.*, 2000; Capranica *et al.*, 2004) are attributed to the combined effects of ageing and a sedentary lifestyle (Capranica *et al.*, 2005). The present results confirm that inter-limb coordination depends on exercise mode and frequency of execution, with better performances shown during isodirectional tasks and at lower execution frequencies (Baldissera *et al.*, 1991; Kelso, 1994; Serrien *et al.*, 2000; Capranica *et al.*, 2004, 2005). According to the literature concerning the age-related 'parabola' of coordinative development and degradation (Meinel and Schnabel, 1998; Capranica *et al.*, 2004; Spirduso *et al.*, 2005), inter-limb coordination performance was best in adults, intermediate in children and worst in older individuals, with age-related performance decrements more pronounced during anti-phase and faster movements. Furthermore, the present results confirm that this test succeeds in discriminating coordination performances between active and sedentary individuals. According to the literature (Capranica *et al.*, 2004), in the anti-phase condition a severe decrement in coordinative performance emerged with increasing frequency of execution, independent of age and activity level. Physical activity exerts a preservative effect particularly during the fastest in-phase performances that are strongly affected by age-related muscle weakness, prolonged reaction times, and changes in stretch reflexes. These differences are also evident in the anti-phase condition at the highest execution frequency that needs increased monitoring and attentional allocation in order to inhibit the natural in-phase mode (Baldissera *et al.*, 1982; Wuyts *et al.*, 1998). However, in disagreement with reports on old rhythmic gymnasts (Capranica *et al.*, 2005), no difference emerged between active and sedentary individuals in the fastest anti-phase conditions. These results might reflect that rhythmic gymnastics requires a massive variety of coordination patterns with high spatial and temporal constraints. This high coordinative component of training helps individuals to overrule the easier movement patterns, which spontaneously emerge under stressful conditions or during the acquisition of new skills. Although physical activity plays an important role in developing and preserving coordination quality across a lifespan, the most challenging coordinative patterns are arduously carried out and rely strongly on fine neuromuscular control that need specific training more than general physical activity.

Although the relationship between strength and inter-limb coordination has been hypothesized (Baldissera *et al.*, 1994), in the present study we did not find any predictor effect of grip strength in inter-limb coordination, indicating that strength per se can not be a relevant factor in determining better coordination performances. On the other hand, the counter-movement jump proved to be a good predictor of coordination in both in-phase and anti-phase mode, explaining around 30% of the variance. Considering that phase transitions from anti-phase to in-phase coordination occur when the foot fails to mirror the movement of the hand, the rhythmical synchrony of hand and foot flexion-extensions might be achieved by a higher control of the lower limb (Baldissera *et al.*, 2000). In the literature (Bobbert and Van Soest, 2001), the fine neuromuscular activation timing has been proved to be central for counter-movement jump performances. Thus, it is possible that timing is the coordinative factor shared by the two capabilities.

The predicted value of CMJ is moderated by age only in the more complicated task. In fact, age acts as a moderator of the relationship between counter-movement jump and anti-phase coordination only, indicating that CMJ does not have the same predictor activity of the anti-phase coordination at all ages. In fact, CMJ is a good predictor of the anti-phase coordination only for adult and old individuals, indicating that the more difficult coordination in adolescents is governed by qualitative factors partially different from in adults. This difference can be explained by a different control of CMJ between adult and adolescent, which may use more attention than proprioception. On the other hand, the modest contribution of activity level might be due to the limited discriminatory power of the questionnaire. Further studies are needed to establish the role played by more specific exercise training on the relationship between coordination, strength and power performances.

References

Baldissera, F., Borroni, P. and Cavallari, P., 2000, Neural compensation for mechanical differences between hand and foot during coupled oscillations of the two segments. *Experimental Brain Research*, **133**, pp. 165-177.

Baldissera, F., Cavallari, P. and Chivaschi, P., 1982, Preferential coupling between voluntary movements of ipsilateral limbs. *Neuroscience Letters*, **34**, pp. 95-100.

Baldissera, F., Cavallari, P. and Tesio, L., 1994, Coordination of cyclic coupled movements of hand and foot in normal subjects and on the healthy side of hemiplegic patients. In: *Interlimb coordination: Neural, Dynamical and Cognitive Constraints*, edited by Swinnen, S., Heuer, H., Massion, J. and Casaer, P., (San Diego: Academic Press), pp. 229-242.

Baldissera, F., Cavallari, P., Marini, G. and Tassone, G., 1991, Differential control of in-phase and anti-phase coupling of rhythmic movements of ipsilateral hand and foot. *Experimental Brain Research*, **83**, pp. 375-380.

Bean, J.F., Leveille, S.G., Kiely, D.K., Bandinelli, S., Guralnik, J.M. and Ferrucci, L., 2003, A comparison of leg power and leg strength within the InCHIANTI study: Which influences mobility more? *Journal of Gerontology: Medical Sciences*, **58A**, pp. 728-733.

Bobbert, M.F. and Van Ingen Schenau, G.J., 1988, Coordination in vertical jumping. *Journal of Biomechanics*, **21**, pp. 249-262.

Bobbert, M.F. and Van Soest, A.J., 2001, Why do people jump the way they do? *Exercise Sport Science Review*, **29**, pp. 95-102.

Bosco, C. and Komi, P., 1980, Influence of aging on the mechanical behavior of leg extensor muscles. *European Journal of Applied Physiology*, **45**, pp. 209-219.

Capranica, L., Tessitore, A., Olivieri, B. and Pesce, C., 2005, Homolateral hand and foot coordination in trained older women. *Gerontology*, **106**, pp. 309-315.

Capranica, L., Tessitore, A., Olivieri, B., Minganti, C. and Pesce, C., 2004, Field evaluation of cycled coupled movements of hand and foot in older individuals. *Gerontology*, **50**, pp. 399-406.

Carson, R.G. and Kelso, J.A.S., 2004, Governing coordination: behavioural principles and neural correlates. *Experimental Brain Research*, **154**, pp. 267-274.

Carson, R.G., Goodman, D., Kelso, J.A.S. and Elliott, D., 1995, Phase transitions and critical fluctuations in rhythmic coordination of ipsilateral hand and foot. *Journal of Motor Behaviour*, **27**, pp. 211-224.

Hager-Ross, C. and Rosblad B., 2002, Norms for grip strength in children aged 4-16 years. *Acta Pediatrica*, **91**, pp. 617-625.

Hansen, L., Bangsbo, J., Twisk, J. and Klausen, K., 1999, Development of muscle strength in relation to training level and testosterone in young male soccer players. *Journal of Applied Physiology*, **87**, pp. 1141-1147.

Jeka, J.J. and Kelso, J.A., 1995, Manipulating symmetry in the coordination dynamics of human movement. *Journal of Experimental Psychology: Human Perception and Performance*, **21**, pp. 360-374.

Kay, B.A., Saltzman, E.L. and Kelso, J.A.S., 1991, Steady-state and perturbed rhythmicall movements: a dynamical analysis. *Journal of Experimental Psychology: Human Perception and Performance*, **17**, pp. 183-197.

Kelso, J.A., Fink, P.W., DeLaplain, C.R. and Carson, R.G., 2001, Haptic information stabilizes and destabilizes coordination dynamics. *Proceedings of the Royal Society B: Biological Sciences*, **268**, pp. 1207-1213.

Kelso, J.A.S., 1984, Phase transitions and critical behaviour in human bimanual coordination. *American Journal of Physiology - Regulatory, Integrative and Comparative Physiology*, **246**, R1000-1004.

Kelso, J.A.S., Scholz, J.P. and Schoner, G., 1986, Non-equilibrium phase transitions in coordinated biological motion: Critical fluctuations. *Physics Letters A*, **134**, pp. 8-12.

Koutedakis, Y., and Bouziotas, C., 2003, National physical education curriculum: motor and cardiovascular health related fitness in Greek adolescents. *British Journal of Sports Medicine*, **37**, pp. 311-314.

Kuo, H-K., Leveille, S.G., Yen, C-J., Chai, H-M, Chang, C.-H., Yeh, Y.-C., Yu, Y.-H. and Bean, J.F., 2006, Exploring how peak leg power and usual gait speed are linked to late-life disability: Data from the National Health and Nutrition Examination Survey (NHANES), 1999-2002. *American Journal of Physical Medicine and Rehabilitation*, **85**, pp. 650–658.

Lafargue, G., Paillard, J., Lamarre, Y. and Sirigu, A., 2003, Production and perception of grip force without proprioception: is there a sense of effort in deafferented subjects? *European Journal of Neuroscience*, **17**, pp. 2741-2749.

Laforest, S., St-Pierre, D.M.M., Cyr, J. and Gayton, D., 1990, Effects of age and regular exercise on muscle strength and endurance. *European Journal of Applied Physiology*, **60**, pp. 104-111.

Larsson, L.G., Grimby, G. and Karlsson, J., 1979, Muscle strength and speed of movement in relation to age and muscle morphology. *Journal of Applied Physiology*, **46**, pp. 451-456.

Meesen, R.L., Wenderoth, N., Temprado, J.J., Summers, J.J. and Swinnen, S.P., 2006, The coalition of constraints during coordination of the ipsilateral and heterolateral limbs. *Experimental Brain Research*, **174**, pp. 367-375.

Meinel, K. and Schnabel, G., 1998, *Bewegungslehre - Sportmotorik.* (Berlin: Sportverlag).

Osness, W.H., Adrian, M., Clark, B., Hoeger, W., Raab, D. and Wiswell, R., 1996, *Functional Fitness Assessment for Adults Over 60 Years.* (Dubuque: Kendall/ Hunt Publishing).

Oxyzogluo, N., Kanioglu, A., Rizos, S., Mavridis, G. and Kabitsis, C., 2007, Muscular strength and jumping performance after handball training versus physical education program for pre-adolescent children. *Perceptual and Motor Skills*, **104**, pp. 1282-1288.

Paper, C.E. and Beek, P.J., 1998, Are frequency-induced transitions in rhythmic coordination mediated by a drop in amplitude? *Biological Cybernetics*, **79**, pp. 291-300.

Rowland, T.W., 2005, *Children's Exercise Physiology.* (Champaign, IL: Human Kinetics).

Salesse, R. and Temprado, J.J., 2005, The effect of visuo-motor transformations on hand–foot coordination: evidence in favor of the incongruency hypothesis. *Acta Psychologica*, **119**, pp. 143-157.

Salesse, R., Temprado, J.J. and Swinnen, S.P., 2005, Interaction of neuromuscular, spatial and visual constraints on hand-foot coordination dynamics. *Human Movement Sciences*, **24**, pp. 66-80.

Sayer, A.A., Syddall, H.E., Martin, H.J., Dennison, E.M., Roberts, H.C. and Cooper, C., 2006, Is grip strength associated with health-related quality of life? Findings from the Hertfordshire Cohort Study. *Age and Ageing,* **35**, pp. 409-415.

Schuit, A.J., Feskens, E.J., Launer, L.J. and Kromhout, D., 2001, Physical activity and cognitive decline, the role of the apolipoprotein e4 allele. *Medicine and Science in Sports and Exercise*, **33**, pp. 772-777.

Serrien, D.J. and Swinnen, S.P., 1998, Load compensation during homologous and non-homologous coordination. *Experimental Brain Research*, **121**, pp. 223-229.

Serrien, D.J., Li, Y., Steyvers, M., Debaere, F. and Swinnen, S.P., 2001, Proprioceptive regulation of interlimb behaviour: interference between passive movement and active coordination dynamics. *Experimental Brain Research*, **140**, pp. 411-419.

Serrien, D.J., Swinnen, S.P. and Stelmach, G.E., 2000, Age-related deterioration of coordinated inter-limb behavior. *Journal of Gerontology: Psychological Science,* **55B**, P295-P303.

Spirduso, W.W., Francis K.L. and MacRae, P.G., 2005, *Physical Dimensions of Aging.* (Champaign, IL: Human Kinetics).

Strong, W.B., Malina, R.M., Blimkie, C.J.R., Daniels, S.R., Dishman, R.K., Gutin, B., Hergenroeder, A.C., Must, A., Nixon, P.A., Pivarnik, J.M., Rowland, T.,

Trost, S. and Trudeau, F., 2005, Evidence based physical activity for school-age youth. *Journal of Paediatrics*, **146**, pp. 732-737.

Swinnen, S.P., Jardin, K., Meulenbroek, R., Dounskaia, N. and Hofkens-Van Den Brandt, M., 1997, Egocentric and allocentric constraints in the expression of patterns of interlimb coordination. *Journal of Cognitive Neuroscience,* **9**, pp. 348–377.

Syddall, H., Cooper, C., Martin, F., Briggs, R. and Aihie Sayer, A., 2003, Is grip strength a useful single marker of frailty? *Age and Ageing*, **32**, pp. 650–656.

Tessitore, A., Meeusen, R., Tiberi, M., Cortis, C., Pagano, R. and Capranica, L., 2005, Aerobic and anaerobic profile, heart rate and match analysis in older soccer players. *Ergonomics*, **48**, pp. 1365-1377.

Tessitore, A., Tiberi, M., Cortis, C., Rapisarda, E., Meeusen, R. and Capranica L., 2006, Aerobic-anaerobic profiles, heart rate and match analysis in old basketball players. *Gerontology*, **52**, pp. 214-222.

Visnapuu, M. and Jurimae, T., 2007, Handgrip strength and hand dimensions in young handball and basketball players. *Journal of Strength and Conditioning Research*, **21**, pp. 923-929.

Wuyts, I.J., Byblow, W.D., Summers, J.J., Carson, R.G. and Semjen, A., 1998, Intentional switching between patterns of coordination in bimanual circle drawing. In: *Motor Behaviour and Human Skill*, edited by Piek, J.P., (Champaign, IL: Human Kinetics), pp. 191-208.

CHAPTER TWENTY-ONE

Body composition of international- and club-level professional soccer players measured by dual-energy x-ray absorptiometry (DXA)

L. Sutton, J. Wallace, M. Scott and T. Reilly

Research Institute for Sport and Exercise Sciences, Liverpool John Moores University, Liverpool, UK

1. INTRODUCTION

The aim of body composition analysis is to differentiate and quantify different body compartments. Given the health implications of overweight and obesity, and the reduction in lean mass and bone density associated with various disease states, assessments of body composition are frequently employed in clinical, research, health and general settings. In applied sport settings, a single assessment may be used alongside other physiological tests as a measure of preparedness for competition. Furthermore, repeat assessments are useful in evaluating the effects of changes to exercise and/or dietary regimens on body composition status.

Whilst body composition measures are determined more by genetic than by environmental factors, body composition is a key consideration in the physical preparation of players for competitive soccer. Excess fat mass acts as a dead weight in activities in which the body is lifted against gravity (Reilly, 1996). This added weight has an adverse effect upon general locomotion and soccer-specific skills such as jumping to contest possession in the air. Muscular development is important for strength and speed production, power output and for specific activities such as contesting and maintaining possession and kicking, amongst others. Muscle mass provides the bulk of the body's lean component, which, together with bone strength, is thought to play an important role in injury-prevention (Karlsson, 2007).

Dual-energy x-ray absorptiometry (DXA) scanning is considered to be the 'gold standard' method of bone densitometry (Lewiecki, 2005) and is further able to divide the body's non-mineral content into fat and lean compartments based on the x-ray attenuation properties of the different molecular masses. The DXA-derived three-compartment (3C) model is the criterion method of body composition analysis currently employed in clinical contexts (Oates, 2007).

Previous comparative studies and reviews of body composition in soccer have demonstrated differences between players based on the national league division in which they play (Bandyopadhyay, 2007), club position within the league (Kalapotharakos *et al.*, 2006), playing position (Davis *et al.*, 1992; Matković *et al.*, 2003; Arnason *et al.*, 2004; Silvestre *et al.*, 2006a; Gil *et al.*, 2007), first team versus substitutes (Kraemer *et al.*, 2004; Silvestre *et al.*, 2006a), and professional versus amateur (Kutlu *et al.*, 2007). Anthropometric comparisons have also been made between professional soccer players and participants in other sports (Strudwick *et al.*, 2002; McIntyre, 2005; Wallace *et al.*, 2006; Bandyopadhyay, 2007). Whilst these studies have highlighted the relevance of body composition for performance among professional soccer players, it is not clear whether small differences in body composition variables are critical for success among relatively homogenous groups of top-level players. To date, there are no direct comparisons of body composition variables between international and non-international professional soccer players in the highest national leagues. It is evident that optimal body composition is important in elite soccer performance, yet it is unknown whether body composition is a key factor in the progression from professional club-level to international-level performance.

Since the introduction of DXA to the field of body composition analysis, investigations have been carried out in which soccer players have been compared to non-players. Such studies enable a three-compartment model to be constructed in which the effects of participation in elite soccer competitions on bone, lean and fat mass can be evaluated. Assuming an adequate diet, development of bone and muscle mass would be anticipated given the loading patterns on the soccer player during training and competition, and the accumulation of body fat should be limited (Shephard, 1999). Wittich *et al.* (1998) reported significantly greater bone area, mineral content and mineral density in Division I soccer players compared to a reference group matched for age and body mass index (BMI). Increased values were attributed to the high-impact loading of the sport. The same research group later conducted another study in which the soft-tissue mass of professional soccer players was examined and compared to a reference group. Fat mass was found to be significantly higher in the reference group, and bone and lean mass higher in the soccer group (Wittich *et al.*, 2001). Similar findings were obtained by Calbet *et al.* (2001) in a comparison of recreational soccer players and non-players. Bone area, BMC, BMD and lean mass were all significantly greater in the soccer players, and both relative and absolute amounts of body fat were lower compared to a sedentary reference group. Results were attributed to long-term participation in the game. As soccer players show superior values to non-players in each of the body compartments, it remains to be determined which variables, or combination of variables, are the more important among soccer players. This question can be addressed by comparing those players at professional level who have progressed to full international representation and those who have not made the progression.

Body composition variables are labile, being subject to dietary, training and other influences. It has previously been reported that the body composition of professional players varies considerably over the course of the season (Casajús, 2001; Kraemer *et al.*, 2004; Egan *et al.*, 2006; Silvestre *et al.*, 2006b). These findings suggest that the time of the competitive season should be taken into account when conducting comparative studies involving soccer players.

The aims of the present study were:
- To investigate differences in body composition between international and non-international soccer players competing at the same club level;
- To compare professional soccer players to a reference group to assess both whole-body and regional differences and determine which body composition variables best distinguish professional soccer players from non-players.

2. METHODS

2.1 Participants

Thirty-one senior international soccer players were recruited from four professional clubs competing in the English Premier League. Three of the clubs finished in the top four of the 2007-8 Premier League, the fourth club also qualifying for European club competitions. Between them the participants represented 12 different countries from Europe, South America, Africa and Australia. The number of international appearances ranged from five to 91 (38.0 ± 23.7 caps). The non-international senior professional players (n = 23) were recruited from the same clubs. The majority of players were of Caucasian ethnicity. Of the international players, six were of African/Caribbean descent and one Asian. Of the non-international players, four were of African/Caribbean origin. A reference group comprising 24 male participants was recruited, matched for age and height to the combined means of the two soccer groups. Members of the reference group did not participate regularly in competitive sport and demonstrated low to moderate levels of physical activity. The participants' characteristics are shown in Table 1. Goalkeepers were excluded from the study due to reported anthropometric differences from outfield players (Davis *et al.*, 1992; Rico-Sanz, 1998; Matković *et al.*, 2003; Arnason *et al.*, 2004).

2.2 Procedure

The investigation was approved by the Research Ethics Committee of Liverpool John Moores University and verbal and written informed consent was obtained. Laboratory sessions were held both prior to and during the competitive season. Each participant attended at least one session during which an adult whole-body DXA scan was carried out using a Hologic QDR Series Discovery A fan-beam scanner (Hologic Inc, Bedford, MA). Participants wore shorts and were asked to remove any metal and jewellery from their persons. Height by stadiometry and scale mass were recorded to the nearest 0.5 cm and 0.1 kg, respectively (Seca 702, Seca GmbH & Co.KG, Hamburg, Germany). The participant then assumed a stationary, supine position on the scan bed. Total scan time was approximately three minutes.

Measured variables included:

- Areal bone mineral density (BMD) – calculated by dividing bone mineral content by projected bone area;
- Z-score – the BMD expressed as a standard deviation above a population mean, taking into account age and sex (NHANES database);
- Fat mass – this variable is reported as both an absolute (kg) and a relative (percent body fat) value;
- Lean mass – this value is the bone-free, fat-free soft tissue mass, and is therefore largely comprised of muscle mass.

Scans were grouped into 'pre-season' and 'competitive season' data to reduce the impact of seasonal variation on the data. Forty-one scans on soccer players were obtained at each time point. The body as a whole was assessed using the DXA subtotal values. The subtotal value, which is the total minus the head, was selected instead of the total value as the addition of the head induces more measurement error due to the inability of DXA to detect mineral-free mass that is overlying pixels containing bone mineral. The lean and fat content of the head is estimated based on the ratio of the lean and fat mass in pixels from the rest of the body that contain no mineral content (Sutcliffe, 1996). Furthermore, some taller participants do not completely fit within the scan field, meaning the head has to be excluded from the analysis. Wallace (2007) reported a lower systematic error, coefficient of variation and technical error of measurement, and greater correlation with the previous reference method of hydrodensitometry, when the subtotal value was utilised in place of the total. In addition, the DXA regional analyses were combined to allow for a comparison of upper-body and lower-body segments between groups. The upper-body segment comprised the left arm, right arm and trunk regions, whilst the lower-body segment comprised the left and right legs.

2.3 Statistical analysis

Differences between group means were analysed using one-way between groups ANOVA with Tukey's *post hoc* test. Separate tests were conducted for 'pre-season' and 'competitive season' upper-, lower- and whole-body data. Stepwise discriminant function analysis was employed to establish which variables were most influential when distinguishing groups and assessing their relative contribution by determining whether groups could be correctly classified using those variables. To improve the results of discriminant function analyses, multicollinearity should be checked (Tabachnick and Fidell, 1996). Where one variable is calculated from another, such as is the case for BMD and Z-score, or relative and absolute body fat values, high correlations occur between variables (>0.9). To reduce the resulting multicollinearity, one variable from each anthropometric compartment was selected (BMD, lean mass and percent body fat). Inclusion of these three variables in the analyses resulted in acceptable levels of collinearity, as assessed by detection tolerance and variance inflation factor values.

Table 1. Characteristics of the participants: International soccer players (n = 31), non-international soccer players (n = 23) and a reference group (n = 24).

Group		N	Age (years)		Height (m)		Mass (kg)	
			Mean±SD	Range	Mean±SD	Range	Mean±SD	Range
International	Season	20	27.2±3.4	21.0-35.0	1.82±0.06	1.69-1.94	83.1±4.8	74.0-90.1
	Pre-season	22	28.0±4.2	17.0-36.0	1.81±0.07	1.69-1.94	81.6±6.2	71.9-94.3
Non-international	Season	21	23.6±4.6	16.0-34.0	1.80±0.08	1.66-1.93	81.2±7.8	68.7-95.0
	Pre-season	19	26.1±4.9	17.0-34.0	1.83±0.04	1.75-1.89	83.8±8.3	70.1-98.2
Reference group	-	24	26.8±5.2	16.0-36.0	1.80±0.06	1.65-1.92	81.3±12.7	62.8-106.0

The alpha level was set at $P \leq 0.05$. All analyses were carried out using SPSS for Windows version 14 (SPSS Inc., Chicago, IL).

3. RESULTS

3.1 Group comparisons

Results are presented for international soccer players (Int), non-international professional players (Club) and the reference group (Ref) as mean ± SD where relevant. The P values represent the significance of the differences between soccer and reference groups. Percent differences from reference values for each variable are shown in Tables 2 and 3, providing insights into the patterning of body composition variables in the phenotype presented by the elite soccer players.

3.1.1 Bone mineral density

The pre-season international and non-international soccer groups both demonstrated significantly greater whole-body BMD than the reference group ($F_{2,62}$ = 36.50, P < 0.001). The same was true of the competitive season data ($F_{2,62}$ = 27.76, P < 0.001). There were no significant differences between international and non-international professional soccer players at either time point (Figure 1).

Upper-body BMD was significantly greater in the soccer groups compared to the reference group at both time points (pre-season: $F_{2,62}$ = 18.26, P < 0.001; season: $F_{2,62}$ = 12.63, P < 0.001), with no significant differences between international and non-international soccer players. The same was true of the lower-body results (pre-season: $F_{2,62}$ = 39.26, P < 0.001; season: $F_{2,62}$ = 32.53, P < 0.001), as shown in Figure 1.

When converted to a Z-score, the pattern of whole-body BMD results remained unchanged, as both soccer groups demonstrated significantly greater values than the reference group, but with no differences between soccer groups (pre-season: Int 3.1 ± 1.0, Club 3.3 ± 1.0, $F_{2,62}$ = 34.10, P < 0.001; season: Int 3.1 ± 1.3, Club 3.1 ± 1.1, $F_{2,62}$ = 23.00, P < 0.001; Ref 1.2 ± 0.9). The overall (combined international and non-international, pre-season and season data) mean increase from the reference whole-body BMD was +16.5 %, and departure from reference Z-score was +172 %

3.1.2 Lean mass

During both pre-season and competitive season time points, whole-body lean mass was significantly greater in soccer groups compared to the reference group (pre-season: $F_{2,62}$ = 6.01, P < 0.004; season: $F_{2,62}$ = 7.98, p < 0.001). No differences were evident between the international and non-international soccer players (Figure 2).

Table 2. Percent difference from reference values of DXA variables measured during the pre-season period in international (n = 22) and non-international (n = 19) professional soccer players.

	International players			Non-international players		
	Upper body	Lower body	Whole body	Upper body	Lower body	Whole body
BMD (g.cm⁻²)	↑12%	↑16%	↑16%	↑16%	↑16%	↑17%
Z-score (SD)	-	-	↑171%	-	-	↑183%
Lean mass (kg)	↑4%	↑16%	↑8%	↑9%	↑16%	↑11%
Fat mass (kg)	↓43%	↓31%	↓41%	↓42%	↓31%	↓40%
Fat mass (%)	↓37%	↓35%	↓37%	↓36%	↓34%	↓38%

Table 3. Percent difference from reference values of DXA variables measured during the competitive season in international (n = 20) and non-international (n = 21) professional soccer players.

	International players			Non-international players		
	Upper body	Lower body	Whole body	Upper body	Lower body	Whole body
BMD (g.cm⁻²)	↑12%	↑20%	↑17%	↑13%	↑16%	↑16%
Z-score (SD)	-	-	↑168%	-	-	↑165%
Lean mass (kg)	↑10%	↑16%	↑12%	↑8%	↑11%	↑9%
Fat mass (kg)	↓47%	↓38%	↓46%	↓44%	↓37%	↓44%
Fat mass (%)	↓42%	↓41%	↓43%	↓38%	↓37%	↓40%

When the upper-body was isolated, the pre-season data for the international soccer players tended to be higher than reference upper-body lean mass, but lacked statistical significance (P = 0.41). Non-international players also showed greater upper-body lean mass values than the reference group during the pre-season period (P ≤ 0.02). During the competitive season, both soccer groups had higher upper-body lean mass values than the reference group ($F_{2,62}$ = 5.23, P = 0.008). When the lower-body alone was considered, both soccer groups showed greater results than the reference group (pre-season: $F_{2,62}$ = 8.42, P < 0.001; season: $F_{2,62}$ = 7.59, P < 0.001), with no significant differences between international and non-international players (Figure 2). The overall mean difference from reference whole-body lean mass was +10 %.

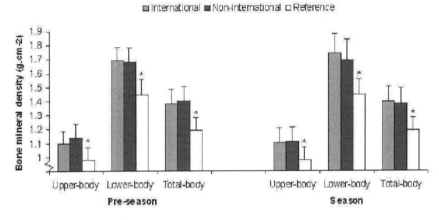

Figure 1. Segmental and whole-body bone mineral density values for international soccer players (n = 31), non-international soccer players (n = 23) and a reference group (n = 24).
* denotes significant difference between soccer and reference groups (P≤0.05)

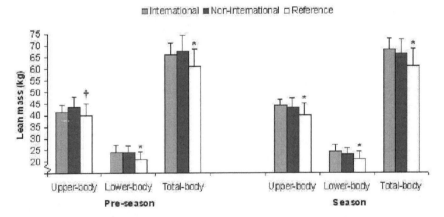

Figure 2. Segmental and whole-body lean mass values for international soccer players (n = 31), non-international soccer players (n = 23) and a reference group (n = 24).
* denotes significant difference between soccer and reference groups (P≤0.05)
† denotes significant difference between non-international soccer players and reference group (P≤0.02)

3.1.3 Fat mass

In terms of absolute whole-body fat values, there were no significant differences between international and non-international soccer players. Both groups demonstrated significantly lower results than the reference group (pre-season: Int 8.5 ± 1.9, Club 8.6 ± 2.1, $F_{2,62}$ = 16.03, P < 0.001; season: Int 7.8 ± 1.5, Club 8.6 ± 1.9, $F_{2,62}$ = 21.28, P < 0.001; Ref 14.4 ± 5.2 kg).

The same trend was true of upper-body (pre-season: Int 4.9 ± 1.0, Club 5.0 ± 1.3, $F_{2,62}$ = 16.05, P < 0.001; season: Int 4.6 ± 1.0, Club 4.8 ± 1.2, $F_{2,62}$ = 18.62, P < 0.001; Ref 8.6 ± 3.8 kg) and lower-body fat values (pre-season: Int 3.6 ± 1.1, Club 3.6 ± 1.0, $F_{2,62}$ = 10.13, P < 0.001; season: Int 3.2 ± 0.7, Club 3.3 ± 0.9, $F_{2,62}$ = 18.06, P < 0.001; Ref 5.2 ± 1.8 kg). Overall, the soccer groups demonstrated 43 % lower absolute fat mass than the whole-body fat value for the reference sample.

The relative amount i.e. percent body fat was also considered (Figure 3). Both soccer groups exhibited consistently lower results than the reference group (P < 0.001); there were no significant differences between soccer groups in any of the body segments neither during pre-season nor the competitive season. The mean whole-body percent fat values for the international soccer players were 10.9 ± 2.2 % during pre-season and 9.8 ± 1.8 % during the season. For the non-international players, these values were 10.6 ± 2.1 % for the pre-season group and 10.3 ± 1.9 % during the season. The reference group averaged 17.3 ± 3.9 % body fat. Overall, the soccer groups demonstrated mean whole-body percent fat values 40 % lower than the reference value.

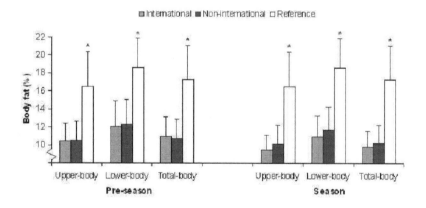

Figure 3. Segmental and whole-body percent fat values for international soccer players (n = 31), non-international soccer players (n = 23) and a reference group (n = 24).
* denotes significant difference between soccer and reference groups (P≤0.05)

3.2 Discriminant analysis

It was not possible to distinguish international from non-international professional soccer players using discriminant function analysis as group similarity meant none of the DXA variables qualified for the stepwise model. Subsequently, these groups were combined to form one 'professional soccer' group for comparison with the reference group using stepwise discriminant function analysis.

3.2.1 Pre-season

The variables best able to differentiate between the professional soccer players and the reference group were, in order of contribution to the stepwise model, percent body fat and BMD ($1 = 0.25$, $F_{2,62} = 93.00$, $P < 0.001$). Using the predictive model derived from these two variables, 96.9 % of participants were correctly assigned to their respective groups. Two of the professional soccer players were misclassified as reference individuals. The predictor variables remained the same when the upper-body data were assessed ($1 = 0.33$, $F_{2,62} = 63.24$, $P < 0.001$), yet the classification strength decreased slightly, with 93.8 % of the participants correctly assigned to their group. The model based on lower-body data also resulted in 93.8 % of cases correctly classified, yet the two discriminating variables were reversed, with BMD being the greater contributor to lowering the overall Wilks' lambda, followed by percent fat ($1 = 0.26$, $F_{2,62} = 89.78$, $P < 0.001$).

3.2.2 Competitive season

The best predictor model for distinguishing the professional soccer players from non-players during the season, taking into account the whole body, included all three variables ($1 = 0.21$, $F_{3,61} = 74.76$, $P < 0.001$). This model was able to classify both the professional soccer and reference groups correctly. Isolating the upper-body data also lead to the inclusion of all three variables ($1 = 0.28$, $F_{3,61} = 53.12$, $P < 0.001$), resulting in 98.5 % correct group classification. The lower-body model included just percent fat and BMD ($1 = 0.25$, $F_{2,62} = 91.62$, $P < 0.001$), similarly classifying 98.5 % of cases.

4. DISCUSSION

When comparing results between studies, actual DXA-reported values should not be compared directly unless the scan equipment is the same, as there are differences between DXA scanners produced by different manufacturers, and between current and older models of the same company (Kistorp and Svendsen, 1998; Tothill and Hannan, 2000). However, it is acceptable to compare and contrast trends in results produced by different scanners. Consequently, percent differences between the soccer and reference groups within the current study are used when comparing results with previous findings.

4.1 Bone mineral density

Differences in BMD between the professional soccer players and the reference sample were evident in both the upper and lower body. It has been reported that changes in bone mass due to physical activity are site-specific, meaning adaptations occur in the areas at which increased loading and mechanical strain occurs, and that impact-loading sports, such as soccer, cause greater improvements in bone mass than unloaded or part-loaded activities (Uzunca *et al.*, 2005). Due to the loading pattern and lower-body dominance of soccer, it would therefore be expected that lower-body BMD increases, as has been found previously in the lower limbs and lumbar spine (Calbet *et al.*, 2001). Wittich *et al.* (1998) reported a mean leg BMD in professional soccer players that was 16 % greater than an age-matched and BMI-matched reference sample. This is comparable to the 17 % greater mean lower-body BMD found across the soccer groups and time points in the present study.

In the current sample, the elevated values in the professional players imply that adaptive changes also occur in the upper body, as both soccer groups demonstrated significantly greater upper-body BMD than the reference group. This effect may be due to the contact nature of the game and additional loading incurred during training. The whole-body BMD values of 16 - 17 % higher than the reference group are slightly greater than previously reported values of 12 % (Wittich *et al.*, 1998) and 11 % (Wallace *et al.*, 2006). These cross-sectional studies suggest that high-level soccer performance has a beneficial effect on the BMD of the whole body. This elevation is reflected in the very high Z-scores evident in both soccer groups. The Z-score is a transformation of the BMD value into a standard deviation value above or below an 'average' person of the same age and sex, and is calculated by the QDR software. The mean difference in Z-score from the reference group in the present study was +172 %, which is lower than the +264 % difference from the reference sample reported by Wittich *et al.* (1998), but the participants forming the reference group in the present study had higher Z-scores than those participating in the aforementioned study, possibly due to being older and heavier, with greater BMI. In the current study, BMD was not identified as an important factor in the transition from professional club-level to international-level soccer.

4.2 Lean mass

The results of the present study imply that soccer is associated with beneficial effects on both upper- and lower-body lean masses. Previous studies have indicated greater whole-body lean mass values of 8 % (Calbet *et al.*, 2001) and 12 % (Wittich *et al.*, 1998) above reference values, which were attributed to participation in soccer. The average whole-body lean mass quantity of 10 % greater than the reference value found in the present study corresponds with these findings. The greater difference from the reference value found in the lower-body compared to the upper-body lean mass supports the statements of Reilly (1996) and Shephard (1999) that muscular development is more pronounced in the lower-body in top soccer players, namely the hip flexors, quadriceps, hamstrings and ankle plantar-flexors and dorsi-flexors. This muscular development is of benefit to power output and performance in activities with a lower-body dominance, such as jumping, tackling and kicking. Neither segmental nor whole-body lean mass differed between the international and non-international professional soccer players, suggesting that muscularity is not a discriminating factor in progressing from elite club-level to international-level soccer.

4.3 Fat mass

Both absolute and relative fat values were lower than reference values in the professional soccer players. This finding was true of the upper-body as well as the lower-body. The reported whole-body percent fat values of 9.8 - 10.9 % for the players concord with Wilmore and Costill (1999)'s recommendations of 6 – 14 % fat for soccer players. Shephard (1999) recommended levels of ~10 % in outfield players, with the more successful players being <10 %, and stressed the importance of limiting the rise in body fat between competitive seasons. It is difficult to draw comparisons with previous reports of percent fat in the literature as a variety of methods has been used to assess body fat. However, comparison with previously reported DXA values for professional soccer players [12.1 ± 3.3 % (Wittich *et al.*, 1998); 12.6 ± 4.1 % (Calbet *et al.*, 2001); 12 ± 3.1 % (Wittich *et al.*, 2001); 12.9 ± 1.3 – 13.5 ± 2.8 % (Egan *et al.*, 2006); 13.2 ± 2.0 % – 13.4 ± 1.3 % (Wallace *et al.*, 2007)] indicates a slightly lower percent body fat in the current sample. This is most likely due to the exclusion of goalkeepers from the present study, but may also be affected by player ethnicity and the standard of the league from which the sample was selected. The use of the DXA subtotal in place of the total value may result in lower percent fat values in lean individuals. In the present study, use of the subtotal would account for ~0.5 % of the difference between this study and other DXA studies employing the total value as the whole-body measure. Neither absolute nor relative amounts of body fat appeared to be critical in distinguishing international-level players from non-internationals in the present study.

4.4 Whole-body results and discriminant analysis

The superior results reported in the professional soccer groups in the present study reflect previous findings of Wittich *et al.* (2001), who also utilised a 3C DXA model of body composition assessment in a sample of professional soccer players. The present study furthered current knowledge by then testing the discriminant function of the measured body composition variables. Percent body fat was found to be the most discriminating variable due to its inclusion in the six stepwise models produced (upper-, lower- and whole-body models for both pre-season and competitive season phases), five of which saw it listed as the first step. Bone mineral density was the second most important variable, featuring in each of the models, but only once as the primary discriminant variable. These findings imply that body fat and BMD may have relatively greater importance for soccer players than the amount of lean mass they carry. Whilst lean mass was not a significant discriminator between the professional groups, its absence in the final stepwise models does not suggest that muscular training is unimportant in the preparation of players for competition. Muscle tissue only makes up part of the DXA lean mass value, as the mineral-free, fat-free compartment includes skin and connective tissue also. Therefore, changes in muscle mass will appear relatively small when the entire lean mass of the body is considered.

The upper-, lower- and whole-body models produced using scan data collected during the competitive season were better able to distinguish professional soccer players from non-players than those produced using pre-season scan data. This finding is likely due to the fact that percent body fat, the best discriminant function, tends to display seasonal variation in professional soccer players, with the highest values being reported at the start of pre-season training (Egan *et al.*, 2006). A higher body fat value at pre-season would mean the professional player is closer in body composition to a reference individual, making it slightly more difficult for a statistical model to classify him correctly. This finding reinforces the argument that the time of season should be taken into account when assessing body composition in professional athletes.

During both pre-season and competitive season phases, the whole-body models were better able to predict group membership than were the segmental analyses, indicating that the body as a whole must be taken into account when assessing body composition in soccer groups. Identification of a player's misclassification using a stepwise discriminant model could prompt closer inspection of the scan report to determine which variable is causing the misclassification and be taken into consideration when planning future interventions such as resistance training for increasing muscle mass, impact loading for promoting bone mineral density or a combination of training and dietary regimens to reduce body fat. The patterning of body composition measures is not necessarily critical in the selection of senior international soccer players from those playing regularly in one of the world's top professional leagues.

5. CONCLUSIONS

The main observations in this study were that professional elite soccer players showed more desirable values in all components of a DXA three-compartment model of body composition than an age- and height-matched, low-activity reference group; differences being evident in upper-, lower- and whole-body analyses. Percent fat and BMD were the body composition variables best able to distinguish the professional soccer players from the reference group. There were no differences in any of the measured variables between the international and non-international professional soccer players

Results from this cross-sectional study suggest that high-level soccer performance is associated with a beneficial effect upon the whole body with regard to bone, lean and fat composition, separating professional soccer players from the normal population on the basis of their body composition. The similarities between the international and non-international players at the top professional clubs imply that appropriate body composition may need to surpass threshold levels rather than attain specific values for participation at this elite level. It appears that body composition is not a critical factor in progressing from elite senior club-level to international-level performance.

References

Arnason, A., Sigurdsson, A., Gudmundsson, I., Holme, L., Engebretsen, L. and Bahr, R., 2004. Physical fitness, injuries and team performance in soccer. *Medicine and Science in Sports and Exercise*, **36**, pp. 278-285.

Bandyopadhyay, A., 2007. Anthropometry and body composition in soccer and volleyball players in West Bengal, India. *Journal of Physiological Anthropology*, **26**, pp. 501-505.

Calbet, J.A., Dorado, C., Díaz-Herrera, P. and Rodríguez-Rodríguez, L.P., 2001. High femoral bone mineral content and density in male football (soccer) players. *Medicine and Science in Sports and Exercise*, **33**, pp. 1682-1687.

Casajus, J.A., 2001, Seasonal variation in fitness variables in professional soccer players. *Journal of Sports Medicine and Physical Fitness*, **41**, pp. 463-469.

Davis, J.A., Brewer, J. and Atkin, D., 1992. Pre-season physiological characteristics of English first and second division soccer players. *Journal of Sports Sciences*, **10**, pp. 541-547.

Egan, E., Wallace, J., Reilly, T., Chantler, P. and Lawlor, J., 2006. Body composition and bone mineral density changes during a premier league season as measured by dual-energy x-ray absorptiometry. *International Journal of Body Composition Research*, **4**, pp. 61-66.

Gil, S.M., Gil, J., Ruiz, F., Irazusta, A. and Irazusta, J., 2007. Physiological and anthropometric characteristics of young soccer players according to their playing position: relevance for the selection process. *Journal of Strength and Conditioning Research*, **21**, pp. 438-445.

Kalapotharakos, V.I., Strimpakos, N., Vithoulka, I., Karvounidis, C., Diamantopoulos, K. and Kapreli, E., 2006. Physiological characteristics of elite professional soccer teams of different ranking. *Journal of Sports Medicine and Physical Fitness*, **46**, pp. 515-519.

Karlsson, M.K., 2007. Does exercise during growth prevent fractures in later life? *Medicine and Sport Science*, **51**, pp. 121-136.

Kistorp, C.N. and Svendsen, O.L., 1998. Body composition results by DXA differ with manufacturer, instrument generation and software version. *Applied Radiation and Isotopes*, **49**, pp. 515-516.

Kraemer, W.J., French, D.N., Paxton, N.J., Häkkinen, K., Volek, J.S., Sebastianelli, W.J., Putukian, M., Newton, R.U., Rubin, M.R., Gómez, A.L., Vescovi, J.D., Ratamess, N.A., Fleck, S.J., Lynch, J.M. and Knuttgen, H.G., 2004. Changes in exercise performance and hormonal concentrations over a big ten soccer season in starters and nonstarters. *Journal of Strength and Conditioning Research*, **18**, pp. 121-128.

Kutlu, M., Sofi, N. and Bozkus, T., 2007. Changes in body compositions of elite level amateur and professional soccer players during the competitive season. *Journal of Sports Science and Medicine*, **6**, S53.

Lewiecki, E.M., 2005. Update on bone density testing. *Current Osteoporosis Reports*, **3**, pp. 136-142.

Matković, B.R., Misigoj-Duraković, M., Matković, B., Janković, S., Ruzić, L., Leko, G. and Kondric, M., 2003. Morphological differences of elite Croatian soccer players according to the team position. *Collegium Antropologicum*, **27**, S167-174.

McIntyre, M.C., 2005. A comparison of the physiological profiles of elite Gaelic footballers, hurlers and soccer players. *British Journal of Sports Medicine*, **39**, pp. 437-439.

Oates, M.K., 2007. The use of DXA for total body composition analysis – part I. *International Society for Clinical Densitometry: SCAN newsletter*, 07/2007, pp. 6-7.

Reilly, T., 1996. Fitness assessment. In *Science and Soccer* (edited by T.Reilly), pp. 25-47. London: E&FN Spon.

Rico-Sanz, J., 1998. Body composition and nutritional assessments in soccer. *International Journal of Sport Nutrition*, **8**, pp. 113-123.

Shephard, R.J., 1999. Biology and medicine of soccer: an update. *Journal of Sports Sciences*, **17**, pp. 757-786.

Silvestre, R., West, C., Maresh, C.M. and Kraemer, W.J., 2006a. Body composition and physical performance in men's soccer: a study of a National Collegiate Athletic Association Division I team. *Journal of Strength and Conditioning Research*, **20**, pp. 177-183.

Silvestre, R., Kraemer, W.J., West, C., Judelson, D.A., Spiering, B.A., Vingren, J.L., Hatfield, D.L., Anderson, J.M. and Maresh, C.M., 2006b. Body composition and physical performance during a National Collegiate Athletic Association Division I men's soccer season. *Journal of Strength and Conditioning Research*, **20**, pp. 962-970.

Strudwick, A., Reilly, T. and Doran, D., 2002. Anthropometric and fitness profiles of elite players in two football codes. *Journal of Sports Medicine and Physical Fitness*, **42**, pp. 239-242.

Sutcliffe, J.F., 1996. A review of in vivo experimental methods to determine the composition of the human body. *Physics in Medicine and Biology*, **41**, pp. 791-833.

Tabachnick, B.G. and Fidell, L.S., 1996. *Using Multivariate Statistics (third edition)*. New York: Harper Collins.

Tothill, P. and Hannan, W., 2000. Comparisons between Hologic QDR 1000W, QDR 4500A, and Lunar Expert dual-energy x-ray absorptiometry scanners used for measuring total body bone and soft tissue. *Annals of the New York Academy of Sciences*, **904**, pp. 63-71.

Uzunca, K., Birtane, M., Durmus-Altun, G. and Ustun, F., 2005. High bone mineral density in loaded skeletal regions of former professional football (soccer) players: what is the effect of time after active career? *British Journal of Sports Medicine*, **39**, pp. 154-157.

Wallace, J., Egan, E., George, K. and Reilly, T., 2006. Body composition in competitive male sports groups. In *Contemporary Ergonomics*. Bust, D., ed. , (London: Taylor and Francis), pp. 516-518.

Wallace, J., 2007. *The impact of exercise on human body composition determined using dual-energy x-ray absorptiometry*. Thesis (PhD). Liverpool John Moores University.

Wallace, J., Egan, E., Lawlor, J., George, K. and Reilly, T., 2007. Body composition changes in professional soccer players in the off-season. In *Kinanthropometry X: Proceedings of the 10th International Society for the Advancement of Kinanthropometry conference, held in conjunction with the 13th Commonwealth International Sport Conference*, Marfell-Jones, M. and Olds, T., eds., (London: Routledge), pp. 127-134.

Wilmore, J.H. and Costill, D.L., 1999. *Physiology of Sports and Exercise (second edition)*. Champaign: Human Kinetics.

Wittich, A., Mautalen, C.A., Oliveri, M.B., Bagur, A., Somoza, F. and Rotemberg, E., 1998. Professional football (soccer) players have a markedly greater skeletal mineral content, density and size than age- and BMI-matched controls. *Calcified Tissue International*, **63**, pp. 112-117.

Wittich, A., Oliveri, M.B., Rotemberg, E. and Mautalen, C., 2001. Body composition of professional football (soccer) players determined by dual x-ray absorptiometry. *Journal of Clinical Densitometry*, **4**, pp. 51-55.

Part V

Expertise and co-ordination

Optimising speed and energy expenditure in accurate visually-directed upper limb movements

Digby Elliott[1], Steve Hansen[1] and Lawrence E.M. Grierson[2]

[1]Research Institute for Sport and Exercise Sciences, Liverpool John Moores University, Liverpool, United Kingdom
[2] Department of Kinesiology, McMaster University, Hamilton, Ontario, Canada

1. INTRODUCTION

In the area of perceptual-motor behaviour, there are very few relations between important performance-defining variables that remain unchanged across different participant groups and environments. The relationship between movement time and movement accuracy in goal-directed aiming/reaching is the exception. The co-variation between these two variables is stable across so many contexts that this relation has come to be known as Fitts' Law. Paul Fitts found that mean movement time increased systematically with the accuracy requirements, as defined by the amplitude of the movement and the size of the target. This relationship was observed for both reciprocal (Fitts, 1954) and discrete aiming movements (Fitts and Peterson, 1964). Specifically:

Mean Movement Time = $a + b$ [$\log_2$(2x Movement Amplitude/Target Width)]
Where a and b are empirical constants that depend on the specific performer and task conditions.

In many sport, work and daily living activities, it is important for the performer to complete goal-directed movements with a high degree of spatial precision in the shortest possible period of time. In some activities, it is also important to maintain speed and accuracy while optimising energy expenditure. This is particularly true for motor performances that need to be repeated over a long period of time where fatigue may be an issue. The premise of our work is that at least some understanding of the processes involved in optimising speed, accuracy and energy expenditure in complex motor skills can be gained by examining the performance and learning processes associated with simple tasks such as goal-directed aiming.

The study of goal-directed aiming has a long history. In classic work conducted over a century ago, R.S. Woodworth (1899) had participants make sliding movements with a stylus on a rotating drum to target lines that were a fixed distance from the starting point. He manipulated the movement time of these actions by having participants perform to the beats of a metronome. Woodworth was the first researcher to demonstrate the systematic increase in end-point error with increasing movement speed (i.e., decreased movement time). Woodworth (1899) identified two unique components within the aiming movements from the tracings marked on the paper that was affixed to the drum. There was an initial ballistic component that he called the 'initial adjustment' that brought the limb into the target region, and also a slower, more graded phase of the movement associated with 'current control'. Woodworth's notion was that during this homing phase of the movement participants used visual feedback about the position of the limb and the target to reduce error and achieve the final target position. Aiming attempts with longer movement times were associated with less spatial error because the performer had more time to spend in the error reducing homing phase.

Woodworth's two-component model of limb control has stood the test of time. During the mid-20th century, his model provided the foundation for a number of other dual process models including the "iterative correction model" (Crossman and Goodeve, 1963, 1983; Keele 1968) and the "single correction model" (Beggs and Howarth, 1970 1972, see Elliott *et al.*, 2001 for a review).   In more recent years, the most influential model of speed-accuracy and limb control has been Meyers and co-workers' (1988) "optimised submovement model" which does a very good job of describing the mathematical relation between a number of kinematic and performance variables associated with movement time and movement accuracy/spatial variability. This model provides a starting point for our current theoretical framework on the optimisation of speed, accuracy and energy in limb control.

2. OPTIMISED SUBMOVEMENT MODEL

This model builds directly on important work by Richard Schmidt and his colleagues in the late 1970s. Abandoning Woodworth's (1899) notion that the relation between speed and accuracy was primarily associated with the time available for feedback utilization, Schmidt *et al.* (1978, 1979) proposed that the endpoint spatial variability of a series of rapid aiming movements was related to the absolute force requirements associated with performing those movements rather than the time required for visually based corrections.   Simply put, movements of greater amplitude or speed require more muscular force than shorter, slower movements. With greater force, there will be greater force variability and this greater force variability translates into greater endpoint spatial variability. Schmidt and his colleagues found that this relationship between the muscular forces associated with the initial impulse and the variability in endpoint error held fairly well for very rapid movements (e.g., less than 200 ms). However, the relationship began to breakdown when there was sufficient time for feedback utilization. This shortcoming of Schmidt's impulse variability model was dealt with nicely by Meyer and colleagues (1988) subsequent model.

Like the Schmidt model, Meyers and co-workers' optimised submovement model holds that the endpoint variability of the initial movement increases with the velocity of the movement. According to Meyer *et al.* (1988), the performer must strike a compromise between an initial submovement that is fast and one that is slower and more likely to hit the target. The idea is that fast movements, by their nature, will be more variable and thus more likely to require time-consuming corrections. The optimised submovement model is a stochastic model and holds that for a given participant and target-aiming situation, the distribution of primary movement endpoints will be centred on the middle of the target. Responses from the tails of the distribution will fall outside of the target boundaries and the individual movements associated with the two ends of the distribution will require corrective submovements. Presumably there will also be variability associated with the first corrective submovement and some of these will also need to be corrected for.

3. WORST CASE SCENARIO

Although the stochastic features of the optimised submovement model are intuitively appealing, Meyers and co-workers' idea about normally distributed error around the target centre are inconsistent with the majority of the empirical evidence. Studies from our laboratory (e.g., Elliott *et al.*, 1999, 2004) and elsewhere (Engelbrecht *et al.*, 2003) have repeatedly shown that primary submovements generally undershoot the target rather than overshoot the target. Thus, second accelerations to achieve the final target position are far more common than movement reversals (see Elliott *et al.*, 2001 for a review). We have taken the position that primary movement endpoint distributions are centred short of the target, because not all errors are equal in terms of temporal and energy costs (Elliott *et al.*, 2004). Specifically, when a primary submovement overshoots the target, a movement reversal is necessary to acquire the target. This type of movement and subsequent correction requires the limb to travel a greater distance than when the limb undershoots the target. Moreover the limb must overcome the inertia of a zero velocity situation at the point of reversal. This requires the increased expenditure of time and biological energy. Also, a reversal requires a change in the role of the primary muscle groups; that is, the agonist becomes the antagonist and vice versa. In the context of manual aiming, participants seem to have implicit knowledge about the spatial variability associated with their aiming movements and plan their movements to avoid the worse case scenario (i.e., a costly target overshoot). It may also be the case that, in many daily activities, there are safety costs associated with overshooting a target that warrant a "play it safe" strategy. One can think about the disadvantages associated with overshooting when reaching for a cup of coffee (e.g., burning the hand or spilling the coffee), working with a circular saw or moving food to the mouth with a fork.

In recent work designed to examine our "cost of missing" hypothesis, we reasoned that the degree of undershooting in an aiming situation would increase with the cost of an overshoot (Lyons *et al.*, 2006). We had participants make goal-directed aiming movements from a central home position to one of two targets in a situation in which the target aiming apparatus was either positioned flat on the table top or positioned vertically. The first situation required movements either away from or towards the body while, when the apparatus was positioned vertically, the aiming movements were made either up or down. We reasoned that in the aiming down situation participants would be particularly careful not to overshoot the target with their primary aiming movement. This is because a corrective submovement, that entails a reversal, would be working against gravity, and thus would require additional time and energy. In Figure 1, we have plotted the mean and within-participant standard deviation of the primary submovement endpoint distributions. Not only were we correct in our hypothesis regarding central tendency, it was also the case that greater undershooting was associated with conditions in which there was greater variability. Once again, this is consistent with the notion that performers are implicitly aware of their own variability and plan their aiming trajectories accordingly (e.g., Trommershaüser *et al.*, 2006).

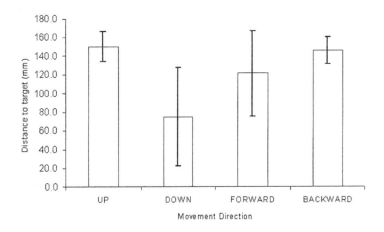

Figure 1. Mean distance covered with the primary submovement as a function of movement direction. The target was 160 mm from the home position. The error bars represent the mean within-participant standard deviations of the primary movement endpoint (Reprinted from Lyons *et al.* 2006, *Experimental Brain Research* with permission).

Even for very simple aiming movements, people become faster and more accurate with practice (e.g., Elliott *et al.*, 1995). In another recent study, we sought to determine how practice influenced both central tendency and variability of primary endpoint distributions (Elliott *et al.*, 2004). We provided participants with feedback and monetary incentives to improve their movement times and introduced a monetary penalty for target misses. Over 4 days of practice participants were able to reduce their movement times from approximately 380 ms to 320 ms with no increase in aiming error. Overall, movements that did not require a correction were fastest (320 ms), movements that required a reversal to hit the target were substantially slower (425 ms), and undershoots that required a second acceleration to hit the target were intermediate (370 ms). However, what is most interesting is how the distribution of primary movement endpoint changed over practice. In Figure 2, we have plotted the proportion of initial undershoots, initial hits and initial overshoots for all the error free trials. What is apparent is that over-practice participants increased their proportion of hits and decreased their proportion of undershoots while maintaining overshoots at a very low and consistent level. The idea is that as participants were able to decrease their aiming variability they started to "sneak-up" on the target while still avoiding costly overshoots. Once again, this finding is consistent with the notion that performers prepare their movements to avoid the worst case scenario.

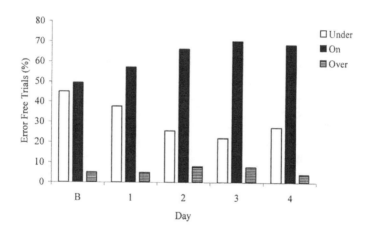

Figure 2. Percentage of error free trials for which the primary submovement undershot, hit and overshot the target as a function of practice day (Reprinted from Elliott *et al.*, 2004, Journal of Motor Behavior with permission).

4. WHAT ABOUT ENERGY MINIMISATION?

Up to this point, we have talked mainly about optimising movement speed while maintaining movement accuracy. Although movement trajectories that undershoot or hit the target are generally more efficient in terms of time and energy, we have attempted to examine the energy minimisation principle independent of time. In order to examine energy expenditure without the involvement of temporal co-variation, we used a slightly different type of targeting task in which participants were required to propel a small disk down a track to hit a spatial target (Oliveira *et al.*, 2005). In what we termed an unassisted condition, participants used one continuous movement to propel the disk down the track. In this condition, more force meant greater distance. For a second condition, a rubber tube was used to assist the movement of the disk down the track. Using a similar arm motion to the first condition, participants applied a force against the assistive force of the rubber tube in order not to propel the disk down the track too far. Thus more force meant less distance. In spite of feedback and a monetary payoff to hit the target, participants still undershoot the target for the first 75 practice trials in the unassisted condition. Their undershooting was especially pronounced for the first 25 trials. Exactly the opposite was true in the assisted condition. That is, participants overshoot the target with a mean performance that was not in the target area until the 4th block of 25 trials. Because time was not an issue in this study, these aiming biases can be attributed to a strategy of energy minimisation. Like the manual aiming studies discussed in the previous section, participants only began to sneak up on the mean target position when within-participant aiming variability began to stabilise. One would suspect that this type of energy minimisation strategy would be even more pronounced under conditions where fatigue is an issue (see Sparrow and Newell, 1998 for a review).

5. BALLISTIC INITIAL IMPULSE?

From the work discussed so far, it is apparent that people get faster with practice partly because they are able to reduce the frequency of costly error-reducing submovements (e.g., Elliott *et al.*, 2004). Certainly this increased efficiency can be achieved by organising and executing a more precise (i.e., less variable) primary submovement (Khan *et al.*, 1998, 2006). Although participants undoubtedly improve in their ability to specify the timing and magnitude of muscular force, over the last decade it has also become apparent that the so-called ballistic initial impulse or primary submovement may not be as preplanned as previously thought (cf. Woodworth, 1899; Meyer *et al.*, 1988). Specifically, well-practiced performers appear to be able to engage in at least some form of current or on-line control prior to the completion of the primary submovement (Elliott *et al.*, 1991).

One way that investigators have attempted to examine early online control is to examine how spatial variability in the movement trajectory changes as the movement unfolds (see Khan *et al.*, 2006 for a review). When visual feedback is available, one often finds that within-participant spatial variability increases until approximately peak deceleration and then drastically diminishes between peak deceleration and the end of the movement (e.g., Khan *et al.*, 2002, 2003) without necessarily any discrete discontinuities in the movement trajectory (e.g., discrete corrections; Elliott *et al.*, 2001). Moreover, Hansen *et al.* (2005) found that participants were able to improve their aiming performance by reducing variability earlier in the movement with practice of the task. Specifically, spatial variability in the primary direction of the movement began to decrease between peak velocity and peak deceleration, long before what is typically deemed the primary submovement was complete.

Until recently most experiments using spatial variability as an index of online control have focused on variability in the primary direction of the movement. We have recently extended this methodology and have started to examine spatial variability in three dimensions. Figure 3 depicts the behaviour of participants aiming at targets that could change position at movement initiation during the experimental conditions. The spatial location and volume of ellipsoids early in the trajectories of movements are quantitatively different when the participants aimed at the farthest target and knew that the target would not move (Figure 3a) compared to when the target unexpectedly moved to the farthest target from the middle position (Figure 3b). Early trajectory differences are particularly noticeable between the condition where the middle target was presented and then the farthest target was presented at movement initiation and the condition where participants aimed at an unperturbed middle target with the potential for the target to change position at movement initiation (Figure 3c). Specifically, differences in the spatial location of the ellipsoids indicate the increases in the velocity of the effector for the perturbed condition. More spatial volume for the early markers also indicates increases in movement variability during the perturbed condition. Although the general pattern remains the same between the conditions without target perturbation, more spatial variability is also evident as the amplitude of the movements increases (Figure 3a compared to Figure 3c; see also Lemay and Proteau, 2001).

Using these three-dimensional methods, we have also compared performance in conditions of full vision to those in which vision is eliminated upon movement initiation, as well as the impact of practice on variability throughout the trajectory (Hansen *et al.*, 2008-a). In both these protocols, we found evidence for very early limb modulation during what Meyer *et al.* would characterise as the primary submovement.

Another way to examine online control is to determine how the spatial position of the limb at an early kinematic event predicts error later in the trajectory. The idea is that if a movement is incorrectly planned and not adjusted on the basis of response associated feedback, early error will be positively related to late error (e.g., a movement that has gone too far at peak velocity will eventually over shoot the target). However, if early error can be detected and rectified, then spatial position early will not necessarily predict spatial position later in the trajectory (Heath, 2005). Under conditions that are ideal for online control, there will be a negative relationship between the physical distance covered early and late in the movement (e.g., Elliott *et al.*, 1999). We have found further evidence for early online control using these types of correlational methods. Specifically, if vision of both the limb and the target are available to participants during movement execution, they begin the corrective process before the end of the primary submovement in a manner that is unassociated with discrete discontinuities in the movement trajectory.

Insight into early and late corrective processes during manual aiming can also be achieved by introducing either real or illusory perturbations to the movement environment designed to change either the perceived goal of the movement (e.g., Heath *et al.*, 1998) or the response-produced feedback associated with achieving

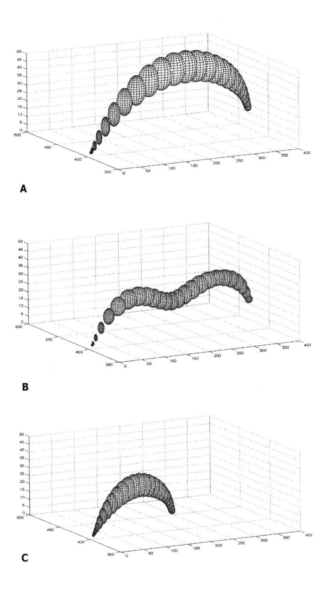

Figure 3. Within-condition trajectory variability ellipsoids averaged over twenty participants during a target perturbation paradigm. Ellipsoids are depicted for movements to the farthest target without perturbation (3A), movements to the far target following presentation of the middle target prior-to movement initiation (3B), and movements to the middle target without perturbation (3C).

that goal (e.g., Hansen *et al.*, 2007). To impact online control processes, these perturbations can be introduced at movement onset, or at a predetermined temporal or spatial location during movement execution. When the technical capability exists for real-time analyses during movement execution, perturbations can also be introduced at a specific kinematic event (e.g., peak velocity; Hansen *et al.*, 2008-b). Of interest is how and when the visual-motor system adjusts the movement trajectory to accommodate these changes to the task or perceived task demands.

In one compelling study, Proteau and Masson (1997) unexpectedly translated the texture elements over which individuals performed goal-directed reaches. Specifically, following movement initiation, the background characteristics of the aiming surface either moved in the direction of, or the direction opposite to, that of the movement. In the case of opposing background translation, participants were faced with an illusory perception of faster than expected effector velocity and, as such, terminated their primary submovements early. Though to a lesser extent, the converse was true for instances in which the texture elements translated in the same direction as the effector. It seems the performers were able to make early adjustments to the primary submovement to meet the new demands imposed by the illusory feedback. Interestingly, the endpoint biases induced by the moving background illusions affected the amplitude of the primary submovement indicating that this portion of the movement is not entirely ballistic (cf. Woodworth, 1899; Meyer *et al.*, 1988).

In a recent study conducted in our laboratory, Hansen *et al.* (2007) used prism-equipped liquid crystal goggles to either introduce or remove a 14.5° perceived lateral shift of target and hand position at, or after, movement initiation. Female participants in particular adjusted their limb trajectories very quickly to accommodate the perceived change in the target goal and the response-produced feedback. When the perturbation was introduced at movement initiation, adjustments to the trajectory were sometimes realised prior to peak velocity of the primary submovement, which provides further evidence for early online control.

6. THE TWO COMPONENT AND OPTIMISED SUBMOVEMENT MODELS REVISITED

Although Woodworth (1899, see also Meyer *et al.*, 1988) was probably correct that there are two essential phases/components to manual aiming movements, our current work suggests that, during both phases, there are processes that provide some degree of online control. Prior to movement initiation the performer prepares a movement designed to get the limb the majority of the way to the target goal. The specific nature of this planned movement depends on a number of factors. As reported earlier, the amplitude of the movement trajectory depends not only on the relative temporal and energy costs of overshooting or undershooting the target, it also depends on the performer's implicit knowledge about what sort of variability to expect in movement execution. Both internal variables (e.g., practice, fatigue) and external variables (e.g., mass of the effector, gravity, environment resistance to the movement) impact variability. What we have not mentioned thus far is that prior knowledge about the availability of sensory feedback will also impact the planning process. For example, Hansen *et al.* (2006) have shown that participants plan their movement trajectories quite differently under conditions in which there is certainty versus uncertainty about the availability of visual feedback over the course of the impending movement. Consistent with some of the ideas presented earlier, under conditions of uncertainty people prepare to avoid the worst case scenario.

Prior knowledge about what to expect is important, because the expected sensory consequences associated with each aiming attempt provide the basis for early online control. This sort of idea, which stems from work by von Holst (1954), has been developed by investigators such as Miall and Wolpert (1996). The notion is that an internal model or representation of the expected sensory consequences of the aiming movement is created concurrent with the movement planning process. This representation then provides the basis for early online control. Specifically dynamic visual and proprioceptive feedback about the movement of the limb in space is compared to this representation. If the afferent feedback and expected feedback are congruent the movement unfolds as planned. However, adjustments can be realised very quickly and early in the movement trajectory in cases where there is a mismatch between feedforward and feedback representations. In terms of feedback, it is the dynamic information about limb velocity and direction that is important. Perturbations that impact these perceived movement parameters (e.g., a moving background for velocity, prismatic displacement for direction) will create this mismatch and thus jump-start this type of early corrective process.

As the movement unfolds and the limb approaches the physical target, there is an opportunity for the second type of online processing that Woodworth (1899), Keele (1968), Beggs and Howarth (1970) and Meyer *et al.* (1988) identified. Specifically, independent of the initial movement plan, the performer is able to compare the position of the limb to the position of the target and make one or more discrete corrective adjustments to the trajectory to reduce error. Although there is some ocular information available about the position of the target (e.g., Binsted and Elliott, 1999) and proprioceptive information available about the position of the limb, visual afference is the primary source of data from this error reducing process (Elliott *et al.*, 2001).

In the most recent work from our laboratory, we have attempted to test these ideas about early and late online control by concurrently introducing manipulations expected to impact one or the other. For example, in one study we used a moving background perturbation to influence the perceived velocity of the limb (e.g., Proteau and Masson, 1997) and the Muller-Lyer illusion (e.g., Mendoza *et al.*, 2006) to impact the perceived position of the target. We found that the two manipulations had additive and independent effects on movement outcome (Grierson and Elliott, in press).[1] More importantly, the moving background manipulation had an impact on early trajectory changes while the Muller-Lyer illusion had an impact on late trajectory changes. This finding is consistent with the notion that there are two types of online control. Specifically, there is early control based on a comparison of dynamic information about limb velocity and direction to an internal model of what is expected, and late, discrete control based on a comparison of the relative positions of the limb and target. Future work will be necessary to determine the degree of independence/covariation between these two corrective processes.

7. THEORETICAL IMPLICATIONS

Most current explanations of Fitts' Law attribute the relation between movement speed and accuracy to some combination of feedforward processes that determine initial movement variability and feedback processes that require time to complete (e.g., Meyer *et al.*, 1988). Our framework for understanding speed-accuracy relations in goal-directed reaching and aiming is consistent with Meyers and co-workers' notion that a performer must learn to strike a compromise between endpoint variability associated with moving fast and the extra time it takes to modify movement error based on response-produced feedback. Our main departure from the optimised submovement model is associated with the precise current control or online processes used to regulate movement. Specifically, we hold that, as well as late control based on feedback about the relative positions of the limb and target, skilled performers[2] can also exercise a form of early control based on a comparison of early dynamic information about the limb's movement in space to an internal representation of the expected sensory consequences associated with the particular movement. This latter type of control appears to be more graded or continuous in nature compared to the type of discrete error reducing regulation associated with most two component models of limb control (e.g., Woodworth, 1899; Keele, 1968; Meyer *et al.*, 1988). Elsewhere we have suggested that the absence of discontinuities in the movement trajectory associated with this type of control could reflect "graded adjustments to the muscles being used to decelerate the movement" (Elliott *et al.*, 2001, p. 348). Alternatively it could reflect "many overlapping discrete adjustments to the movement trajectory giving the movement the appearance of continuity (i.e., pseudo-continuous control)" (Elliott *et al.*, 1991, p. 415).

One of our other contributions to the speed-accuracy literature has been to document the flexibility with which the human performer faces particular movement tasks in order to maximise performance. Certainly, prior knowledge about what to expect influences the movement planning process and, as a result, how movements are ultimately executed and regulated. Generally, participants prepare for the worst case scenario in an attempt to minimise the occurrence of errors that will be costly in terms of time and energy (Elliott *et al.*, 2004; Hansen *et al.*, 2006). This type of strategic behaviour indicates that the performer has a great deal of implicit knowledge about what to expect under a variety of movement circumstances. Clearly if we are correct in speculating that early limb regulation is based on a comparison of afferent information to feedforward information about the impending movement, then it is important that movement planning processes consider what afferent information will be both available and reliable during movement execution. In addition, the movement goal (i.e., central tendency) will be constrained by implicit knowledge about the variability associated with the particular movement situation and the practice history of the performer in that situation given that some errors are more costly than others over a set of trials. Over practice, shifts in central tendency toward the precise movement goal will occur with reductions in variability. In contrast to most models of speed-accuracy, our theoretical framework highlights the flexibility, learning history and trial-to-trial strategic behaviour of the performer (Elliott *et al.*, 2001; 2004; Lyons *et al.*, 2006).

8. PRACTICAL IMPLICATIONS

From a purely technical perspective, we are excited that some of the techniques we have developed to examine goal-directed reaching will also serve researchers in the ergonomics community who are interested in designing work-friendly and play-friendly environments (Hansen *et al.*, 2008-a). In addition to the quantification of individual movement patterns and the flexibility of the sensory-motor system following perturbations, the ellipsoid methodology (e.g., Figure 3), based on the within-condition variability of a performer over a number of movements, has a utility for movement rehabilitation, and equipment or workstation design. When investigating the movement patterns of individuals with movement dysfunction such as observing prehension in individuals following stroke or spinal cord injury, identifying the spatial and temporal location of increased movement variability can facilitate the movement intervention selected by the therapist. Recognising changes in the spatial variability of movement trajectories can also enable the interventionist to provide both spatial and temporal information to the participant about their progression through the therapy.

Utilization of the ellipsoid technique can also assist designers to build an environment around the individual's space requirements. Often equipment and workstations are designed without regard to human interaction. Many awkward designs are created when the designer accounts only for the amount of volume the person will physically occupy. Designers can avoid the potentially costly mistake of creating a workstation that humans find awkward to interact with by determining the location and volume of space that the person's movements, accurate or inaccurate, will occupy. In other words, designers can build the workstation to cater to the requirements of the individual rather than having the user conform to an uncomfortable and potentially hazardous environment.

Although we have focused on simple goal-directed aiming movements, many of the movement planning and online regulation processes examined in this paper can be applied to more complex targeting (e.g., throwing, kicking, shooting) and interceptive motor skills (e.g., catching, hitting). For example, contrary to traditional wisdom, recent research work on basketball shooting (de Oliveira *et al.*, 2006, 2007) indicates that both jump shot and foul shooting success depends on very late sensory information available just before the ball is released. Moreover degradation of visual information during basketball shooting results in shifts in both central tendency and variability that are consistent with both the "play it safe" and "energy minimisation" principles outlined earlier in this paper (de Oliveira, 2007). What remains to be seen with this research on basketball is whether late visual regulation is related to afferent processing only, or processes that involve the comparison of visual and proprioceptive feedback to sensory expectations.

In research work that has proceeded in parallel to some of the experiments described here, we have been examining specific problems with goal-directed behaviour in a number of special populations including the elderly (Welsh *et al.*, 2007) and persons with Down syndrome (Obhi *et al.*, 2007). Based on findings which indicate that young adults with Down syndrome judge the temporal occurrence of self-generated movements on response-associated feedback (i.e., a late judgement) as opposed to premovement planning processes (i.e., an early judgement), we are currently exploring the hypothesis that some of the judgement differences in this population is related to their inability to either form or maintain an internal model of goal-direct behaviour against which to evaluate feedback. A diminished representation of the expected movement outcome would result in their aiming behaviour being very slow and inefficient (Elliott *et al.*, 2006). Research work such as this contributes to our understanding of specific movement pathologies, and also provides an appropriate model for studying typical goal-directed behaviour by examining the circumstances under which it breaks down.

References

Beggs, W. D. A. and Howarth, C. I., 1970, Movement control in man in a repetitive motor task. *Nature*, **221**, pp. 752-753.

Beggs, W. D. A. and Howarth, C. I., 1972, The accuracy of aiming at a target: Some further evidence for a theory of intermittent control. *Acta Psychologica*, **36**, pp. 171-177.

Binsted, G. and Elliott, D., 1999, Ocular perturbations and retinal/extraretinal information: The coordination of saccadic and manual movements. *Experimental Brain Research*, **127**, pp. 193-206.

Crossman, E.R.F.W. and Goodeve, P.J., 1963/1983, Feedback control of hand movement and Fitts' law. *Quarterly Journal of Experimental Psychology*, **35** (**A**), pp. 251-278.

de Oliveira, R.F., et al., 2007, Basketball jump shooting is controlled online by vision. *Experimental Psychology*, **54**, pp. 180-186.

de Oliveira, R.F., Oudejans, R.R.D. and Beek, P., 2006, Late information pic-up is preferred in basketball jump shooting. *Journal of Sports Sciences*, **24**, pp. 933-940.

Elliott, D., Binsted, G. and Heath, M., 1999, The control of goal-direct limb movements: Correcting errors in the trajectory. *Human Movement Science*, **18**, pp. 121-136.

Elliott, D., et al., 1991, Discrete vs. continuous visual control of manual aiming.

Human Movement Science, **10**, pp. 393-418.

Elliott, D., et al., 1995, Optimizing the use of vision in manual aiming: The role of practice. *Quarterly Journal of Experimental Psychology: Human Experimental Psychology*, **48(A)**, pp. 72-83.

Elliott, D., et al., 2004, Learning to optimize speed, accuracy, and energy expenditure: A framework for understanding speed-accuracy relations in goal-directed aiming. *Journal of Motor Behavior*, **36**, pp. 339-351.

Elliott, D., Helsen, W. F. and Chua, R., 2001, A century later: Woodworth's two-component model of goal directed aiming. *Psychological Bulletin*, **127**, pp. 342-357.

Elliott, D., et al., 2006, The visual regulation of goal-directed reaching movements in adults with Williams syndrome, Down syndrome and other developmental disabilities. *Motor Control*, **10**, pp. 34-54.

Engelbrecht, S.E., Berthier, N.E. and O'Sullivan, L.P., 2003, The undershooting bias: Learning to act optimally under uncertainty. *Psychological Science*, **14**, pp. 257-261.

Fitts, P. M., 1954, The information capacity of the human motor system in controlling the amplitude of movement. *Journal of Experimental Psychology*, **47**, pp. 381-391.

Fitts, P.M. and Peterson, J.R., 1964, Information capacity of discrete motor responses. *Journal of Experimental Psychology*, **67**, pp. 103-112.

Grierson, L.E.M. and Elliott, D., in press. Goal-directed aiming and the relative contribution of two online control processes. *American Journal of Psychology*.

Hansen, S., Elliott, D. and Khan, M.A., 2008-a, Quantifying the variability of three dimensional human movement using ellipsoids. *Motor Control*, **12**, pp. 241-251.

Hansen, S., Elliott, D. and Tremblay, L., 2007, Online control of discrete action following visual perturbation. *Perception*, **36**, pp. 268–287.

Hansen, S., et al., 2006, The influence of advance information about target location and visual feedback on movement planning and execution. *Canadian Journal of Experimental Psychology*, **60**, pp. 200-208.

Hansen, S., Tremblay, L. and Elliott, D., 2005, Part and whole practice: Chunking and online control in the acquisition of a serial motor task. *Research Quarterly for Exercise and Sport*, **76**, pp. 60-67.

Hansen, S., Tremblay, L. and Elliott, D., 2008-b, Real time manipulation of visual displacement during manual aiming. *Human Movement Science*, **27**, pp. 1-11.

Heath, M., 2005, Role of limb and target vision in the online control of memory-guided reaches. *Motor Control*, **9**, pp. 281-309.

Heath, M., et al., 1998, On-line control of rapid aiming movements: Effects of target characteristics on manual aiming. *Canadian Journal of Experimental Psychology*, **52**, pp. 163-173.

Keele, S.W., 1968, Movement control in skilled motor performance. *Psychological Bulletin,* **70**, pp. 387-403.

Khan, M. A., et al., 2002, Optimal control strategies under different feedback schedules: Kinematic evidence. *Journal of Motor Behavior*, **34**, pp. 45-57.

Khan, M. A., et al., 2006, Inferring online and offline processing of visual feedback in target-directed movements from kinematic data. *Neuroscience and Behavioral Reviews*, **30**, pp. 1106-1121.

Khan, M.A., Franks, I.M. and Goodman, D., 1998, The effect of practice on the control of rapid aiming movements: Evidence for an interdependence between programming and feedback processing. *Quarterly Journal of Experimental Psychology*, **51(A)**, pp. 425-444.

Khan, M.A., et al., 2003, The utilization of visual feedback in the control of movement direction. Evidence from a video aiming task. *Motor Control*, **7**, pp. 290-303.

Lemay, M. and Proteau, L., 2001, A distance effect in a manual aiming task to remembered targets: A test of three hypotheses. *Experimental Brain Research,* **140**, pp. 357-368.

Lyons, J., et al., 2006, Optimizing rapid aiming behaviour: movement kinematics depend on the cost of corrective modifications. *Experimental Brain Research,* **174**, pp. 95-100.

Mendoza, J. E., et al., 2006, The effect of the Muller-Lyer illusion on the planning and control of manual aiming movements. *Journal of Experimental Psychology: Human Perception and Performance*, **32**, pp. 413-422.

Meyer, D.E., et al., 1988, Optimality in human motor performance: Ideal control of rapid aimed movements. *Psychological Review*, **95**, pp. 340-370.

Miall, R. C. and Wolpert, D. M., 1996, Forward models for physiological motor control. *Neural Networks*, **9**, pp. 1265-1279.

Obhi, S.S., et al., 2007, The perceived time of voluntary action for adults with and without Down syndrome, *Down Syndrome Quarterly*, **9**, pp. 4-9.

Oliveira, F.T.P., Elliott, D. and Goodman, D., 2005, The energy minimization bias: Compensating for intrinsic influence of energy minimization mechanisms. *Motor Control*, **9**, pp. 101-114.

Proteau, L. and Masson, G., 1997, Visual perception modifies goal-directed movement control: Supporting evidence from a visual perturbation paradigm. *Quarterly Journal of Experimental Psychology*, **50A**, pp. 726-741.

Schmidt, R.A., Zelaznik, H.N. and Frank, J.S., 1978, Sources of inaccuracy in rapid movement. In *Information Processing in Motor Control and Learning*, edited by Stelmach, G.E., (New York: Academic Press), pp. 183-203.

Schmidt, R. A., et al., 1979, Motor output variability: A theory for the accuracy of rapid motor acts. *Psychological Review*, **86**, pp. 415-451.

Sparrow, W.A. and Newell, K.M., 1998, Metabolic energy expenditure and the regulation of movement economy. *Psychonomic Bulletin & Review*, **5**, pp. 173-196.

Trommershaüser, J., et al., 2005, Optimal compensation for changes in task-relevant movement variability. *The Journal of Neuroscience*, **25**, pp. 7169-7178.

Von Holst, E., 1954, Relations between the central nervous system and the peripheral organs. *British Journal of Animal Behaviour*, **2**, pp. 89-94.

Welsh, T.N., Higgins, L., and Elliott, D., 2007, Are there age-related differences in learning to optimize speed, accuracy and energy expenditure? *Human Movement Science*, **26**, pp. 892-912.

Woodworth, R.S., 1899, The accuracy of voluntary movement. *Psychological Review*, **3**, (Monograph Supplement), pp. 1-119.

Footnotes

1.   The moving background illusion impacts the perceived velocity of the limb movement through the physical space between the home position and the target.   This dynamic information is important early, before the limb reaches the vicinity of the target.   The Muller-Lyer illusion affects the perceived position of the target. This information becomes important as the limb approaches the target. However, because any adjustments made early in the trajectory (e.g., adjustments based on dynamic information about limb velocity) will ultimately affect the position of the limb relative to the target late in the movement, statistical independence is not a necessary feature of processing independence.

2.   We use the term "skilled performer" to indicate that the individual must have enough experience with the task to form a viable model or internal representation of the expected sensory consequences associated with a set of motor commands.

CHAPTER TWENTY-THREE

Tracing the process of expertise in a simulated anticipation task

A.P. McRobert[1], A.M. Williams[1], P. Ward[2] and D.W. Eccles[2]

[1]Research Institute for Sport and Exercise Sciences, Liverpool John Moores University, Liverpool, UK
[2]Center for Expert Performance Research, Florida State University, USA

1. INTRODUCTION

Over the past few decades, the study of perceptual-cognitive expertise has expanded rapidly in a range of disciplines (Ericsson *et al.*, 2006). Perceptual and cognitive skills have been identified as crucial to performance in many domains, such as driving (e.g. McKenna and Horswill, 1999), aviation (e.g., Russo *et al.*, 2005), medicine (e.g., Patel *et al.*, 1990), and sport (e.g., Williams *et al.*, 1999). In such situations, the ability to anticipate future events based on information arising early in the visual display is crucial. For example, skilled drivers are able to anticipate potentially hazardous traffic situations more effectively than novice drivers, resulting in a reduction in accident liability (McKenna and Horswill, 1999). Similarly, in many sports, whether externally paced such as tennis and cricket (Penrose and Roach, 1995; Williams *et al.*, 2002) or self paced such as orienteering (Eccles *et al.*, 2002a, 2002b), the ability to anticipate upcoming events based on relevant prior information can provide a performance advantage. Experts identify relevant information early and make use of domain-specific knowledge to facilitate performance.

Ericsson and Smith (1991) proposed the expert performance approach as a descriptive and inductive framework for the study of expertise. Using this approach, researchers first observe superior performance *in situ,* and then recreate a task that is representative of the domain such that reliable individual differences in performance can be objectively measured under controlled laboratory conditions. Second, researchers attempt to determine the mechanisms underlying performance using process-tracing measures such as eye-movement recordings, verbal reports of thinking, and/or representative task manipulations. Finally, researchers detail the process of skill acquisition during the development of expertise, while considering the practical implications for guiding practice and instruction (for a detailed review, see Williams and Ericsson, 2005). In this study, we developed a laboratory-based simulation to try and capture representative performance in the sport of cricket. The anticipation task contained deliveries from fast and spin bowlers in order to assess how the mechanisms underpinning skilled performance altered with the task constraints. We extended previous research by collecting both eye-movement data and think-aloud verbal reports simultaneously in an attempt to identify better the mechanisms underpinning skilled performance.

Researchers examining advance cue utilisation in cricket have previously identified a strong correlation between anticipation and skill level (Penrose and Roach, 1995). Visual temporal occlusion techniques have been used to present film displays of bowlers' pre-release movement patterns with limited or no ball flight information. Skilled batters have consistently demonstrated superior response accuracy compared to less skilled batters when presented with only pre-release movement information from fast bowlers (Abernethy and Russell, 1984; Penrose and Roach, 1995), and when facing different types of deliveries from left- and right-hand fast (McRobert and Tayler, 2005) and spin bowlers (Renshaw and Fairweather, 2000). The importance of advance information has been further demonstrated using realistic protocols with life-size film images and movement-based response measures or in the natural setting using liquid crystal occlusion glasses (e.g. Tayler and McRobert, 2004; Müller and Abernethy, 2006). Superior speed and/or accuracy of anticipatory response in skilled batters is consistent with observations made on expert performers in other fastball such as soccer (Williams and Davids, 1998) and tennis (Williams *et al.*, 2002), as well in domains such as driving (McKenna and Horswill, 1999) and aviation (Russo *et al.*, 2005).

An abundance of evidence exists to indicate that the extraction of advance cues is a critical component for skilled cricket batting, yet few have examined specifically 'what' sources of relevant information are extracted from the bowler's movement pattern. Spatial occlusion techniques have been employed to present batters with a consistent time course of events while selectively removing specific sources of information for the duration of the action. The information extracted from the bowling arm, hand, and ball locations have been reported to be primary sources for skilled and less-skilled batters when facing fast bowlers, whereas when facing spin bowlers skilled batters appear to rely mainly on information from the bowling arm (Tayler and McRobert, 2004; Müller *et al.*, 2006).

Analyses of visual search behaviours have provided information on individual differences in point of gaze, duration time, and search order. In general, skilled performers use different search strategies and fixate on more informative cues compared to less-skilled counterparts (Williams and Davids, 1998; Williams *et al.*, 2002). However, only a few researchers have recorded gaze patterns in cricket. Barras (1990) reported that skilled batters fixated on the bowling hand and ball until the bowler assumed the 'side-on' delivery position. At this point, they altered their point of gaze to the location above the bowler's right shoulder in anticipation of ball release. Land and McLeod (2000) adopted a field-based approach and demonstrated that, when facing a bowling machine, the skilled batter fixated the early portion of ball flight only. This batter subsequently made a 'saccadic' eye movement to the predicted bounce location then tracked the ball for 200 ms after it had bounced. However, this study was limited due to the slow ball speeds (i.e., 25 m/s) compared to the match situation, the very small sample size (n = 3), and the use of a bowling machine instead of real bowlers.

Think aloud and retrospective verbal reports have been used to gather information about the underlying knowledge structures and in-event thought processes across various domains (for a detailed review, see Ericsson and Simon, 1993). McPherson (1993) examined conceptual knowledge in skilled and less-skilled baseball batters. After being instructed to act as if they were the fourth in the batting order, batters viewed a videotape of the first three batters from each team from a game containing a variety of events (e.g., strike ball count, in-fielders catching balls, types of pitches, outfielders making plays, score). The skilled batters' verbal reports collected while observing the videotaped batters included sophisticated content (e.g., detailed physical cues, abstract tactical conditions) and structure (e.g., pattern of condition-actions rules) that allowed them to monitor changes in game situation and plan for possible actions. Less-skilled batters used a more global approach and monitored a wider variety of events when compared to skilled batters. Although this work provided direct insight into batters' planning strategies, it did not indicate how these strategies were implemented during subsequent action neither did it examine the relationship between these planning activities and anticipation performance.

Typically, perceptual and cognitive skills are inferred from the quality, speed, and accuracy of an individual's performance, with minimal attempt to explain the cognitive processes involved during anticipation. According to Ericsson and Kintsch's (1995) theory of long-term working memory (LTWM), experts have acquired the necessary skills to index and encode information into an elaborate representation stored in long-term memory (LTM). This information remains accessible via the use of retrieval cues in short-term memory (STM). Ward, Ericsson, and Williams (submitted; see also Ward *et al.*, 2003) proposed that these skills and underlying representations provide a dual function; they provide memory support for performance, in the form of planning, monitoring and evaluations, while simultaneously enabling retrieval structures to be built and updated 'on the fly' that promote direct access to task pertinent options. This process allows experts to predict the occurrence and consequences of future events and anticipate the retrieval demands likely to be placed on the system.

In this study, we collected eye movements and verbal reports simultaneously while skilled and less-skilled batters responded to near-life-size video simulations of cricket bowler deliveries. The test film contains trials from fast and spin bowlers, providing a manipulation of task constraints. We hypothesised that skilled batters are significantly more accurate at anticipating ball location than less skilled batters due to their task specific knowledge and previous exposure to a variety of fast and spin bowlers (Renshaw and Fairweather, 2000).

An eye-movement registration system was used to examine the visual search behaviours employed when making anticipation judgements. Previous literature suggests that skilled performers employ more efficient and effective search behaviours when confronted with such tasks in sport and other domains (Williams *et al.*, 2004). We hypothesised that skilled batters employ more fixations to extract information from both distal sources (i.e., bowling arm) and central body features (i.e., head-shoulders, trunk-hips), whereas less skilled batters rely on more distal cues (i.e., ball-hand). Researchers have indicated that manipulations to task constraints have a marked impact on the cognitive and visual search processes underpinning successful performance (e.g., see Williams and Davids, 1998;

Vaeyens *et al.*, 2007). Visual search behaviours were also expected to alter as a function of observing fast or spin bowlers. We hypothesised the primary source of information extraction when facing spin bowlers is the ball-hand, compared to additional central features when facing fast bowlers (see Müller *et al.*, 2006). Finally, protocol analysis on think-aloud verbal reports was employed to provide some insight into the underlying representations held by the batters in LTWM. We predicted that skilled batters would engage in more verbal evaluation, prediction, and deep planning statements, compared to their less-skilled counterparts.

2. METHODS

2.1 Participants

Twenty male cricket batters were recruited. Skilled participants (n = 10; mean age = 25.2 years, SD = 6.8) were professional, county-level batters, and three of these had played at an international level. They had a mean of 14 years (SD = 4) playing experience and had, on average, taken part in 363 (SD = 91) competitive matches. The less-skilled participants (n = 10; mean age = 23.7 years, SD = 4) had a mean of 11 years (SD = 3) playing experience at a recreational level (e.g., local club) and had taken part in 89 (SD = 47) competitive matches. Participants signed an informed consent form and reported normal or corrected to normal levels of visual function. The study was approved by the institution's ethics committee and carried out under its ethical guidelines.

2.2 Materials and apparatus

A simulated task environment (STE) was developed to assess participants' batting performance on a representative cricket task (see Figure 1). The STE, which was managed by LabVIEW (National Instruments Corporation, Newbury, Berkshire, UK), allowed the participants to interact with a life-size, digital video simulation that was back-projected onto a large screen (3 m ´ 3 m). The screen was placed 6 m directly in front of the participant and the angle subtended by the opponent was 18° in the vertical plane and 28° in the horizontal plane, ensuring that the image was representative of the real situation.

Ten male (mean age = 19.5 years, SD = 2.5), county-level cricket bowlers (6 fast; 4 spin) were recruited to create the video-based test stimuli. A digital video camera (Canon 3CCD Digital Video Camcorder XM2 PAL, Tokyo, Japan) was used to record the video stimuli from a first-person perspective. The camera was positioned on the crease at 1.7 m in height and in line with middle stump so that it represented an individual's normal viewing perspective while batting. Each stimulus video included the bowler's preparatory phase, run-up, delivery action, and follow-through (80 ms of ball flight after ball release) and was approximately 8 s in length. In total, 36 video stimuli were created, of which there was a minimum of two and a maximum of six deliveries from each of the bowlers, resulting in 20 fast and 16 spin trials. To record the actual location of the ball when it passed the hitting zone, a second digital video camera was placed 3 m directly behind middle stump and at a height of 0.90 m. This perspective provided the experimenter with the actual x and y position of each bowl in the video stimuli as it passed through a known plane.

Participants' visual search data were recorded using the Applied Science Laboratories (ASL) 5001 eye-movement registration system (Bedford, MA). This is a video-based, monocular corneal reflection system that records eye point-of-gaze with respect to a helmet-mounted scene camera. The system measures the relative position of the pupil and corneal reflection in relation to each other. These features are used to compute point-of-gaze by superimposing a crosshair onto the scene image captured by the head-mounted camera optics. The image is transferred to DVD format and analysed frame-by-frame using a DVD player (Panasonic DMR-E50, Osaka, Japan) with a sampling frequency of 50 Hz. System accuracy was ± 1° visual angle, with a precision of 1° in both the horizontal and vertical fields.

A lapel microphone, telemetry radio transmitter (Sennheiser EW3, High Wycombe, UK), and telemetry radio receiver (Sennheiser EK100 G2, High Wycombe, UK) were employed to collect verbal reports. Verbal reports were recorded onto miniDV tape using a digital video camera, converted into computer audio .wav files and then transcribed prior to analysis.

Figure 1. An example of the experimental set-up used to collect data.

2.3 Procedure

Prior to testing participants were given an overview of the experiment, a demonstration of the STE, and a biographical information sheet. Ericsson and Kirk's (2001) adapted directions for giving think-aloud verbal reports (based on the original instructions from Ericsson and Simon 1993, pp. 375-379) were combined with domain specific warm-up tasks. Participants practiced giving verbal reports with feedback by solving generic and sport specific tasks for approximately 30 minutes ensuring that the criteria for providing level I or II verbal reports were met (see Ericsson and Simon, 1993).

For each trial, participants were instructed to take up their normal batting stance holding a cricket bat and to play a stroke that would intercept the ball's anticipated flight path based on line and length of the delivery observed. After playing the stroke, participants marked in pen the anticipated location of the ball when it passed the strike zone on a paper response sheet that depicted a scaled representation of the view from behind the stumps.

Prior to commencing the actual trials the lapel microphone, telemetry radio transmitter, and head-mounted optics were fitted to the participant and checked that they were comfortable and did not interfere with performance. The ASL eye-movement registration was calibrated using a 9-point grid so that the fixation mark corresponded precisely to the participant's point-of-gaze. A simple eye calibration was performed for each participant to verify point-of-gaze, and periodic calibration checks were conducted before and during testing.

After calibration, participants were presented with six practice trials in the STE. These trials allowed participants to respond to simulated scenarios and provide retrospective verbal reports of their thoughts from the start of the bowlers' approach run to the completion of their response. Additional feedback on giving verbal reports was given when necessary. Participants then viewed all 36 video stimuli in the STE in randomized order. They were instructed to record the ball location on the response sheet after each trial and give retrospective verbal reports on every third trial and 8 additional trials selected at random. The practice and test trials took approximately 90 minutes in total.

2.4 Data analysis

2.4.1 Outcome data

Absolute radial error was measured to determine anticipation response accuracy for each participant. This measure was defined as the distance between the predicted ball location relative to the actual location of the ball in the strike zone. Performance on the simulated anticipation task was analysed using a factorial analysis of variance (ANOVA), with group as the between-participant factor and bowler type as the within-participant factor.

2.4.2 Visual search data

The data were analysed on two separate occasions by the lead investigator with a random sample being analysed by an independent investigator to determine reliability. Inter/intra observer agreement formulas were used to determine the percentage of agreement for percentage viewing time and search rate data (see Thomas and Nelson, 2001). Two skilled and two less-skilled participants were removed from the analysis due to calibration and equipment failure.

Percentage viewing time was the percentage of time spent fixating on each area of the display. The display was divided into eight fixation locations: ball-hand, bowling arm, non-bowling arm, head-shoulders, trunk-hips, legs, predicted ball release, and unclassified. An 'unclassified' category was included to account for all the fixations that did not fall within any of the other areas (i.e., stumps, site screen, or background areas). Percentage viewing time data were analysed using a factorial three-way ANOVA with group as the between-participant factor and bowler type and fixation location as within-participant factors. The intertester and intratester agreement for percentage viewing time were 91% and 93%, respectively.

Search rate comprised the mean number of fixation locations, the mean number of fixations, and the mean fixation duration of each fixation location per trial. A fixation was defined as the period of time (^{3} 100 ms) when the eye remained stationary within 1.5° of movement tolerance (see William and Davids, 1998). The variables were analysed separately using a two-way ANOVA, with group as the between-participant factor and bowler type as the within-participant factor. Search rate data reached an intertester percentage agreement of 94% and an intratester percentage agreement of 97%

2.4.3 Verbal report data

According to Ericsson and colleagues (Ericsson and Smith, 1991; Ericsson and Simon, 1993) the trials that show the largest skill-based difference across anticipation performance on the STE should be the subject of protocol analysis. Consequently, the combined z scores for each trial across all of the performance data was used to identify trials 15, 18, 21 and 24 as the four trials that discriminated most clearly between skill groups.

The transcriptions of retrospective verbal reports were segmented using natural speech and other syntactical markers. The data were initially encoded based on an adapted predicate calculus notation system (e.g., relation [argument 1, argument 2]) using various task-specific relations (e.g., hit, leave, delivery type, foot movement) (also see Ericsson and Simon, 1993; Ward, 2003). Arguments were related to a specific subject, object, or element to which they referred (e.g., bowler, ball pitch). The coding system was developed primarily as a result of the inductive task analysis process.

Verbal reports were categorically coded based on a structure originally adapted from Ericsson (Ericsson and Simon, 1993) and further developed by Ward (2003) to identify statements made about cognitions, evaluations, and planning (including predictions and deep planning). Ward (2003) conceptualized cognitions as all statements representing current actions or recalled statements about current events, and evaluations as some form of positive, neutral or negative assessment of a prior statement. Planning statements were divided into predictions and deep planning. Predictions reflected statements about what would and could occur next and deep planning statements concerned information about searching possible alternatives beyond the next move by strategically developing potential action outcomes.

The verbal report data were encoded on two separate occasions by the primary investigator, with a random sample being encoded a third time by an independent investigator. Inter/intra observer agreement formulas were used to determine the percentage of agreement for verbal report data (see Thomas and Nelson, 2001). The intertester and intratester agreement values were 87% and 89%, respectively. One skilled and one less-skilled participant were removed due to audio failure. Statistical analysis was conducted using a three-way ANOVA. Group was the between-participant factor and bowler type and types of verbal statement were the within-participant factors.

Partial eta squared (h_p^2) values were provided as a measure of effect size for all main effects and interactions. Cohen's *d* measures were reported when making comparisons between two means. Posthoc Bonferroni corrected pairwise comparisons were employed as follow-ups where appropriate.

3. RESULTS

3.1 Outcome data

The ANOVA revealed significant main effects for group ($F_{1,18}$ = 40.88, P < 0.0005, h_p^2 = 0.85) and bowler type ($F_{1,18}$ = 4.44, P = 0.049, h_p^2 = 0.20). Skilled batters (M = 37.3 mm, SD = 2.8) recorded lower error scores when anticipating ball location compared to the less-skilled batters (M = 48.9 mm, SD = 5.9). Anticipation error scores revealed that all batters were more accurate when viewing spin (M = 41.4 mm, SD = 9.5) compared to fast (M = 44.5 mm, SD = 6.9) bowlers. No significant interaction between group and bowler type was observed ($F_{1,18}$ = 0.33, P = 0.57, h_p^2 = 0.02).

3.2 Visual search data

Percentage viewing time data showed no significant main effects for group ($F_{1,14}$ = 3.03, P = 0.10, h_p^2 = 0.18) and bowler type ($F_{1,14}$ = 0.80, P = 0.39, h_p^2 = 0.05). A significant main effect for fixation location ($F_{7,98}$ = 81.7, P < 0.0005, h_p^2 = 0.85) was observed (see Figure 2). Pairwise comparisons demonstrated that more time was spent fixating the ball-hand location compared to all other fixation locations (P < 0.05). In addition, participants spent more time viewing the bowling arm compared to the non-bowling arm and legs (P < 0.05). Finally, participants spent less time viewing the legs than the ball-hand, bowling arm, head-shoulders, trunk-hips, predicted ball release, and unclassified fixation locations (P < 0.05). There were no significant differences between any of the other fixation locations (P > 0.05).

The ANOVA showed a significant Group ´ Fixation Location interaction ($F_{7, 98}$ = 2.64, P = 0.02, h_p^2 = 0.16). These data are illustrated in Figure 3. Less-skilled batters spent more time fixating the ball-hand and unclassified locations compared to the skilled batters (d = 0.57 and 0.51, respectively). In contrast, skilled batters spent additional time viewing the bowling arm (d = 0.72), head-shoulder (d = 0.90), trunk-hips (d = 0.74), and predicted ball release area (d = 0.68) when compared to the less skilled batters.

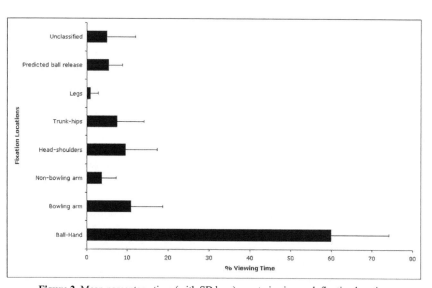

Figure 2. Mean percentage time (with SD bars) spent viewing each fixation location.

The Bowler Type ´ Fixation Location interaction ($F_{7, 98}$ = 2.06, P = 0.056, h_p^2 = 0.13) for percentage viewing time was sufficiently close to conventional levels of significance to warrant discussion (see Figure 4). Participants spent more time viewing the ball-hand fixation location during the spin compared to fast trials (d = 0.59). In comparison, participants spent more time viewing the head-shoulder (d = 0.57) and trunk-hip (d = 0.42) areas during the fast compared with spin trials. No further interactions for Group ´ Bowler ($F_{1, 14}$ = 0.45, P = 0.51, h_p^2 = 0.03) and Group ´ Fixation Location ´ Bowler Type ($F_{7, 98}$ = 0.62, P = 0.74, h_p^2 = 0.04) were observed.

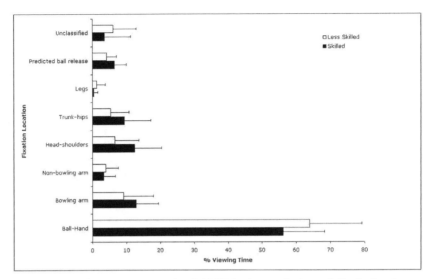

Figure 3. Mean percentage time (with SD bars) spent viewing each fixation location
for skilled and less skilled batters.

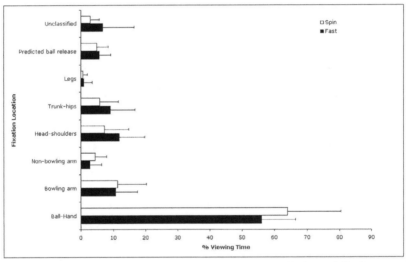

Figure 4. Mean percentage time (with SD bars) spent viewing each fixation
location for fast and spin trials.

ANOVA revealed a significant main effect for group in relation to the number
of fixation locations ($F_{1, 14}$ = 5.05, P = 0.04, h_p^2 = 0.27). Skilled batters (M = 5.1,
SD = 0.5) viewed significantly more fixation locations then their less-skilled
counterparts (M = 4.2, SD = 0.9). However, there was no significant main effect for
bowler type ($F_{1, 14}$ = 0.09, P = 0.78, h_p^2 = 0.01) and no Group ´ Bowler Type
interaction ($F_{1, 14}$ = 0.28, P = 0.61, h_p^2 = 0.08).

A significant bowler type main effect was observed for the mean fixation duration ($F_{1, 14}$ = 9.47, P = 0.01, h_p^2 = 0.40). Participants demonstrated fixations of longer duration when viewing spin (M = 385.1 ms, SD= 100.7) compared to fast bowlers (M = 339.3 ms, SD= 108.5). There was no significant main effect for group ($F_{1, 14}$ = 1.05, P = 0.32, h_p^2 = 0.07) and no Group ´ Bowler Type interaction ($F_{1, 14}$ = 0.07, P = 0.79, h_p^2 = 0.01) for mean fixation duration. Similarly, for the mean number of fixations there were no significant main effects for group ($F_{1, 14}$ = 0.95, P = 0.35, h_p^2 = 0.06) and bowler type ($F_{1, 14}$ = 0.45, P = 0.52, h_p^2 = 0.03) and no Group ´ Bowler Type interaction ($F_{1, 14}$ = 1.83, P = 0.19, h_p^2 = 0.12).

3.3. Verbal report data

The frequency with which the categorical codes (e.g., cognitions, evaluations, predictions, and deep planning) occurred for the four most discriminating trials was assessed for each participant group. There were no significant differences for group ($F_{1, 16}$ = 0.03, P = 0.86, h_p^2 = 0.002) and bowler type ($F_{1, 16}$ = 0.001, P = 0.98, h_p^2 = 0.000) in the total number of verbal statements. A significant main effect for type of verbal statement was observed ($F_{3, 48}$ = 88.13, P < 0.0001, h_p^2 = 0.85). Bonferroni corrected pairwise comparisons showed that participants made more verbal statements coded as cognitions (M = 48.48%, SD = 25.95) than evaluations (M = 39.39%, SD = 16.08), predictions (M = 4.13%, SD = 6.65), and deep planning (M = 7.98%, SD = 8.81) statements (P < 0.05). Secondly, more evaluation statements were coded compared to prediction and deep-planning statements (P < 0.05). There was no significant difference in the number of prediction and deep-planning statements coded (P > 0.05).

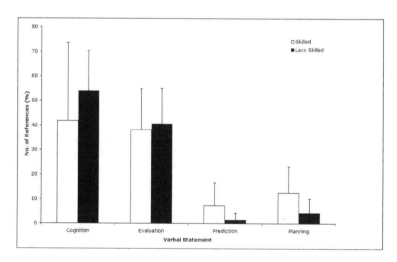

Figure 5. No of references (%)(with SD bars) for type of verbal statement between skilled and less skilled.

There was a significant Group ′ Type of Verbal Statement interaction ($F_{3, 48}$ = 6.52, P = 0.001, h_p^2 = 0.29) in relation to the number of verbal references (see Figure 5). Skilled batters made more statements that were coded as prediction (d = 0.86) and deep planning (d = 0.95) compared to the less-skilled batters. In contrast, less skilled batters made reference to more statements that were coded as cognitions (d = 0.48). The Group ′ Bowler Type ($F_{1, 16}$ = 0.07, P = 0.79, h_p^2 = 0.004), Bowler ′ Type of Verbal Statement ($F_{3, 48}$ = 2.10, P = 0.11, h_p^2 = 0.12), and Group ′ Bowler ′ Type of Verbal Statement ($F_{3, 48}$ = 0.89, P = 0.45, h_p^2 = 0.05) interactions were not significant.

4. DISCUSSION

As an index of the perceptual cognitive mechanisms that underpin expert performance, we collected eye movement and verbal report data of skilled and less-skilled batters during a simulated cricket batting task. Skill-based differences in visual search behaviours and thinking were assessed. A secondary aim was to examine how these indices alter as a function of task constraints. We hypothesised that skilled batters demonstrate superior anticipation accuracy, fixate on more locations and extract information from both distal (i.e., bowling arm) and central body features (i.e., head-shoulders, trunk-hips) compared to less-skilled batters. The latter would focus primarily on the ball-hand of the bowler prior to ball release. We further predicted that visual search behaviours are altered as a function of the task constraints. Batters would fixate more on the ball-hand location when observing spin bowlers compared to central body features when viewing fast bowlers. Finally, we hypothesised that skilled batters make more evaluation, prediction, and deep planning statements.

As expected, skilled players demonstrated superior anticipation skill compared to less skilled performers on the simulated cricket batting task. These findings support previous research where representative real-world tasks have been used to measure anticipation skill in cricket (Penrose and Roach, 1995; Renshaw and Fairweather, 2000). The two groups of batters were less accurate when viewing fast compared to spin bowlers. This finding was somewhat unexpected. We had expected the reverse effect because when facing a spin bowler accurate prediction of where the ball passes through the strike zone typically requires that the batter extracts information from the final portion of ball flight due to movement of the ball 'off the pitch' (Müller and Abernethy, 2006). Moreover, anecdotally the perception in professional cricket is that batters have less exposure to spin compared with fast bowlers, a factor likely to influence the ability of batters to anticipate the outcome successfully (Williams *et al.*, 1999). A possible explanation may be that the spin bowlers employed were less effective than the fast bowlers at disguising the critical cues underpinning anticipation judgements by the batters.

The eye-movement data revealed a number of systematic differences in visual search behaviour between skilled and less-skilled batters. Less-skilled batters extracted information primarily from the ball-hand location. In contrast, skilled batters used a more systematic search strategy, presumably using prior knowledge and experience to direct their gaze towards additional task-relevant sources of information. Skilled batters adopted a more exhaustive search pattern, fixating on more locations so that the primary cue source (i.e., ball-hand) was supplemented with more subtle-task specific information from the bowling arm and central body features (i.e., head-shoulders and trunk-hips). The search for information from more central body cues supports previous research in tennis (Williams *et al.*, 2002). However, skilled cricket batters did spend more time than skilled tennis players extracting information from distal cues (i.e., ball-hand compared to racket). Task-specific task constraints might account for these differences, with the ball-hand being a pertinent cue area for cricket batters because it provides information on how the bowler grips the ball as well as seam location.

The eye-movement data demonstrated that all batters altered their visual search strategy as a function of observing fast and spin bowlers. For the spin bowlers, batters used fixations of longer duration and spent more time extracting information from the ball-hand location compared to fast bowlers. In comparison, all batters spent time fixating on the ball-hand and central body locations (i.e., head-shoulders, trunk-hips) when viewing fast bowlers. These results coincide with those derived from spatial occlusion techniques that have identified the relationship between the bowling hand and arm as being crucial when anticipating a fast bowler's intentions (Tayler and McRobert, 2004; Müller *et al.*, 2006). In addition, our findings suggest that focusing on the shoulder and trunk region provides additional information about the direction of delivery. Müller *et al.* (2006) found that removing the bowler's arm and hand had the greatest decrement on skilled batters' performance when viewing spin deliveries. In conclusion, the ball and hand seem to be the primary source of information when facing spin deliveries, whereas, more central regions provide additional information when viewing fast deliveries.

Protocol analysis of think-aloud verbal reports was conducted to gain some insight into the nature of the batters' cognitive representations. Due to qualitative differences in the nature of the representation between experts and novices, it was predicted that skilled batters make more evaluation, prediction, and deep-planning statements. There was no difference between batters in the number of evaluation statements made. The lack of differences might be due to the lack of sensitivity of the generic evaluation coding category to distinguish between task-relevant and task-irrelevant evaluation statements. Follow-up analysis of this would be needed for confirmation. As expected, the skilled batters made more prediction and deep-planning statements than their less skilled counterparts. The retrospective verbal reports of thinking indicated that skilled batters were able to engage in systematic deep planning of potential outcomes resulting in a higher percentage of correct anticipations.

The current findings have implications for those interested in developing anticipation skill across domains. A growing body of research has highlighted the potential benefits that may be gleaned from using simulated task environments, coupled with instruction as to the important information guiding anticipation, and feedback on task performance (for a review, see Williams and Ward, 2003; Ward *et al.*, 2006). In order to develop such training programmes effectively, knowledge of the mechanisms underpinning successful performance and how these may alter as a function of task constraints and participant skill level is essential (e.g., see Williams *et al.*, 2002). The findings in this report provide an essential step toward establishing a foundation for the development of systematic training programmes that facilitate the acquisition of perceptual-cognitive skills in cricket. Moreover, the systematic approach employed provides a framework for performance enhancement in other, related domains such as the military, aviation, driving, and law enforcement.

In summary, the expert performance approach was adopted to determine differences in expert performance in a representative task domain, and to examine differences in process measures by collecting eye-movements and verbal reports. A novel approach was employed by recording visual search behaviours and verbal reports simultaneously during a simulated batting task. Secondly, the representative cricket protocol examined the effects of task-specific constraints (i.e., fast compared to spin bowlers) on perceptual and cognitive processes. Skilled batters' superior anticipation skill was characterised by more refined and effective visual search strategies and by differences in thinking. These processes allow information to be processed at a deeper level in order to plan an appropriate response strategy. The batters adopted different visual search patterns when facing fast compared to spin bowlers. The potential implications extend beyond the domain of cricket and sport. The development of methods to effectively capture expert performance in different domains helps promote understanding of the general and specific mechanisms underpinning expertise and its development (Ericsson and Ward, 2007; Ericsson and Williams, 2007).

References

Abernethy, B. and Russell, D.G., 1984, Advance cue utilisation by skilled cricket batsmen. *The Australian Journal of Science and Medicine in Sport,* **16**, pp. 2-10.

Barras, N., 1990, Looking while batting in cricket: What a coach can tell batsmen. *Sports Coach,* **13**, pp. 3-7.

Eccles, D. W., Walsh, S. E., and Ingledew, D. K., 2002a, A grounded theory of expert cognition in orienteering. *Journal of Sport and Exercise Psychology,* **24**, pp. 68-88.

Eccles, D. W., Walsh, S. E., and Ingledew, D. K., 2002b, The use of heuristics during route planning by expert and novice orienteers. *Journal of Sports Sciences,* **20**, pp. 327-337.

Ericsson, K.A. and Kintsch, W., 1995, Long-term working memory. *Psychological Review,* **102**, pp. 211-245.

Ericsson, K.A., and Kirk, E., 2001, Instructions for giving retrospective verbal reports. Department of Psychology, Florida State University, Tallahassee, Florida, USA. *Unpublished manuscript.*

Ericsson, K.A. and Simon, H.A., 1993, *Protocol Analysis: Verbal Reports as Data (Revised Edition),* (Cambridge, MA: Bradford books/MIT Press).

Ericsson, K.A., Charness, N., Hoffman, R. and Feltovich, P., 2006, *Cambridge Handbook of Expertise and Expert Performance*, (Cambridge: Cambridge University Press), pp. 243-262.

Ericsson, K.A. and Smith, J., 1991, *Toward a General Theory of Expertise: Prospects and Limits*, (Cambridge: Cambridge University Press).

Ericsson, K.A., and Ward, P., 2007, Capturing the naturally-occurring superior performance of experts in the laboratory: Toward a science of expert and exceptional performance. *Current Directions in Psychological Science*, **16**, pp. 346-350.

Ericsson, K.A. and Williams, A.M., 2007, Capturing naturally-occurring superior performance in the laboratory: Translational research on expert performance. *Journal of Experimental Psychology: Applied,* **13**, pp. 115-123.

Land, M.F. and McLeod, P., 2000. From eye-movements to action: How batsmen hit the ball. *Nature*, **3**, pp. 1340-1345.

McKenna, F.P. and Horswill, M.S., 1999, Hazard perception and its relevance for driving licensing. *Journal of International Association of Traffic and Safety Sciences,* **23**, pp. 36-41.

McPherson, S.L., 1993, The influence of player experience in problem solving during batting preparation in baseball. *Journal of Sport and Exercise Psychology,* **15**, pp. 304-325.

McRobert, A. and Tayler, M., 2005, Perceptual abilities of experienced and inexperienced cricket batsmen in differentiating between left and right hand bowling deliveries. *Journal of Sports Sciences,* **23**, pp. 190-191.

Müller, S. and Abernethy, B., 2006, Batting with occluded vision: an in situ examination of the information pick-up and interceptive skills of high- and low-skilled cricket batsmen. *Journal of Science and Medicine in Sport,* **9**, pp. 446-458.

Müller, S., Abernethy, B. and Farrow, D., 2006, How do world-class cricket batsmen anticipate a bowler's intention? *The Quarterly Journal of Experimental Psychology,* **59**, pp. 2162-2186.

Patel, V.L., Groen, G.J. and Arocha, J.F., 1990, Medical expertise as a function of task difficulty. *Memory and Cognition,* **18**, pp. 394-406.

Penrose, J.M.T. and Roach, N.K., 1995, Decision making and advanced cue utilisation by cricket batsmen. *Journal of Human Movement Studies,* **29**, pp. 199-218.

Renshaw, I. and Fairweather, M.M., 2000, Cricket bowling deliveries and the discrimination ability of professional and amateur batters. *Journal of Sports Sciences,* **18**, pp. 951-957

Russo, M.B., et al., 2005, Visual perception, psychomotor performance, and complex motor performance during an overnight air refueling simulated flight. *Aviation, Space, and Environmental Medicine*, **76**, C92-103.

Tayler, M.A. and McRobert, A., 2004, Anticipatory skills and expertise in cricket batsmen. *Journal of Sport and Exercise Psychology,* S26, **185**.

Thomas, J.R. and Nelson, J.K., 2001, *Research Methods in Physical Activity*, (Champaign, IL: Human Kinetics).

Vaeyens, R., et al., 2007, The effects of task constraints on visual search behaviour and decision-making skill in youth soccer players. *Journal of Sport and Exercise Psychology*, **29**, pp. 147-169.

Ward, P., 2003, *The development of perceptual-cognitive expertise*. Unpublished doctoral thesis. Liverpool John Moores University.

Ward, P., Ericsson, K.A., and Williams, A.M., (submitted), Identifying mechanisms of perceptual-cognitive expertise in an applied domain using think-aloud protocols.

Ward, P., Williams, A.M., and Ericsson, K.A., 2003, Underlying mechanisms of perceptual-cognitive expertise in soccer. *Journal of Sport and Exercise Psychology*, **25**, S136.

Ward, P., Williams, A.M. and Hancock P., 2006, Simulation for performance and training. In *Handbook of Expertise and Expert Performance*, edited by Ericsson, K.A., Hoffman, R.R., Charness, N. and Feltovich, P.J., (Cambridge: Cambridge University Press), pp. 243-262.

Williams, A.M. and Davids, K., 1998, Visual search strategy, selective attention, and expertise in soccer. *Research Quarterly for Exercise and Sport,* 69 (2), 111-128.

Williams, A.M., Davids, K. and Williams, J.G., 1999, *Visual Perception and Action in Sport*, (London: E. & F.N. Spon).

Williams, A.M. and Ericsson, K.A., 2005, Perceptual-cognitive expertise in sport: Some considerations when applying the expert performance approach. *Human Movement Science*, **24**, pp. 283-307.

Williams, A.M., Janelle, C.M. and Davids, K., 2004, Constraints on the search for visual information in sport. *International Journal of Sport and Exercise Psychology*, **2**, pp. 301-318.

Williams, A.M. and Ward, P., 2003, Developing perceptual expertise in sport. In *Expert Performance in Sports: Advances in research on sport expertise*, edited by Starkes, J.L. and Ericsson, K.A., (Champaign, Illinois: Human Kinetics), pp. 220-249.

Williams, A.M., et al., 2002, Perceptual skill in real-world tasks: Training, instruction, and transfer. *Journal of Experimental Psychology: Applied,* **8**, pp. 259-270.

CHAPTER TWENTY-FOUR

Determination of a technical learning line for "Big Air" snowboarding

Evert Zinzen[1], Jerom Pannier[1] and Peter Clarys[2]

[1]Department of Movement Education and Sportstraining,
Vrije Universiteit Brussel, Brussel, Belgium
[2]Department of Human Biomechanics and Biometry,
Vrije Universiteit Brussel, Brussel, Belgium

1. INTRODUCTION

Snowboarding is a relatively new discipline in snow sport. It evolved from a ski-like sport, which in the beginning mainly entailed the "slalom" discipline, but now includes other disciplines such as "snowboardcross", "big air", "half pipe" and "slope style". In this short period, a world circuit was developed with World Cups, World Championships and even Olympic disciplines (FIS, 2007). The evolution towards the more recent snowboard disciplines was only possible due to a fast evolution of the snowboard material (boots, fixations and boards) and snowboard accommodation in snowparks e.g. jumps (different in height and slope), half pipes (U-curved area for jumps and tricks) and handrails (see Figure 1).

The positioning on the board indicates that riding is an asymmetrical skill with, according to individual preference, either the left foot in the front fixation ("regular" position) or the right foot in the front fixation ("goofy" position). Riding "switch" means that the non-preferred foot is in the front position.

Big Air is one of the most recently developed snowboard disciplines with a competition circuit. It is a very popular and spectacular discipline often performed on artificial slopes (Figure 2). In the Big Air discipline the riders perform one jump. Besides amplitude (distance) and altitude (height) of the jump, the number and type of rotations and the landing are scored (FIS, 2007). The scoring system is similar to that used in gymnastics.

The importance of the snowboard circuit is constantly growing (from commercial aspects to Olympic honours) and professionalism is required at different levels. This raises the need to organize a professional framework in which athletes can develop from beginner to world top levels. Within this framework coaches and trainers should be able to spot talented youngsters. Talent identification (TID) has been described for several sports e.g. soccer (Williams and Reilly, 2000), swimming (Richards, 1999; Hoare, 2001), athletics (Henning *et al.*, 2006), gymnastics (Irwin *et al.*, 2005), physical education (Bailey and Morley, 2006) and more in general by Vrijens *et al.* (2001), Wylleman *et al.* (2004), Loko (2007), Malina (2007) and others.

The above cited authors agree that talent is composed of different aspects and evolves with age. Talent should therefore be identified at an early age, followed by a selection of the talented youngsters and further developed into excellence (Hoare, 2001; Vrijens *et al.*, 2001). This way, a talented youngster will start in a "participation programme" where all levels of athletes are present. In this stage he or she should be detected and move from participation programmes into a "talent programme" where only youth with talent is selected. From there on, the real elite athletes can be chosen and followed up to excellence (Hoare, 2001). Wylleman *et al.* (2004) described the evolution of an average top athlete on four levels:- athletic, psychological, psychosocial and academic all of which have a different evolution according to the age of the athlete (Figure 3).

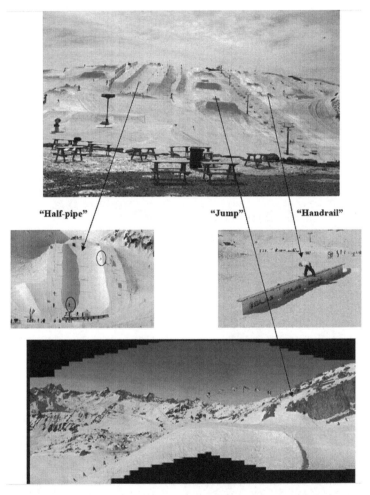

Figure 1. Overview of a snowpark (top) with a half pipe (middle left), handrails (middle right) and a jump (used also in Big Air).

Figure 2. An artificial jump, typical for World-Cup Big Air competitions.

AGE	10	15	20	25	30	35
ATHLETIC LEVEL	INITIATION	DEVELOPMENT	MASTERY		DISCONTINUATION	
PSYCHO-LOGICAL LEVEL	CHILD-HOOD	ADOLESENCE	ADULTHOOD			
PSYCHO-SOCIAL LEVEL	PARENTS/ SIBLINGS: PEERS	PEERS/ COACH/ PARENT	PARTNER/ COACH		FAMILLY (coach)	
ACADEMIC/ OCCUPATION-AL LEVEL	PRIMARY EDUCATION	SECUNDARY EDUCATION	HIGHER ERDUCATION	VOCATIONAL TRAINING/ PROFESSIONAL OCCUPATION		

Figure 3. Talent development model according to Wylleman, Alferman and Lavallee (2004, with permission).

It is on the athletic level that the Flemish (Belgian) elite sport coordinator defined a *technical learning line* as *"an average evolution in technical performances of top-level athletes"* (Van Aken, 2005). He added that the fastest and slowest evolving athletes have to be taken into account to derive a *technical learning zone*. He considered this technical learning zone together with learning zones on the tactical, conditional and mental evolution of an athlete as indispensable in the prediction for elite success.

Although substantial research is done in the area of talent detection in different sports, publications concerning technical learning lines in snowboard are still lacking. It was the aim of this study to try to determine a technical learning line (and zone) and thereby provide Big Air snowboard coaches with information on what age their talented youngsters should start and how they should evolve technically in time. This study also questions if this learning line could be influenced by:

- the evolution in snowboard equipment: the older elite Big Air athletes started their training with less developed snowboard materials and accommodations,
- by different starting ages and
- the availability of natural snowboard conditions in the home country.

2. METHODS

To determine this technical learning line, the world's top Big Air Snowboard riders were questioned retrospectively. Therefore, a classification of all available snowboard (Big Air) techniques into ten levels was developed in collaboration with an international expert panel (coaches of world cup riders, ex-world cup riders and decision makers in selection). The scoring system of the FIS (Fédération Internationales de Ski, International Ski Federation) was used as basis for the classification.

The 10 levels were:

1. basic snowboarding techniques (able to descend every slope)
2. freestyle techniques (ollie, 180° spin and flat tricks); Doing an ollie – using the tail of your board as a spring – gets you in the air without using a jump.
3. straight air, 180° or 360° spins over a flat distance of 3 m, front or backside
4. **180°, 360° or 540° spins over a flat course of less then 10 m, front or backside**
5. 180°, 360° or 540° spins in switch backside or cab over a flat section less then 10 m: Riding switch means that you are riding with the opposite foot forward than you normally would; switch backside = riding switch followed by a backside rotation after take-off. Cab = riding switch followed by a front side rotation after take-off.
6. **180°, 360° or 540° spins over ± 15 m in front and/or backside**
7. 180°, 360°, or 540° spins in switch backside or cab over a flat section of ±15 m
8. 720° or 900° spins over a flat course of ±15 m, front or backside
9. **720° or 900° spins switch backside or cab over ± 15 m**
10. (absolute top) 1080°, 1260° spins front and/ or backside, cab and/or switch backside, over a flat section of ± 15 m

The different levels were explained to the riders who filled out the questionnaire. The levels mentioned in **bold** were chosen by the expert panel as landmarks in the evolution of a world athlete and will be used in further analysis. These levels were chosen as landmarks because of the important differences between them in the distance jumped, the amount of spins, and regular or switch techniques.

Before introducing the questionnaire in the World Top Big Air riders, feasibility was checked in five Belgian low-level Big Air riders. Riders were questioned during different FIS (International Ski Federation) World-Cup events each time before participating in a contest. These events were at the end of the Big

Air competition season. Due to practical (time) and coaching reasons (not all were prepared to reveal their history to a fellow competitor) only 34 of 95 FIS World Cup Big Air riders participated. These riders (all males, since a female World Cup Big Air competition does not exist) were asked, besides some general snowboard demographics (age, years in world cup, nationality, regular/goofy and so on), to indicate the age at which they reached the different defined levels. Age was defined as the age they were at the time of the competition season (winter time).

Descriptive statistics (percentiles) were used for the construction of the learning line and learning zone. A t-test was performed to evaluate differences in time needed to attain the landmark levels between junior (<20 years old) and senior (≥ 20 years old) riders. Possible differences should give an indication of the influence of the evolution of the snowboard materials. Juniors started snowboarding with much more sophisticated material (e.g. better designed jumps, improved softboots, boards and fixations) than the seniors did (i.e. often self-made jumps, hardboots, first generation snowboards). Differences in time needed to attain the landmark levels between riders coming from a country with or without natural snowboard conditions, were examined with a t-test and indicate if there is an advantage (or not) to living in an alpine country. Finally, differences in time needed to achieve the landmark levels between starting ages, were all checked with t-tests or Kruskall-Wallis (not normal or too little data) tests. This will give an indication whether starting at a younger age is beneficial or not. Prior to all statistical tests, normality was checked and proven by a Kolmogorov-Smirnov one-sample goodness of fit test. All significance levels were set on 5% and SPSS 15.0[©] was used to perform these analyses.

3. RESULTS

The world cup standings (rankings) of the Big Air riders were: 1, 2, 3, 4, 5, 7, 9,10, 13, 25, 26, 27, 30, 34, 35, 38, 40, 43, 46, 48, 56, 61, 63, 64, 69, 72, 74, 81, 85, 87. Four participants were not ranked.

On average, Big Air riders questioned had started snowboarding at 11.8 ± 2.6 years of age, participated in their first competition at the age of 15.3 ± 2.7, became professional at 17.0 ± 3.4, became a member of the national team at 18.2 ± 3.1 and participated for the first time in a FIS World cup at the age of 18.9 ± 2.8 years. At the time of questionnaire administration, the riders were 21.1± 3.2 years old. Numbers after ± are standard deviations.

By calculating the 50 percentile (median) age for every level we were able to construct the base of the **learning line**. When adding the fastest and slowest learner the **learning zone** is provided (Figure 4).

In general, the top level Big Air riders started around the age of 12 and needed on average 4.2 years to reach landmark level 4. To attain level 6 they needed on average 5.7 years, for level 9 this figure was 8.4 years and it took on average 9.7 years to reach level 10 (Table 1).

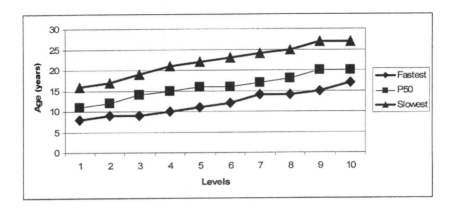

Figure 4. Fastest, slowest and median Technical Learning Line for Big Air Snowboard techniques. The area between the fastest and slowest learner is considered the Technical Learning Zone.

Table 1. Years needed to reach critical levels.

	N	Minimum	Maximum	Average	Std. Dev.
Level 4	34	1	8	4.2	1.6
Level 6	33	2	10	5.7	2.0
Level 9	27	3	13	8.4	2.4
Level 10	12	3	16	9.7	3.2

In order to answer the first question arising from Table 1: "Are these needed times influenced by the starting age of the riders", we divided the snowboarding group into two starting age categories: 8-12 years (N=19 for level 4 and 6; N=17 for level 9 and N= 9 for level 10) and 10-14+ years (N=15 for level 4; N=14 for level 6; N=10 for level 9 and N=3 for level 10). The results are shown in Figure 5. Significantly shorter times (P<0.05) were needed to reach the higher levels when the athletes started snowboarding at a later age. To attain level 4 or 6, no differences in time could be noted for the different starting ages.

As mentioned in the methodology, possible evolution influences of the snowboard equipment were measured by dividing the snowboarders into juniors (< 20 years of age: N=20) and seniors (≥20 years of age, N=14). Results can be seen in Figure 6. Again some absolute differences were found but the t-test did not reveal anything significant.

Finally the influence of living in a country with natural snowboard conditions was investigated. Again the experimental group was split according to natural snowboard conditions (N= 26) versus artificial conditions (N=8). Again no statistical differences could be found (Figure 7).

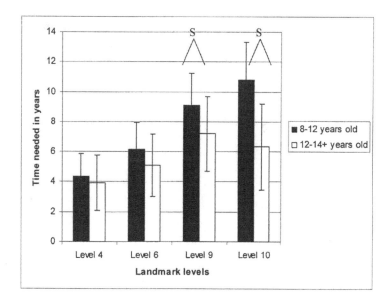

Figure 5. Time to reach landmark levels as a function of starting to snowboard age.

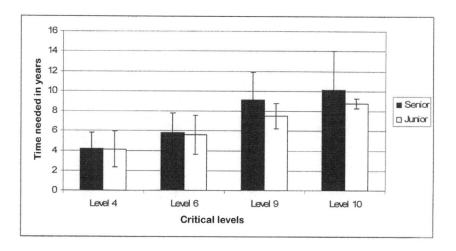

Figure 6. Time to reach landmark levels for senior and junior riders.

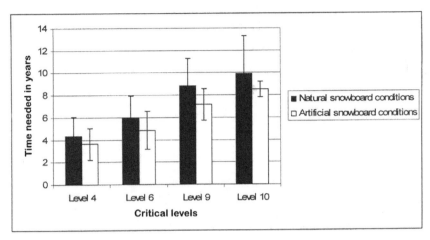

Figure 7. Time to reach different landmark levels: natural versus artificial snowboard conditions.

4. DISCUSSION

The construction of the levels can be considered as subjective. Partially they might be, but care was taken in constructing them in conjunction with snowboard Big Air experts. They also reflect the criteria (at least the higher levels) of the scoring system which is used in Big Air. Since no other ranking of Big Air techniques is available to our knowledge, the levels are considered as a guideline for Big Air techniques with a gradually increasing difficulty. No comments on the choice of techniques for the different levels were made by the sampled riders, indicating that they could agree. The choice of the landmarks may be even more subjective. They also were carefully chosen by Big Air experts but were, post hoc, mainly chosen for statistical purposes in this study. Since this study is considered as a pilot, it was believed to be a good choice to display the results with the landmarks as a guide.

Figure 8 shows that results found in the Big Air riders fit the model of Wylleman *et al.* (2004) although the "mastery" phase seems to start earlier (around 17 in stead of 20 years) for these Big Air riders. In comparison with other models such as Vrijens *et al.* (2001) it was noted that the Big Air riders started to snowboard in the second phase (initiation phase according to Wylleman *et al.*, 2004) of their talent development. They started to participate in different competitions in the "third" phase. This confirmed the theory that in the first phase of talent development (before the age of 10) attention is paid to development of general motor skills. Later on, their talent was selected in the second phase and developed in the third, in accordance with the statements of Hoare (2001).

It is notable that, when the Big Air riders start to snowboard at a later age (12 - 14+) they need less time to reach the top levels. Until level six, the improvement is quite similar but, from there, significant less time is required to reach the higher levels (7.2 in stead of 9.1 years to reach level 9 and 6.3 instead of 10.8 years to reach level 10). The data for level 10 should be interpreted with caution since only three riders started at a later age than 12. These three were however normally distributed but in order to reach statistical soundness a Kruskall-Wallis test was used instead of a t-test.

Starting age does seem to influence the learning time. It could be assumed that starting at a younger age lengthens the "mastery" phase or performances at top-level. Hence, data suggest that, to reach the top, the young riders lose partly their advantage of starting young by needing more time to reach the top levels. This indicates that the techniques in the higher levels require motor skills learned at a later age.

No data were available in this study concerning the "discontinuation" age of top-level Big-Air riders. The FIS points (VSSF, 2006) achieved by the top ten level Big Air riders indicate that, in the season 2005-2006, some riders continued to the age of 28 (Figure 9).

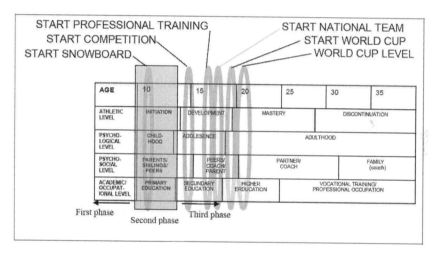

Figure 8. Evolution of the questioned World Cup level Big Air placed in Wylleman, Alferman and Lavallee (2004) model. Phases according to Vrijens, Bourgeois and Lenoir.

The riders of the VSSF study entered the top-ten FIS ranking at the age of 17 which is consistent with the results found in this study. From age 17 up to 22 the FIS points gathered increase continuously and, from then on, display a plateau up to the age of 28, the oldest active Big Air snowboarders at that time. This could indicate that the majority of the riders questioned for this study were rather young and still have about seven years to go in the top level Big Air circuit. The results from our study together with the FIS ranking indicate that the "mastery" level, as brought forward by Wylleman *et al.* (2004), lasts at least until the age of 28 which is completely in accordance with their proposed model.

As mentioned in the introduction, snowboard competition is evolving in a very fast way. Therefore it was assumed that the older riders did not start with the same qualitative materials as the younger ones. Since these new materials (boards, boots, take-offs and do on) made it possible to increase the degree of difficulty and therefore the spectacle of the jumps performed, it was hypothesized that the junior riders have a faster learning curve. However, a decrease in learning time could not be statistically proven. This could be due to a fast adaptation to the new material by the more experienced senior riders.

Being themselves from a country with no natural snowboard conditions, the authors wondered if this fact had a negative influence on the learning line or on the time the riders need to perform at different landmark levels. This seems not to be the case since no statistical differences between riders from countries with natural snowboard conditions and countries with artificial slopes, could be found. Big Air competitions are mostly performed on artificial slopes, which can be the reason for a lack of differences. On the other hand, this leads to the conclusion that natural snowboard conditions do not decrease the learning line; it appears that the basic snowboard skills and Big Air skills in particular, can be acquired in a comparable way in artificial snow conditions as in natural areas for the world-cup talented riders.

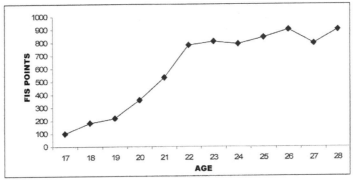

Figure 9. Average amount of FIS points of the top ten Big Air riders according to age during the snowboard season 2005-2006 (VSSF, 2006).

Although these results provide a convenient tool for Big Air snowboard coaches, caution is needed when interpreting them. First of all every athlete should be considered as an individual with his own individual learning line. When this line does not fit in the presented learning zone, it will not necessarily mean that this rider is not talented enough to reach the world top. This should always be a cautious decision taken by the coach, the athlete and his acquaintances. On the other hand, if the learning line fits the proposed learning zone, it does not guarantee that the athlete will reach the top since talent is much more than acquiring the required skills. A fitted learning line, however, indicates that the acquisition of the technical skills follow the expected curve and, if the other talent determining factors are favourable, it is likely that a top level athlete is developing.

Another reason to be careful when interpreting these results is the retrospective nature of the study. This approach is always subject to memory effects. Therefore a prospective study should confirm this learning line.

Finally caution must be expressed with respect to the small sample size. The lack of statistical differences could be explained by the low number of subjects in certain subgroups. Hopefully, these results will lead to an increasing readiness of the Big Air riders to participate in a follow-up study.

5. CONCLUSION

The snowboard Big Air techniques were divided into ten levels. The time needed to reach the top-level is approximately ten years, regardless of the starting age. An evolution over the different levels was made, taking into account the fastest and slowest learning line of the riders. By doing so, a technical learning zone can be presented (Figure 4). This zone is not influenced by the evolution of equipment design nor by the fact that the riders have their roots in a country with natural snowboard conditions or not. The zone is influenced by the starting ages of the athletes where it is noted that the delay in time is compensated by a faster learning line at the higher technique levels.

Users of these results should take in account the individual characteristics of the athletes (riders), the retrospective aspect of this study and the relatively small sample size. Further research on a larger sample, and preferable prospectively, should confirm the data reported here.

References

Bailey R. and Morley, D., 2006, Towards a model of talent development in physical education. *Sport, Education and Society*, **11**, pp. 211-230.

FIS (2007), Rules for the FIS snowboard continental cups 2007/2008, http://www.fis-ski.com/data/document/cocrule08.pdf . Accessed March 2007.

Henning, J.M., Weidnert, T.G. and Jones J., 2006, Peer-assisted learning in the athletic training clinical setting. *Journal of Athletic Training*, **41**, pp. 102-108.

Hoare, J., 2001, SISA Talent Identification and Selection Manual, http://www.srsa.gov.za/ClientFiles/SISA%20Swimming.doc. Accessed March 2007.

Irwin, G., Hanton, S. and Kerwin, D.G., 2005, The conceptual process of skill progression development in artistic gymnastics. *Journal of Sport Sciences*, **23**, pp. 1089-1099.

Loko, J., Talent Selection Procedures, http://www.athleticscoaching.ca/UserFiles/File/Sport%20Science/Youth%20Development/Talent%20Identification/Loko%20Talent%20Selection%20Procedures.pdf Accessed March 2007.

Malina, R.M., Talent Identification and Selection in Sport, http://road.uww.edu/road/jonesd/Coaching-463/Readiness%20for%20Sport/Talent%20Identification%20and%20Selection%20in%20Sport%20--%20Spotlight,%20Spring,%201997_files/Talent%20Identification%20and%20Selection%20in%20Sport%20--%20Spotlight,%20Spring,%201997.htm. Accessed March 2007.

Richards, R., 1999, Talent identification and development. *ASCTA Convention 1999*, pp. 1-17.

Van Aken, I., 2005, Topsportstructuur sportfederaties (Structure of topsport in the sportfederations), press presentation in October 2005.

Vrijens J., Bourgeois, J. and Lenoir, M., 2001, Basis voor verantwoord trainen (Fundamentals for responsible training), Gent, PVLO, pp. 308-358.

VSSF, 2006, Topsportactieplan Snowboard VSSF, KBSF (Action plan for topsport snowboard by the Flemish (VSSF) and Belgian (KBSF) ski and snowboard federations), report of the Flemish ski and snowboard federation (Brussels: VSSF).

Williams, A.M. and Reilly, T., 2000, Talent identification and development in soccer. *Journal of Sports Sciences*, **18**, pp. 657-667

Wylleman, P., Alfermann, D. and Lavallee, D., 2004, Career transitions in sport: European perspectives. *Psychology of Sport and Exercise*, **5**, pp. 7-20.

CHAPTER TWENTY-FIVE

Fundamental movement skill development of four to six year-old pre-school children in Flanders

W. Cools[1], K. De Martelaer[1], B. Vandaele[1,3], C. Samaey[1] and C. Andries[2]

[1]Department of Movement Education and Sport Training, Vrije Universiteit Brussel, Belgium
[2]Department of Developmental and Lifespan Psychology, Vrije Universiteit Brussel, Belgium
[3]Department of Physical Education Teacher Education, Erasmus Hogeschool Brussel, Belgium

1. INTRODUCTION

Early childhood is a sensitive period for the development and refinement of fundamental movement skills [FMS] in children, including object control, locomotion and stability movement skills (Gallahue and Ozmun, 2006). This development is an interactional process affected by constraints within the task, the child and the environment (Haywood and Getchell, 2005).

The importance of well mastered FMS is multiple. On the one hand, FMS are required for proper day-to-day living (Davies, 2003). Even during early childhood, children use and refine their FMS by exploring their surroundings, playing voluntary and engaging in movement activities, while later in adolescence and adulthood well mastered FMS are still of significant importance e.g. keep one's balance on a bus, use a screwdriver, climb a ladder, engage in physical activity [PA] (with the children) and so on. On the other hand, lifelong recreational and / or competitive sports and participation in physical activity demand that a number of FMS (preceding the development of specialized movement skills) have developed towards their mature pattern (Gallahue and Ozmun, 2006).

The development of FMS is interrelated directly with PA (to type as well as to amount) and indirectly with health and socio-affective factors. Children who are more physically active are more skilful than their sedentary peers (Ketelhut et al., 2005; Medeková et al., 2000; Wrotniak et al., 2006). 'Rough and tumble' play in young children and vigorous activities which children generally participate in, are decisive factors for the development of FMS (Sääkslathi et al., 1999). In early childhood, participation in PA is negatively related to cardiovascular risk factors (Sääkslathi et al., 1999; Ketelhut et al., 2005) and postural diseases (Weiss et al., 2004). The mastering of FMS is also related to higher rates of self-esteem and self-image (Bunker, 1991). In addition, lifetime wellness is strongly correlated to regular PA (Le Masurier and Corbin, 2006).

There is overwhelming evidence that time spent in sedentary activities among young children is too much and that many young children do not meet daily PA recommendations (Certain and Kahn, 2002; Nelson and Gordon-Larssn, *et al.*, 2006; Reilly *et al.*, 2006). Therefore, there is reason to be concerned about an increase of hampered and delayed FMS development during early childhood and so the development of FMS among pre-school children should be closely monitored. So far, only a few studies on the development of FMS of typically developing pre-school children have been published (Lamon, 1989; Krombohlz, 1997; Sigmundsson and Stølan, 2003; Booth *et al.*, 2006). Studies on the development of FMS mostly focus on specific target groups, e.g. children with developmental delays or disorders (Van Waelvelde, 2000), children born after Intercytoplasmic Sperm Injection or In Vitro Fertilization (Ponjaert-Kristoffersen *et al.*, 2005) and so on. Vanvuchelen, Mulders and Smeyers (2003) concentrated on the scarcity of scientifically-founded psychomotor assessment tools which include Flemish normative data. Normative data on FMS developmental status of typically developing children, however, are important to assess a child's natural and nurtured developmental process. The data can be used by parents, caregivers, teachers and so on, to guide them in their choice of appropriate activities, e.g. curriculum, education and activity planning and to enhance practical-oriented diagnosis. In summary, having a thorough knowledge of FMS development of pre-school children is indispensable to provide the children with appropriate opportunities to be physically active.

Currently, empirical research addressing the issue of movement skill development of pre-school children is rather limited in number as well as in scope (Jongmans, 2004). Therefore, the first aim of this study is to report on FMS development among a clustered sample of Flemish pre-school children. In addition, we also aim to report prevalence of gender, age and anthropometric differences in FMS development. As there is no general agreement on the presence of gender differences (Krombholz, 1997; Sigmundsson and Rostoft, 2003; Simons and van Hombeek, 2003; Kroes *et al.*, 2004; Van Waelvelde *et al.*, 2008), we aim to report on gender differences in FMS development. Because of rapid growth and development during the pre-school years, some authors believe that it is important to use age category classification in ½ year categories or smaller (Zimmer and Volkamer, 1987; Follio and Fewell, 2000; Ulrich, 2000). In this study we aim to provide supporting evidence for this classification. Obesity is believed to be a restricting factor in developing FMS in young children (D'Hondt *et al.*, 2007), this study aimed to report on supporting evidence on this issue.

2. METHODS

2.1 Subjects

The study was conducted in Flanders (the northern part of Belgium) and Brussels.

A representative sample (Cools *et al.*, 2007a) was drawn from Dutch-speaking *preschools* in this area. A clustered sample of pre-school children (of 40 preschools) and their parents participated in the study. In Flanders the preschool has three official pre-primary school classes Pre-K 1, 2 and 3. Children are allowed starting from the age of 3 to enter Pre-K 1 and most of the children enter elementary school at the age of 6 years. In this study children were recruited in Pre-K 2 and 3.The assumption of valid sample drawing through preschool sampling was based on the fact that, in Belgium, almost every 4-year-old child [>99%] (Eurydice, 2005) is enrolled in pre-primary education. We included 6-year-old children who still attended preschool in the sample, because the focus of this study is pre-school children's FMS development. Further, this study has focused on the majority group of Dutch-speaking respondents. Children of whom both parents did not master the Dutch language were not included in the study. The collected data are part of a larger study on the relation between FMS development of pre-school children aged 4 to 6 and the opportunities to be physically active in their home and school environment, which included the prerequisite that at least one of the child's parents mastered the Dutch language.

The study was approved by the medical ethical committee of the Vrije Universiteit Brussel. First, consent for participation in the study was obtained from the school's principals and teachers. Then, written informed consent was obtained from the custodial caregiver(s) for each child participating in the study. Verbal assent was obtained from each child individually by asking the pre-school child if he or she would like to participate in similar activities to physical education lessons. Response rates for participation (based on parental consent) ranged between 70% and 82% with an average of 77% (See Table 1).

Table 1. Participating pre-school children and corresponding response rates.

Province	Invited	Received Informed Consent	Response Rate
Antwerpen	408	304	0.82
Vlaams Brabant / Brussels Hoofdstedelijk gewest	325	245	0.75
West Vlaanderen	345	242	0.84
Oost Vlaanderen	257	215	0.70
Limburg	302	248	0.75
Total	1637	1254	0.77

Detailed information on the pre-school children revealed that 46 children did not meet the test criteria (e.g. children younger than 4 years of age, absence on the test day, unwillingness to participate), so these data were not included for further data analysis. Finally, 1208 pre-school children between 4 to 6 years of age (654 boys and 554 girls) were included for data analysis in the study. In Table 2 anthropometric and gender data are presented. The age of pre-school children was calculated by means of their birth day in relation to the day of the assessment. There were no corrections for children born preterm or low birth weight, because there were no significant differences in movement skill development for preterm birth or low birth weight (n = 57) in this sample (U = 69649,5, ns).

Table 2. Anthropometric and gender sample data.

Age [years]	5.18 (±0,61)
Mass [kg]	20.02 (±3,45)
Length [m]	1.111 (±0.067)
Gender	654 male (54%) 554 female (46%)

2.2 Instrumentation

The pre-school children were weighed and measured preceding the assessment conforming to standard procedures (Roelants and Hauspy, 2004). Body mass was measured using a digital balance (SOEHNLE 7311, accuracy 100 grams), length was measured with a standard tape measure (accuracy 1 mm). Pre-school children were weighed barefoot wearing light clothes, i.e. t-shirt and shorts. Flemish reference data were used (Roelants and Hauspy, 2004) to place the children in five categories [extreme underweight, underweight, normal, overweight and obese] (Table 3).

Table 3. Under- and overweight prevalence among pre-school children.

BMI categories	n
Extreme underweight	14
Underweight	109
Normal weight	803
Overweight	191
Obese	84

Development of FMS was assessed by the Motoriktest für Vier- bis Sechs-jährige Kinder [MOT 4-6] (Zimmer and Volkamer, 1987). The MOT 4-6 is an assessment tool on FMS development and specifically designed for children of preschool age (Cools *et al.*, 2007b). The test features 18 test items, including: jumping forward in a hoop, keeping ones balance (forward), placing dots, grasping tissue with the toes, jumping sideward, caching a stick, carrying balls from box to box, keeping one's balance (reverse position), throwing at a target disk, collecting matches, passing through a hoop, jumping on one foot in a hoop and standing in balance on one leg, catching a tennis ring, jumping jacks, jumping over a rope, rolling around the length axis, standing up holding a ball on the head, and jumping and turning in a hoop. A detailed overview on the FMS items that are measured by MOT 4-6 is presented in appendix 1. The first test item is not rated because it is used to accustom the children to the assessment. The test is product oriented and refers to a norm. Although qualitative aspects of movement are not scored separately, they are somehow included in the test protocol. Test administration time is about 15 to 20 minutes. The scoring system of the MOT 4-6 has a three-point range for each item. The raw performance scores range from 0 (skill not mastered) to 2 (skill mastered) on each of the test items. The total motor score (raw score) is calculated by summing the item scores, generating a possible total of 34, with higher scores indicating better movement skill development. Most importantly, conversions of the total score used, are percentiles and Motor Quotients. The Motor Quotients are conversions of the total raw scores that level out age differences.

2.3 Selection and training of testers

Twenty-two examiners were involved in the data collection. All examiners were students, enrolled in their final year of physical education, physiotherapy or early childhood teacher education.

All examiners received a standardised training on the administration procedure of the MOT 4-6. The standardised training consisted of two meetings. First there was an information meeting on the aim and process of the research project and the MOT 4-6. In preparation of the training meeting the examiners had to study the manual of the MOT 4-6 in detail. During this second meeting a video modelling correct test administration was used. Correct installation of the test battery was practised (in an unfamiliar room, to approach the real research situation). This practise was filmed and discussed in group afterwards. Additionally an overview of frequently committed errors with the corresponding corrections was given in a written document. After this training the examiners were asked to score 51 individual performances of the different test parts (17 items, each 3 times). Incorrect scoring was corrected and discussed. During the initial data collection a test expert visited each examiner once, to verify correct test installation, appropriate test use and to score the test simultaneously. No differences in scoring appeared.

The training procedure was tested for inter-rater reliability. Scores of one examiner, trained as previously described, were correlated with those of a test expert for a group of 42 pre-school children (20 boys, 22 girls). The pre-school children's mean age was 4.9 years. There was a significant relationship between the scoring of the examiner and the scoring of the test expert, $\tau = 0.93$, P< 0.01. The correlation coefficients (τ, Table 4) of item scores ranged from 0.62 - 1.0 (P<0.01).

Table 4. Inter-rater reliability, Kendall's Tau correlation coefficients for MOT 4-6 test items and total score.

Item	τ	Item	τ
2	0.84**	11	0.94**
3	1.00**	12	0.97**
4	0.77**	13	0.88**
5	0.97**	14	0.62**
6	0.90**	15	0.94**
7	0.97**	16	0.85**
8	1.00**	17	0.89**
9	0.97**	18	0.93**
10	0.97**		
		Total	0.93**

** significance P< .01

2.4 Procedure

Data were collected between January 2005 and October 2006. Examiners visited the school for 2 or 3 days, depending on the number of pre-school children to be tested. Pre-school children who were absent on the test day, were not re-invited for participation. The examiners worked in pairs, so each participating preschool was visited by two examiners. In every school there was a separate room available with a minimum size of 4 by 6 m. Precautions were taken to limit distraction of the children to a strict minimum during the assessment. The children were picked-up and brought back to their classrooms two by two. Before starting the assessment, the pre-school children were asked for their oral consent by putting down the question: "Would you like to participate in a PE-class like activity?" Each examiner assessed one child at a time.

2.5 Data analyses

Descriptive statistics were used to present FMS development of 4 to 6 year old pre-school children. In agreement with the test manual the data of children were presented in half-year age categories. Assumptions for parametric tests were not met to compare groups on age category differences nor on BMI category differences, nor on gender differences, so non-parametric tests were used for further analysis of these data. Kruskal-Wallis tests were used, followed up with Mann-Whitney tests and corrected for type I error rate using the Bonferroni method. We selected a 5% level for statistical significance. Approximate effect sizes were calculated using the equitation, $r = Z/\sqrt{N}$ (from Field, 2005, p.532). Raw scores where used for analysis when age category of the children was considered, MQ were used when the whole sample was considered.

3. RESULTS

3.1 General movement skill development and age differences

Conforming to the test manual's subdivision by ½ year categories, the pre-school children's total scores are presented in Table 5. The pre-school children achieved higher scores with increasing age. The pace of FMS improvement was higher in the first three age categories. (mdn =10-13-17 compared to mdn =19.5-21-22).

Table 5. Pre-school children's raw scores on MOT 4-6.

Age category		4 - 4 ½ 1	4 ½ -5 2	5 - 5 ½ 3	5 ½ - 6 4	6 - 6 ½ 5	6 ½ - 7 6
n	1208	207	291	307	298	95	10
Median (IQR)		10(6)	13(6)	17(8)	19,5(7)	21(8)	22(7,75)
Percentiles	3	2	4	6	9,97	6,88	16
	10	5	7	9,8	13	13,6	16,1
	25	7	10	12	16	17	20
	50	10	13	17	19,5	21	22
	75	13	16	20	23	25	27,75
	90	16,2	19	23	26	27	30,9
	97	20	22	26	28,03	31,12	31

IQR: interquartile range

Movement skill score was significantly affected by age (H (4) = 411.76, P< 0.001). Children in following age categories showed significantly improvements in movement skill (age category 1-2, U= 20370.50, P < 0.001, r = 0.28; age category 2-3, U = 27748.50, P < 0.001, r = 0.33; age category 3-4, U = 31076, P < 0.001, r = 0.27; age category 4-5, U= 11600, P < 0.01, r = 0.13) The effect sizes ranged from small to medium, explaining between 1% and 11% of the variance. The Bonferroni method was applied to correct the type I error rate, so all effects were reported at a 0.01 level of significance. Between age category 5-6, the age difference no longer appeared (U = 379.5, ns). These results are shown in Figure 1.

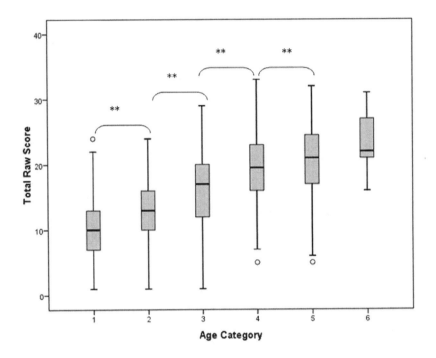

Figure 1. Boxplots of Total raw scores for the age categories of the MOT 4-6.
The box plots represent the distribution of the children's total raw scores over the age categories. The lowest and highest scores are shown by the bottom and top horizontal lines. The interquartile range or the middle 50% of scores is shown by the box, while the bottom and top quartiles are the distances between the boxes and the lower and upper horizontal lines. The middle horizontal line (in the box) represent the value of the median.
** P < 0.01

3.2 Motor Quotients [MQs]

In Table 6, descriptive results on Motor Quotients are presented. The largest group of pre-school children (59%, n= 714) were found in the 'normal' category, they were referred to as 'typically developing' pre-school children. A group of 422 (35%) pre-school children had lower scores, while a group of 72 (6%) had higher scores than the 'normal' category. When the two extreme categories were focused on separately, the category 'very good' FMS development only included a group of 4 children, while the category 'weak' included 92 pre-school children.

3.3 Gender differences in fundamental movement skill development

More girls (7.6%) than boys (4.6%) scored higher than average FMS development and more boys (36.7%) than girls (33.4%) had lower performances than average FMS development (see Table 7). Four to six year-old girls (Mdn = 93) significantly attained higher levels of movement skill competence than four to six year-old boys (Mdn = 91) (U = 165 760, P < 0.05, r = 0.07). The effect size, however, referred to a small effect explaining less than 1% of the variance (See Figure 2). When each age category was considered separately this gender difference appeared to be in age category 4, the age category in which children advance from preschool to elementary school (5 ½ and 6 years). After correction with the Bonferroni method, this difference appeared to be not significant.

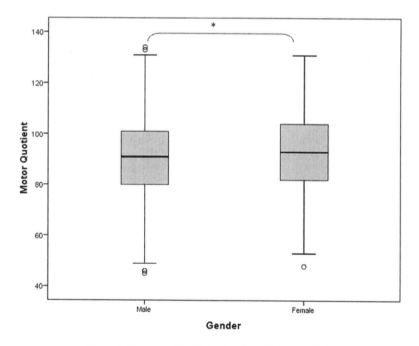

Figure 2. Boxplots of the Motor Quotients for boys and girls.

* P < 0.05

Table 6. Classification of movement skill development with MOT 4-6 in percentages.

MQ classification	Total (n=1208)	4 - 4½ (n=207)	4½-5 (n=291)	5 - 5½ (n=307)	5½-6 (n=298)	6 - 6½ (n=95)	6½-7 (n=10)	Boys (n=664)	Girls (n=554)
Weak	7.6	5.8	7.2	9.8	6.7	9.5	0.0	8.1	7.0
Under average	27.3	32.9	32.6	27.0	21.1	20.0	20.0	28.6	25.8
Normal	59.1	56.0	57.0	54.7	65.4	66.3	60.0	58.7	59.6
Good	5.6	4.8	3.1	7.8	6.4	4.2	20.0	4.1	7.4
Very good	0.3	0.5	0.0	0.7	0.3	0.0	0.0	0.5	0.2

Table 7. Gender differences in Fundamental Movement skill development.

	N	U	p	r
Total sample [N]	1208	165760.00*	.011	-0.07
4 - 4½ (cat 1)	207	5099.50	.621	-0.03
4½-5 (cat 2)	291	9825.00	.328	-0.06
5 - 5½ (cat 3)	307	11070.00	.360	-0.05
5½ - 6 (cat 4)	298	9406.00	.045	-0.12
6 - 6½ (cat 5)	95	814	.068	-0.19

* $P<0.05$, r = effect size

Bonferroni method was used to correct for type I error, so reported differences are significant only at $P<0.0125$. The analysis of data was based on the Motor Quotients for the total sample and based on the total raw scores for the age categories.

3.4 Anthropometric differences in general fundamental movement skill development

The distribution of MQ for Body Mass categories is shown in Figure 3. There was a significant difference of FMS development between children of different BMI categories (H(4) = 13.786, P<0.008). Pre-school children categorised as obese (mdn = 85, n = 84) scored significantly lower on FMS than pre-school children who have normal BMI (mdn = 93, n = 803) (U = 26 259.5, P<0.001, r = -0.11). A Bonferroni correction was applied and so the effect was reported at a 0.0125 level of significance. No differences in movement skill development were found between children in the category of 'normal weight' and children in the remaining categories [extreme underweight (n = 14, U = 5357.5, ns), underweight (n = 109,U = 39813.5, ns), overweight (n = 191 , U = 70703, ns)].

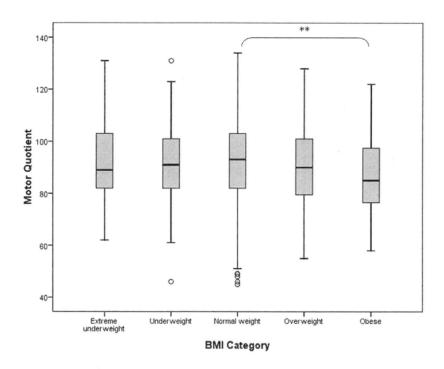

Figure 3. Boxplots of Motor Quotients for BMI category groups.
** P < 0.01

4. DISCUSSION

In this report, data were presented on the performance of Flemish children on the MOT 4-6 test (Zimmer and Volkamer, 1987). The sample was drawn in order to create a movement skill development profile of children attending preschool in

Flanders. As specific standardisation of the test in Belgium or Flanders has not been carried out yet; the normative data on German children were assumed to be equally valid for Flemish children (Cools *et al.*, 2007b). In view of possible differences between Flemish and German pre-school children, there is a need for recent German normative data which will address the question of whether the norms are valid for Flemish children or not. This study focused on pre-school children aged 4 to 6 years, so we did not attempt to present a FMS profile of the 6-year old child in general.

4.1 Motor Problems

Bös (2003), Gaschler (1999, 2000, 2001) and Zimmer (2003) reported prevalence of weaker FMS development among pre-school children and reported shifts towards weaker FMS development of at least 10% of the children. Based on the assumption of a comparable framework with these German data, our data show a similar prevalence. The results showed an unfavourable balance towards a weaker development of FMS. More pre-school children were assigned to categories 'under average'. An interesting and worrying finding from this study was that already a relative large group of children (n= 92, 8%) was classified as 'weak' or 'developmentally disturbed'. A remarkable issue was that many children are identified as being 'under average' or 'borderline' (n=330, 27%). This borderline phenomenon expresses the number of children that are at risk for further constraints in their FMS development. The large number of children in this category might be the result of a continuous change in their leisure activities: an increasing amount of opportunities for sedentary activities competing with the time spent in PA, spontaneous play and movement. Supporting evidence was found in studies reporting low PA levels among pre-school children (Cardon and De Bourdeaudhui, 2007). When this behaviour is continued for a longer period of time, there is a substantial risk of strong delay or incomplete development of mature FMS. A vicious circle threatens: by being less physically active, the rate of progress also declines and might result in more restricted FMS competence with even less participation in PA as a result. It is recommended that special attention is paid to this issue through maximising the number of appropriate developmental FMS opportunities. Governmental support and other initiatives that encourage pre-school children to be more physically active in school and at home have already proven their benefits (Sallis *et al.*, 1997; 1999; Booth *et al.*, 2006). Therefore, this support and these initiatives should be promoted, distributed and further developed on a wide scale.

4.2 Age differences

The FMS of pre-school children improved significantly with age. The chronological age of children was a dependent factor for the development of FMS. The small to medium effect sizes, however, show limited variance explained by chronological age. Further research including more anthropometric measures and expressing biological age of preschool children might explain more of the variance in movement skill development (Van Den Bosch and de Cocq, 2006).

4.3 Gender differences

The girls slightly outperformed boys on FMS development. These observations support findings of other studies (Gaschler, 1998; Sigmundsson and Rostoft, 2003). The small effect size (ES = 0.07) on the significant outcome, however, showed that only little variance was explained by gender. The absence of significant gender differences in age category groups might be due to the smaller number of children in the samples.

Controversy exists as to how these differences were induced. Some authors believed that gender differences prior to puberty were environmentally induced (Thomas and French, 1985), while other authors claimed that differences were partially induced by the earlier maturation of girls (Eaton and Yu, 1989, Berk, 2003). We believe that, according to Newell's model of motor development (Haywood and Getchell, 2005), both aspects induced these differences. More gender differences, both favourable for girls or boys, existed on item level (Duger *et al.*, 1999; Bös, 2003). Although most motor assessment tools purport to be gender neutral, comparing sums of test items for gender differences will always be an issue of doubt. The finding that boys are more often affected than girls (i.e. more boys fall into the motor impaired and borderline groups than girls) has been supported by other studies, although the outcome was not to the same extent. The reported gender range varied from 2:1 to 10:1 (boys to girls) (Gaschler, 1998; Sigmundsson and Rostoft, 2003) and was even lower (1.12:1) in our findings.

4.4 Prospects and limitations

The data of this study can be valuable for pre-school- and PE teachers as well as for parents and other care givers. The data provide a guideline to assess the present level of FMS development of pre-school children and enables specific provision of appropriate movement opportunities for pre-school children. The data also form a basis for evaluation and legitimization of additional attention, help or treatment if necessary. Based on the results which showed a strong shift towards weaker FMS development of pre-school children, we recommend that adequate curriculum time, resources, and professional development is increased in schools and that appropriate initiatives are set up to promote more physical activity among pre-school children in their home environment.

Further research on development of FMS in a longitudinal setting is needed to follow up the evolution of FMS development. Additional data that include minority groups in Flanders (e.g. migrants, underprivileged children and so on) would complete the sample. We also recommend further collaboration between European countries in order to broaden the range of knowledge on the evolution of FMS development in a European perspective. Additional to this study, evaluation more sensitive to qualitative aspects of FMS development can further complete the holistic view on FMS development among pre-school children. Finally, in our society with growing concern about expanding sedentary lifestyles, we suggest that movement skill development will be studied in relation to the type and amount of physical activity in and out of school.

5. CONCLUSION

These results showed that a large group of children were slowed down in the development of their FMS patterns further through the elementary and mature stage. This might have implications such as hampered, delayed or blocked progress in further specialized movement skill development, with a possibility of drop-out in life-long PA participation. Knowing that FMS are cornerstones in the development of the child, this delayed FMS development can affect the children's health and personal well being. Therefore, it is strongly recommended that PA interventions are planned in pre-schools and the children's home environments, with special focus on parental involvement. In addition, we recommend a close follow-up of the evolution of FMS development among pre-school children by repeating FMS assessments on a regular basis. In doing so, it might be possible to counter further unidirectional declines of FMS development among pre-school children.

Acknowledgments

The authors would like to thank all school principals, school teams and parents for their voluntary participation in the study. We would like to thank the pre-school children in particular for their almost inexhaustible enthusiasm and expression of joy during the assessments. We also want to express our gratitude to all the students and their instructors of the different institutions for higher education (Karel De Grote hogeschool, Hogeschool Limburg, Hogeschool West Vlaanderen, Hogeschool Gent and Erasmus hogeschool Brussel) for their contribution in the data collection. The authors also would also like to thank Carla Cammaert (Wesco – Erhard Sport) for her logistic support.

	Locomotion movement skill	Object control movement skill	Stability movement skill	Theme
Forward jump in a hoop			X	Jumping
Forward balance			X	Balancing
Placing dots on a sheet		X		Fine Movement skill: manual dexterity
Grasping a tissue with the toes		X	X	Fine Movement skill: bilateral coordination: grasping, Balancing
Sideward jump	X		X	Jumping, Balancing
Catching a stick		X		Trapping
Carrying balls from box to box	X	X	X	Running, Stopping, Fine movement skill: grasping
Reverse balance			X	Balancing
Throw at a target disk		X		Throwing
Collecting matches		X		Fine movement skill: manual dexterity
Passing to a hoop	X			Crawling
Jumping in a hoop on 1 foot, Standing on 1 leg	X		X	Balancing, Jumping
Catch a tennis ring		X		Catching
Jumping Jacks	X			Bilateral coordination, Jumping
Jump over a rope	X		X	Jumping, Balancing
Roll around the length axe of the body			X	Body rolling
Stand up holding a ball on the head			X	Balancing, Rising
Jump and turn in a hoop	X		X	Turning, Jumping

Appendix 1. Movement skills of MOT 4-6 test items.

References

Berk, L. E., 2003, *Child Development*. (Boston: Allyn and Bacon).

Booth, M., Okely, A., Denney-Wilson, E., Hardy, L., Yang, B., and Dobbins, T., 2006, *NSW Schools Physical Activity and Nutrition Survey (SPANS) 2004 Full Report*. (Sydney: NSW Department of Health).

Bös, K., 2003, *Motorische Leistungsfähigkeit von Kindern und Jugendlichen*, edited by Schmidt, W., Hartmann-Tews, I. and Brettschneider, W.D. (Schorndorf, Germany: Verlag Karl Hofmann), pp. 85-108.

Bunker, L. K., 1991, The role of play and motor skill development in building children's self-confidence and self-esteem. *The Elementary School Journal*, **91**, pp. 467-471.

Cardon, G. and De Bourdeaudhuij, I., 2007, Comparison of pedometer and accelerometer measures of physical activity in preschool children. *Pediatric Exercise Science*, **19**, pp. 205-214.

Certain, L. K. and Kahn, R. S., 2002, Prevalence, correlates, and trajectory of television viewing among infants and toddlers. *Pediatrics,* **109**, pp. 634-642.

Cools, W., De Martelaer, K., Samaey, C., and Andries, C., 2007a, Participating in physical activities among preschoolers: the home environment as socializing agent. (Under review)

Cools, W., De Martelaer, K., Samaey, C., and Andries, C., 2007b, Evaluating the movement skill development of typically developing preschoolers: review of seven movement skill assessment tools. (Under review)

Davies, M., 2003, *Movement and Dance in Early Childhood*. (London, California, New Delhi: Paul Chapman Publishing).

D'Hondt, E., Van Hoorne, G., Van Hoye, K., Deforche, B., De Bourdeaudhuij, I. and Lenoir, M., 2007, Relationship between motor skill and BMI in Flemish primary Schoolchildren. In *12th VK-Symposium: Bewegen in extreme omstandigheden*, edited by Derave, W. and Phillipaerts, R., (Gent: Universiteit Gent), pp. 15.

Duger, T., Bumin, G., Uyanik, M., Aki, E. and Kayihan, H., 1999, The assessment of Bruininks-Oseretsky test of motor proficiency in children. *Pediatric Rehabilitation,* **3**, pp. 125-131.

Eaton, W.O. and Yu, A.P., 1989, Are Sex differences in child motor activity level a function of sex differences in maturational status? *Child Development,* **60**, pp. 1005-1011.

Eurydice, 2005, *European Glossary on Education. Volume 2 - Second edition: Educational Institutions,* 2nd edition, (Brussel: Eurydice).

Field, A., 2005, *Discovering Statistics Using SPSS*, 2nd Ed., (London, Thousand Oaks, New Delhi, SAGE Publications Ltd).

Folio, M.R. and Fewell, R.R., 2000, *Peabody Developmental Motor Scales. Examiners Manual*. (Pro-ED. Inc., Austin-Texas).

Gallahue, D. L. and Ozmun, J. C., 2006, *Understanding Motor Development. Infants, Children, Adolescents, Adults,* 6th ed., (New York: McGraw-Hill).

Gaschler, P., 1998, Motorische Entwicklung und Leistungsfahigkeit von

Vorschulkindern in Abhängigkeit von Alter und Geslecht. *Haltung und Bewegung*, **18**, pp. 5-18.

Gaschler, P., 1999, Motorik von Kindern und Jugendlichen Heute, Teil 1. *Haltung und Bewegung,* **19**, pp. 5-16.

Gaschler, P., 2000, Motorik von Kindern und Jugendlichen Heute, Teil 2. *Haltung und Bewegung,* **20**, pp. 5-16.

Gaschler, P., 2001, Motorik von Kindern und Jugendlichen, Teil 3. *Haltung und Bewegung,* **21**, pp. 5-17.

Haywood, K. and Getchell, N., 2005, *Life Span Motor Development*, 4th ed., (Leeds: Human Kinetics).

Jongmans, M., 2004, *Vroege motorische ontwikkeling: verloop, problemen, interventie*, (Baarn: HB uitgevers), pp. 52-65.

Ketelhut, K., Mohasseb, I., Gericke, C. A., Scheffler, C., and Ketelhut, R. G., 2005, Regular exercise improves risk profile and motor development in early childhood. Verbesserung der Motorik und des kardiovaskulären Risikos durch Sport im frühen Kindesalter. *Deutsches Ärzteblatt*, **102**, A1128-A1136.

Kroes, M., Vissers, Y.L.J., Sleijpen, F.A.M., Feron, F.J.M., Kessels, A.G.H., Bakker, E., Kalff, A.C., Hendriksen J.G.M., Troost, J., Jolles, J. and Vles, J.H.S., 2004, Reliability and validity of a qualitative and quantitative motor test for 5- to 6-year-old children. *European Journal of Pediatric Neurology*, **8**, pp. 135-143.

Krombholz, 1997, Physical performance in relation to age, sex, birth order, social class, and sports activities in kindergarten and elementary school in kindergarten and elementary school. *Perceptual & Motor Skills*, **84**, pp. 1168-1170.

Lamon, A., 1989, Motoriek bij kleuters. *Tijdschrift voor de lichamelijke opvoeding*, **119**, pp. 35-42.

Le Masurier, G. and Corbin, C. B., 2006, Top 10 reasons for quality physical education. *Journal of Physical Education Recreation and Dance*, **77**, pp. 44-53.

Medeková, H., Zapletalová, L. and Havlíček, I., 2000, Habitual physical activity in children according to their motor performance and sports activity of their parents. *Gymnica*, **30**, 21-24.

Nelson, M.C. and Gordon-Larsen, P., 2006, Physical activity and sedentary behavior patterns are associated with selected adolescent health risk behaviors. *Pediatrics*, **117**, pp. 1281-1290.

Ponjaert-Kristoffersen, I., Bonduelle, M., Barnes, J., Nekkebroek, J., Loft, A., Wennerholm, Tarlatzis, B., Peters, C., Hagberg, B., Berner, A. and Sutcliffe, A.., 2005, International collaborative study of intercytoplasmic sperm injection-conceived, in vitro fertilization, and naturally conceived 5-year-old child outcomes: Cognitive and motor assessments. *Pediatrics,* **115**, pp. 283-298.

Reilly, J.J., Kelly, L., Montgomery, C., Williamson, A., Fisher, A., McColl, J.H., Lo Conte, R., Paton, J.Y. and Grant, S., 2006, Physical activity to prevent obesity in young children: cluster randomised controlled trial. *British Medical Journal*, **333**, pp. 1041-1046.

Roelants, M. and Hauspie, R., 2004, Groeicurven, Vlaanderen 2004, 2007(08-08). http://www.vub.ac.be/groeicurven.

Sääkslathi, A., Numminen, P., Niinkoski, H., Rask-Nissilä, L., Viikari, J., Tuominen, J. and Välimäki, I., 1999, Is physical activity related to body size, fundamental motor skills, and CHD risk factors in early childhood? *Pediatric Exercise Science*, **11**, pp. 327-340.

Sallis, J. F., McKenzie, T. L., Alcaraz, J. E., Kolody, B., Faucette, N. and Hovell, M. F., 1997, The effects of a 2-yaer physical education program (SPARK) on physical activity and fitness in elementary school students. *American Journal of Public Health*, **87**, pp. 1328-1334.

Sallis, J.F., McKenzie, T.L., Kolody, B., Lewis, M., Marshall, S. and Rosengard, P., 1999, Effects of health-related physical education on academic achievement: project SPARK. *Research Quarterly for Exercise and Sport*, **70**, pp. 127-34.

Sigmundsson, H. and Rostoft, M. S., 2003, Motor development: exploring the motor competence of 4-year-old Norwegian children. *Scandinavian Journal of Educational Research*, **47**, pp. 451-459.

Thomas, J. R. and French, K. E., 1985, Gender Differences Across Age in Motor Performance: A Meta-Analysis. *Psychological Bulletin*, **98**, pp. 260-282.

Ulrich, D.A., 2000, *Test of Gross Motor Development, 2nd ed. Examiner's manual*, (Pro-ED. Inc., Austin, Texas).

Van Den Bosch, J. and de Cocq, C., 2006, *Sportief talent ontdekken: scouten, meten en testen van jonge kinderen voor diverse sportdisciplines*, (Leuven: Acco).

Van Waelvelde, H., 2000, Motorische ontwikkelingsproblemen. *Signaal*, **31**, pp. 19-29.

Van Waelvelde, H., Peersman, W., Lenoir, M., Smits Engelsman, B. C. M. and Henderson, S. E., 2008, The Movement Assessment Battery for Children: Similarities and differences between 4- and 5-year-old children from Flanders and the United States. *Pediatric Physical Therapy*, **20**, pp. 30-38.

Vanvuchelen, M., Mulders, H. and Smeyers, K., 2003, Onderzoek naar de bruikbaarheid van de recente Amerikaanse Peabody Development Motor Scales-2 voor vijfjarige Vlaamse kinderen. *Signaal*, **45**, pp. 24-41

Weiss, A., Weiss, W., Stehle, J., Zimmer, K., Heck, H. and Raab, P., 2004, The influence of a psychomotor training program on the posture and motor skills of children in pre-school age. *Deutsche Zeitschrift für Sportmedizin*, **55**, pp. 101-105.

Wrotniak, B.H., Epstein, L.H., Dorn, J.M., Jones, K.E. and Kondilis, V.A., 2006, The relationship between motor proficiency and physical activity in children. *Pediatrics*, **118**, pp. 1758-1765.

Zimmer, R., 2003, Zu wenig Bewegung – zu viel Gewicht. In: Deutsche Liga für das Kind (Hrsg.): *Frühe Kindheit*. Berlin , **4**, S. 15-17.

Zimmer, R., and Volkamer, M., 1987, *Motoriktest für vier- bis sechsjärige Kinder (manual)*, (Weinheim: Betltztest).

CHAPTER TWENTY-SIX

Inter-limb coordination, strength, and jump performances following a senior basketball match

Cristina Cortis[1,2], Antonio Tessitore[2,3], Caterina Pesce[4], Maria Francesca Piacentini[2], Maurizio Olivi[2], Romain Meeusen[3] and Laura Capranica[2]

[1]Department of Health Sciences, University of Molise, Campobasso, Italy
[2]Department of Human Movement and Sport Science, IUSM, Rome, Italy
[3]Vrije Universiteit, Brussel, Belgium
[4]Department of Education in Sport and Human Motion, IUSM, Rome, Italy

1. INTRODUCTION

Basketball is one of the most popular sports in the world, played both at professional and amateur levels. Being an open-skill interval-activity sport, basketball strongly depends on the players' capability to move quickly, jump, hold the ball and coordinate lower and upper limb movements. Basketball players need high levels of speed, agility, muscular strength, and anaerobic power (Hoffman *et al.*, 1999; Kalinski *et al.*, 2002), and a moderate aerobic endurance to maintain the level of the numerous all-out performances throughout the match (Apostolidis *et al.*, 2004; Laplaud *et al.*, 2004). In the literature (Hoffman *et al.*, 1999; Apostodolis *et al.*, 2004; Cook *et al.*, 2004; Balciunas *et al.*, 2006; Ostojic *et al.*, 2006; Tessitore *et al.*, 2006; Visnapuu and Jurimae 2007), anaerobic profiles of basketball players has been assessed by means of sprints over 10 to 30m distances, vertical jumps, and handgrip strength, while technical skills were evaluated by requiring players to cover a given distance as fast as possible whilst bouncing the ball (Apostodolis *et al.*, 2004; Tessitore *et al.*, 2006). Recently, the measurement of in-phase and anti-phase rhythmical synchrony of hand and foot flexion-extensions has been proposed to evaluate the inter-limb coordination of team sport players (Lupo *et al.*, 2006). The movement constraints of in-phase and anti-phase coordination patterns, represent a phenomenological framework to explain how the central nervous system manages the control of complex tasks and strategies (Turvey, 1990; Kelso, 1994; Meesen *et al.*, 2006) and to evaluate the attentional control of the individual in maintaining the less stable anti-phase movement patterns (Baldissera *et al.*, 2000).

Since 2007, the Italian Basketball Federation organizes a championship for senior players older than 40 years of age. Studying the aerobic and anaerobic profiles of senior basketball players and the physiological load during matches, Tessitore and colleagues (2006) reported that they have good anaerobic characteristics and a moderate aerobic capacity when compared to co-aged individuals but lower than young and elite basketball players. Furthermore during their match, senior team sport players spend long periods at low-intensity activities (i.e., walking and standing) to recover from periods of high-intensity running (Tessitore *et al.,* 2005, 2006) and show a decrease in heart rates (HR) exceeding 85% of the individual's HRmax toward the end of the competition (Tessitore *et al.,* 2005), which might indicate the development of fatigue. Since it might be hypothesized that during its later part the physical abilities of players are compromised (Mohr *et al.,* 2003), the primary purpose of this study was to examine changes in jump, sprint, handgrip and inter-limb coordination performances of senior basketball players before and after their match. Considering that the development of fatigue is strictly related to the previous exercise stress, in the present study a multiple system approach of measurement (performance, physiological and psychological) is applied to the evaluation of the load imposed by the match relating HR and technical-tactical parameters to the initial, middle, and final periods of their match, and ascertaining the subjective ratings of exertion (RPE).

2. METHODS

2.1 Participants

Ten senior basketball players (age: 51 ± 7 years; range 42-61 years) gave their written consent to participate in this study which was approved by the local ethical committee. The players declared that they underwent a basketball training regimen consisting of a weekly 1.5 hour session for the previous ten years. The experimental session was organized during their regular training, which included a 50-minute friendly match with no breaks.

2.2 Procedures

To avoid any learning effect, one week before the experimental session the players were familiarized with the test procedures. The performances were measured indoors to prevent external interferences with the measurement system and to ensure that the compliance of the surface on which the tests were executed did not affect the scores. For each anaerobic test, participants were allowed two trials, with a 3-min recovery period in between, and the best performance was used for further analysis. Before the test, players underwent a 10-min warm-up period (jogging and strolling locomotion). Before and after their match, the basketball players were continuously recommended to perform at their best sprint (10 m sprint-10 m; 10 m sprint bouncing ball-$10m_{BB}$), jump (countermovement jump-CMJ), handgrip and inter-limb coordination tests. Post-match decrements, if any, were attributed to fatigue.

2.3 Sprint tests

In team sports the velocity of running is often tested over 10-30 m distances. Considering that basketball players generally do not engage in sprinting for the whole pitch length (28 m) and that 10-m distance proved to be a relevant test variable to provide significant information in team sports (Cometti *et al.*, 2001, Chamari *et al.*, 2004, Tessitore *et al.*, 2005, 2006), running speeds (i.e., 10 m and 10m$_{BB}$) were measured on the basketball pitch with infrared photoelectric cells (Polifemo, Micogate, Bolzano, Italy) positioned 10 m apart. Players began from a standing position start with the front foot 0.5 m from the first timing gate. For the 10m $_{BB}$, participants were instructed to keep the ball as close as possible to their feet and to bounce the ball at least three times. Any test that deviated from the required instruction was repeated. According to the literature (Tessitore et al., 2006), the ratio between 10m$_{BB}$ and the 10-m sprint performances has been used to evaluate the technical difficulty of running while bouncing the ball.

2.4 Vertical jump test

The jump performances were assessed measuring with 10^{-3} s precision the player's flight time (i.e., from the instant of take-off to the instant of contact upon landing) by means of an optical acquisition system (Optojump, Microgate, Bolzano, Italy) that automatically calculated the height of the jump. Starting from the standing position with hands kept on the hips, players were instructed to perform standardized two-legged countermovement jumps consisting of a fast downward movement to a freely chosen angle immediately followed by a fast maximal vertical thrust. Any jump that was perceived to deviate from the required instruction was repeated.

2.5 Handgrip test

An adjustable mechanical handgrip dynamometer (Lode, Groningen, The Netherlands) was used to record maximum isometric grip force with the participant standing comfortably with his hand close to the body with a downward direction holding the measuring beam without support. The digital strain-gauge dynamometer displays force measurements to the nearest 1 N to a maximum of 1000 N. During the test the players were verbally encouraged to exert a maximal voluntary contraction on the measuring beam. Isometric force was measured bilaterally and peak values were recorded for each trial from the digital readout on the dynamometer. The best performance was used for further analysis.

2.6 Inter-limb coordination test

According to the literature (Capranica *et al.*, 2004, 2005), to evaluate the inter-limb coordination performance players were seated shoeless on a table with elbows and knees flexed at 90° (Figure 1). The position allowed independent motion of the hands and lower limb in the sagittal plane. Players had to perform flexion and extension movement around the wrist and ankle joints with a 1:1 ratio. They were instructed to make the cyclic homolateral hand and foot movements in a continuous fashion for the total duration of a trial (60 s maximum) and to preserve spatial and temporal requirements of the movement patterns. Two homolateral coordination modes were tested: in-phase mode (i.e., association of hand extension with foot dorsal flexion and hand flexion with foot plantar flexion) and anti-phase mode (i.e., association of hand flexion with foot dorsal flexion and hand extension with foot plantar flexion). Each test condition was performed at three frequencies (80, 120 and 180 cycles.min^{-1}) dictated by a metronome. During the 2-min rest between test trials, the players were allowed to stand. The order of trial was counterbalanced among coordination modes and execution frequencies. Following 15 s at the required metronome pace, a "ready-go" command indicated the start of a trial. Using a stopwatch an observer measured the time of correct execution (s) of the homolateral hand and foot, that is the time from the beginning of the movement up to when the individual failed to meet either the spatial and/or the temporal task requirements. To avoid disagreement among observers, performances were scored by a single competent observer who showed high intra-individual reliability coefficients (range 0.89-0.95) assessed through video-recorded evaluations.

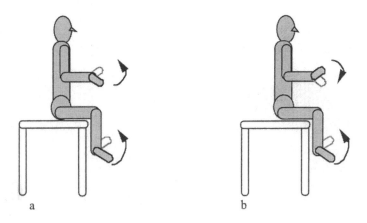

a   b

Figure 1. Schema of the in-phase (a) coordination mode (associations of wrist extension with the ankle dorsal flexion and wrist flexion with the ankle plantar flexion) and anti-phase (b) coordination mode (association of wrist flexion with the ankle dorsal flexion and wrist extension with the ankle plantar flexion).

2.6 Cardiac load and subjective ratings of the basketball match

During the friendly basketball match, players' HR was continuously recorded (Polar team system, Kempele, Finland) as average values over 5 s. According to the literature on team sports (Billows *et al.,* 2004), intensities of effort were subsequently calculated and expressed as percentages of individuals' predicted (220-age) HR_{max} to estimate the amount and duration of moderate (HR<85% of HR_{max}) and intense (HR>85% of HR_{msx}) work carried out. To provide more meaningful information on senior players (Tessitore *et al.,* 2005, 2006), their heart rate responses were also grouped in different percentages of the predicted (220-age) individual's HR_{max} (≤70%, 71-80%, 81-90% >90%). Several authors reported different declines of aerobic and anaerobic parameters in older individuals (Wiswell *et al.,* 2000; Marcell *et al.,* 2003; Stathokostas *et al.,* 2004). Furthermore, Borg's scales (Borg, 1998) of perceived exertion for the whole body (RPE) between 6 (no exertion at all) and 20 (maximal exertion) and of perceived muscle pain for lower limbs (RMP) between 0 (nothing at all) and 11 (maximum pain) were administered about 30 min after each training session to ensure that the player's perceived effort was referred to the whole session rather than the most recent exercise intensity (Impellizzeri *et al.,* 2004).

2.7 Match analysis

The senior basketball match was recorded by means of a video camera (JVC DL 107), positioned at a side of the pitch, at the level of the midfield line, at a height of 10 m and at a distance of 10 m from the touchline. The videotape was later replayed (VHS "JVC BR 8600") to evaluate the following technical-tactical parameters: 1) the number of offensive fouls; 2) the number of defensive fouls; 3) the number of offensive rebounds; 4) the number of defensive rebounds; 5) the percentages of 2-point shots made; 6) the percentages of 3-point shots made; 7) the number of turn-over; 8) the number of assists; and 10) the number of blocked shots (Tessitore *et al.,* 2006). To avoid inter-observer variability, a single observer scored the technical-tactical parameters of the competition.

2.8 Statistical analysis

Throughout the study a 0.05 level of probability was selected. Data are expressed as the mean ± standard deviation. Three equal periods (33.3%) of playing time (i.e., initial: T1; middle: T2; and last: T3), were identified to estimate the exercise load (i.e., intensity of efforts and technical-tactical parameters) imposed by the different phases of the competition. Differences in frequencies of occurrence of technical-tactical parameters were verified by means of a Chi-square test. An ANOVA for repeated measures was used to assess differences between the duration of intensity of efforts expressed in seconds, and between pre-match and post-match sprint, jump, and strength performances. Time of correct execution of synchronized hand and foot movements was submitted to a 2 (Experimental Session: pre-post) x 2 (Coordination Mode: in-phase vs anti-phase) x 3 (Execution Frequency: 80 vs 120 vs 180 cycles.min^{-1}) repeated measures ANOVA (P<0.05). Post-hoc comparisons were performed by means of Tukey's test. To eliminate the problem of an inflated Type 1 error risk for multiple comparisons, the alpha level was adjusted depending on the number of pairwise comparisons by means of Bonferroni's technique. For example, if 3 pairwise comparisons were done, the threshold of significance would be: 0.05/3 = 0.017. Furthermore, to provide meaningful analysis for comparisons from small groups, Cohen's effect sizes (ES) with respect to pre-game values were also calculated (Cohen, 1988). An ES ≤ 0.2 was considered trivial, from 0.3 to 0.6 small, <1.2 moderate and >1.2 large.

3. RESULTS

During the friendly match, the frequency distributions of HR was 35±32% for HR ≤85% of HR$_{max}$ and 65±32% for HR>85% of HR$_{max}$, respectively. Since during the second match period no occurrence of HR ≤70% of HR$_{max}$ was found, the analysis was performed pooling the lowest intensity categories. Significant differences emerged for match periods (F$_{2, 32}$<0.0001, P<0.0001) and for the match period by intensity interaction (F$_{4, 32}$=6.219, P=0.0008). Post hoc analysis showed that players significantly increased the intensity of efforts after the first phase of the match (Figure 2).

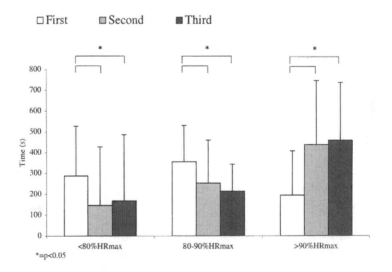

Figure 2. Means and standard deviations of the time spent at HR<80%, 80-90% and >90% of individual's HRmax during the first, second and third portions of the senior basketball match.

Although at the end of the match low RPE ratings (13 ± 2 pt, ranging from "light" to "hard") were reported, these values were different ($F_{1,8}=21.78$, $P=0.0016$, ES=-0.78) from pre-match ratings (8 ± 2 pt). Low RMP values (1.3 ± 1.6 pt, ranging from "nothing at all" to "weak") were also found at the end of the match, with no difference from pre-match (1.0 ± 0.4 pt, ranging from "nothing at all" to "weak") ones.

Regarding the technical aspects of the match, a limited number of fouls (offensive: $n = 2 \pm 2$; defensive: $n = 2 \pm 2$) and blocked shots ($n = 1 \pm 1$) was observed. Players caught more defensive ($n = 5 \pm 4$) than offensive rebounds ($n = 2 \pm 3$) and performed more 2-point shots (n=104; effectiveness = $36 \pm 14\%$) than 3-point shots (n=6; effectiveness = $10 \pm 14\%$). Overall lost possession was 45 balls, 15 of which were intercepted by the opponents. No difference emerged for the technical-tactical parameters of the two teams during the three match periods.

Significantly better performances ($F_{1,8}=29.63$, $P=0.0006$) were shown for the CMJ after the match (28.5 ± 3.7 cm) than before the match (25.0 ± 4.6 cm). In contrast, post-match handgrip performances (434.9 ± 93.9 N) were significantly lower ($F_{1,8}=36.43$, $P=0.0003$) than pre-match (503.5 ± 80.6 N). No difference emerged between experimental sessions for 10-m sprint (pre: 1.97 ± 0.18 s; post: 1.98 ± 0.16 s), 10mBB (pre: 2.02 ± 0.11 s; post: 2.00 ± 0.01 s) and 10mBB/10-m (pre and post: 1.04 ± 0.05) performances. Post-match to pre-match ratios approached 1.0 for sprint and jump tests (10 m: 0.99 ± 0.03; $10m_{BB}$: 1.01 ± 0.1; CMJ: 1.15 ± 0.08), while lower values were found for handgrip performances (0.87 ± 0.07). Generally, small ES with respect to pre-match values were found (range: r = 0.04-0.4).

For inter-limb coordination performances (Figure 3), main effects were found for Coordination Mode ($F_{1,16}=39.1$; $P=0.0002$) and Execution Frequency ($F_{2,16}=55.73$; $P<0.0001$). Significant interactions emerged for Coordination Mode x Experimental Session ($F_{1,16}=19.90$; $P=0.002$), Frequency of Execution x Experimental Session ($F_{2,16}=358.36$; $P=0.013$) and for Coordination Mode x Frequency of Execution x Experimental Session ($F_{22,16}=5.24$; $P=0.018$). Only the three-way interaction was analyzed further, because it includes the former two-way interactions. Post-hoc analysis showed different interaction patterns of Coordination Mode x Execution Frequency for pre-match and post-match performances. In the in-phase coordination mode, the basketball players showed a ceiling effect for the slowest frequencies of execution (i.e. 80, 120 cycles.min^{-1}) and performance decrements at the highest execution frequency, which showed significantly ($F_{1,8}=13.01$, $P=0.0069$, ES=0.62) better post-match performances (50 ± 14 cycles.min^{-1}) with respect to pre-match scores (27 ± 15 cycles.min^{-1}). Instead in the anti-phase coordination mode, performance decrements with increasing execution frequency were always present and post-match performances were similar to pre-match.

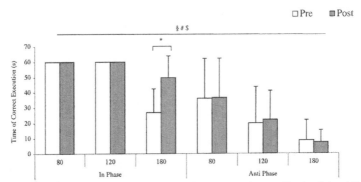

Figure 3. Means and standard deviations of the time of correct execution of homolateral hand and foot coordination in the in-phase (a) and anti-phase (b) coordination mode at three execution frequencies (80, 120, and 180 cycles.min^{-1}) before and after the senior basketball match.

4. DISCUSSION

The present study was carried out to investigate the acute effects of a senior basketball match on all-out and coordinative performances of players. The increasing number of senior players regularly involved in team sport championships, in recreational tournaments, and in friendly matches highlights the necessity to study the age-related changes in athletic profiles and match play of older players. Some authors (Kohno *et al.*, 1991; Tessitore *et al.*, 2005, 2006) reported that senior team sport players showed better aerobic and anaerobic performances than sedentary co-aged individuals although significantly worse than those of younger players. Although Dawson *et al.*, (2005) claimed that young team sport players might be able to produce "one-off" efforts close to their maximum following matches, it remains to be established whether this holds true also for senior players despite their older age and lower fitness level.

Another area that needs investigation is the study of coordinative abilities of the older individuals (Spirduso *et al.*, 2005), also in relation to sports where athletes are often challenged by movement tasks requiring complex coordination (Capranica *et al.*, 2005; Tessitore *et al.*, 2005, 2006). In field settings, task difficulty has been studied in senior team sport players comparing their sprinting performances with and without the ball (Tessitore *et al.*, 2005, 2006) and in older rhythmic gymnasts manipulating the movement pattern (i.e., in-phase and anti-phase) and execution frequency (i.e., 80, 120, 180 cycles.min^{-1}) of homolateral hand and foot cyclic synchronized movements (Capranica *et al.*, 2005). While the first test is sport specific, the second one is often used to investigate the neural control needed for the maintenance of a stable phase relation between the two moving segments that have different mechanical characteristics (Baldissera *et al.*, 2000) and to evaluate the continuous attentive effort required to inhibit conflicting neural commands to maintain the non-isodirectional coordination (Baldissera *et al.*, 1994; Sadato *et al.*, 1997; Swinnen and Jardin, 1997; Goerres *et al.*, 1998; Stephan *et al.*, 1999). Although fatigue is considered detrimental for coordinative performances, no information is available on the acute effects of a match on coordination of basketball players. Thus, the present investigation was carried out to provide evidence regarding the senior basketball players' capability of all-out and coordinative performances at the end of their match using a multiple system approach of measurement and an experimental setting with a high ecological validity. In fact, matches stress players both physiologically and psychologically and there is no laboratory setting that reproduces the exercise stress of the match or a single aspect able to reflect fully the status of the players.

In the present study, the intensities of efforts during the match were evaluated by means of the players' HR recordings expressed as the percentages of each individual's predicted HR_{max} and rating of perceived exertion. The use of the theoretical prediction of HR_{max} was due to the fact that none of the players was available for any further maximal exercise testing where invasive methods are applied or some risks might be present. In agreement with the literature (Tessitore *et al.*, 2006), the senior basketball match frequently (i.e., 65% of total time) elicited heart responses exceeding 85% of the individual HRmax. On the other hand, the high standard deviations confirm the high variability of efforts reported during senior team sport matches (Tessitore *et al.*, 2005, 2006), reflecting the intermittent exercise associated with these matches that provide plenty opportunities for active recovery from high-intensity actions. In agreement with what was reported in young players of other team sports (Castagna *et al.*, 2007; Tessitore *et al.*, 2007), the senior players rated their match as between "very light" and "hard", underestimating their actual cardiac load.

Basketball matches would likely involve less body contact than other team sports (i.e., soccer, rugby and Australian football) and friendly games less likely involve heavy contact with the opponents, which might cause higher rating of muscle soreness. Furthermore, older players tend to adopt a playing strategy to reduce the occurrence of contact with the opponent, which might cause injuries (Tessitore *et al.*, 2005). In the past decades basketball matches have become considerably faster and more spectacular, enhancing the technical and tactical aspects of man-to-man play, the senior players adopted the zone marking tactic, which reflects more the basketball they played in their younger years. This factor facilitates the alternation of short high-intensity and long low-intensity activities, increases the opportunity to get more defensive rebounds, and reduces the contact with the opponent. Therefore, it is not surprising that in the present study the players' RMP ranged between "nothing at all" and "moderate". Furthermore, the relatively low percentage of effectiveness in 2-point (36%) and 3-point (10%) shooting indicates that the basketball regimen of 1.5-hour·week^{-1} with no technical support is not sufficient to maintain a high level of shooting capacity.

In the literature (Tessitore *et al.,* in press), a decrement of CMJ and 10-m sprint performances was reported in young futsal players at the end of their friendly matches. Futsal and basketball teams have the same number of players who have both offensive and defensive roles, and pitches have comparable dimensions (i.e., basketball: 28x15 m; futsal: 38x18 m). Surprisingly, the main findings of the present study were that during the post-match tests the senior basketball players significantly increased their jump and 180 cycles.min^{-1} in-phase coordination performances, maintained their sprint capability, anti-phase coordination capabilities and 10mBB/10-m performance, and significantly decreased only their grip strength.

Handgrip strength is an important measure of general health in older individuals (Sayer *et al.,* 2006), and is measured in sport activities that use hands as tools such as basketball (Visnapuu and Jurimae, 2007). In the present study, the senior basketball players showed higher values than co-aged sedentary individuals (Sayer *et al.,* 2006) and young basketball players (Visnapuu and Jurimae, 2007). At the end of the match the significant strength reduction (14%) could be attributed to fatigue of the hand muscles continuously engaged in catching, holding and throwing the ball during playing. Since, the effects of a match on handgrip performances of young basketball players has not been investigated, at present it is not possible to attribute the present results to an age-related effect.

Although the margin for improvement might be expected when the baseline performance is rather low, in the present study the senior players showed high pre-match CMJ and coordination values. The good jumping capacity of basketball players, especially required to get rebounds and to block shots, was confirmed in the present study. In fact, the mean height of their jumps was higher than that reported for co-aged physically active men (Izquierdo *et al.,* 1999) and older soccer players (Tessitore *et al.,* 2005), and comparable to senior basketball players (Tessitore *et al.,* 2006). Studying the effects of different warm-up protocols on jump and sprint performances of young men, Vetter (2007) showed that the duration, speed, intensity, and type of exercises performed during warm-up strongly affect only the jump capabilities of young individuals. Since jump performances necessitate a fine timing of muscle activation (Bobbert and Van Soest, 2001), it is reasonable to hypothesize that a 10-minute warm-up period is not sufficient for older individuals to reach their optimum condition for jump tests. Conversely during the match, the prolonged intermittent physical exercise alternating all-out performances with sufficient active recovery could have enhanced post-match CMJ performances of the senior athletes. Also the results of sprints with and without bouncing the ball was comparable to that reported by Tessitore *et al.* (2006), confirming that bouncing the ball does not significantly affect the velocity of senior players while running short distances. On the other hand, the non-significant post-match 10-m performances indicate that further studies are necessary to elucidate fully the underlying mechanisms that differentially influence jump and sprint performances.

The different effects of the basketball match on coordination are intriguing. Despite no difference in technical or tactical variables emerging between match periods, the match had a significant effect only on the in-phase inter-limb coordination performances at 180 cycles.min^{-1}. In providing a tentative explanation of the differences between coordination performances it is necessary to take into consideration the pre-match coordination level of the senior players and the overall possible margin of improvement. In the pre-match condition the senior basketball players showed values higher than those reported for co-aged and young sedentary individuals (Capranica *et al.*, 2004), confirming that the practice of sports with high coordinative demands nullifies the regression of motor coordination performance normally observed in older individuals (Capranica *et al.*, 2005). However, as compared to closed-skills master athletes specifically trained in complex rhythmic performances (Capranica *et al.*, 2005), senior basketball players showed worst performances in the pre-match in-phase 180 cycles.min^{-1} condition. Therefore, if a margin of improvement was expected, it was more likely to be observed in this specific condition. The high time constraints of this condition stress the timing capacities of the individual. The fact that senior basketball players showed enhanced post-match performances both in CMJ and the in-phase inter-limb coordination test at 180 cycles.min^{-1} suggests that timing might represent a common factor, which is facilitated by intermittent and prolonged exercise bouts. On the other hand, the lack of post-match improvements in the anti-phase condition might be due to the fact that the pre-competition values observed in senior basketball players match the highest performance level of older individuals trained on complex rhythmic tasks (Capranica *et al.*, 2005), thus reflecting a functional ceiling effect opposite to the floor effect observed in older sedentary individuals that tend to zero with increasing velocity of execution (Capranica *et al.*, 2004). The general similarity between inter-limb coordination of older rhythmic gymnasts and basketball players contrasting with the worst performances of sedentary counterparts suggests that it is not only the specific coordinative component of a closed-skill rhythmic training that is responsible for the preservation of their synchronized coordination in older individuals. Some authors claim that aerobic training probably exerts a preservative effect on the efficiency of those executive and attentional processes involved in the control of complex motor behaviour (Kramer *et al.*, 1999; Hollman and Strüder *et al.*, 2000; Barnes *et al.*, 2003). Thus, the aerobic component of basketball training might contribute to the maintenance of complex inter-limb coordination thanks to its beneficial effects on the efficiency of executive and attentive control functions involved in complex motor behaviours. This might be a critical factor responsible for the observed transfer of acyclic open skills on the cyclic inter-limb coordination task (Barnett and Ceci, 2002). Furthermore, the positive effect of the basketball match on coordination performance offers a new insight on the additional role played by acute exercise in the complex interplay between the effects of ageing and the counteracting effects of chronic exercise training. The facilitating effect observed at the end of the match might also be due to the arousing effect of acute exercise (Tomporowski, 2003), and so more research is needed, particularly focused on the linked effects of chronic and acute exercise on both coordinative and cognitive aspects in older individuals (Pesce *et al.*, 2007).

In conclusion, the results of this study indicate that despite a senior basketball match imposing a high load on players, it does not hamper their overall post-match performances but has a positive effect on their coordinative capabilities. The limited generalizability of the present data due to the small sample size and the lack of a control group could be overcome by further research on players of different ages and sports to differentiate the practical implications of this study.

References

Apostolidis, N., Nassis, G.P., Bolatoglou, T. and Geladas, N.D., 2004, Physiological and technical characteristics of elite young basketball players. *Journal of Sports Medicine and Physical Fitness,* **44**, pp. 157-163.

Balciunas, M., Stonkus, S., Abrantes, C. and Sampaio, J., 2006, Long term effects of different training modalities on power, speed, skill and anaerobic capacity in young male basketball players. *Journal of Sports Science and Medicine,* **5**, pp. 163-170.

Baldissera, F., Cavallari, P. and Tesio, G., 1994, Coordination of cyclic coupled movements of hand and foot in normal subjects and on the healthy side of hemiplegic patients. In *Inter-Limb Coordination: Neural, Dynamical and Cognitive Constraints,* edited by Swinnen, S., Heuer, H., Massion, J. and Casaer, (San Diego, Ca: Academic Press Inc), pp. 230-241.

Baldissera, F., Borroni, P., and Cavallari, P., 2000, Neural compensation for mechanical differences between hand and foot during coupled oscillations of the two segments. *Experimental Brain Research,* **133**, pp. 165-177.

Barnes, D.E., Yaffe, K., Satariano, W., and Tager, I.B., 2003, A longitudinal study of cardiorespiratory fitness and cognitive function in healthy older adults. *Journal of the American Geriatrics Society,* **51**, pp. 459-465.

Barnett, S.M., and Ceci, S.J., 2002, When and where do we apply what we learn? A taxonomy for far transfer. *Psychological Bullettin,* **128**, pp. 612-637.

Billows, D., Reilly, T., and George, K., 2004, Physiological demands of match-play on elite adolescent footballers. *Journal Sports Sciences,* **22**, pp. 524-525.

Bobbert, M.F., and Van Soest A.J., 2001, Why do people jump the way they do? *Exercise Sport Science Review,* **29**, pp. 95-102.

Borg, G., 1998, *Borg's Perceived Exertion and Pain Scales.* (Champaign, IL: Human Kinetics), pp. 2-16.

Capranica, L., Tessitore, A., Olivieri, B. and Pesce, C., 2005, Homolateral hand and foot coordination in trained old women. *Gerontology,* **51**, pp. 309-315.

Capranica, L., Tessitore, A., Olivieri, B., Minganti, C. and Pesce, C., 2004, Field evaluation of cycled coupled movements of hand and foot in older individuals. *Gerontology,* **50**, pp. 399-406.

Castagna, C., Belardinelli, R., Impellizzeri, F.M., Abt, G.A., Coutts, A.J. and D'ottavio, S., 2007, Cardiovascular responses during recreational 5-a-side indoor-soccer. *Journal of Science and Medicine in Sport,* **10**, pp. 89-95.

Chamari, K., Hachana, Y., Ahmed, Y.B., Galy, O., Sghaier, F., Chatard, J.C., Hue, O. and Wisloff, U., 2004, Field and laboratory testing in young elite soccer players. *British Journal of Sports Medicine,* **38**, pp. 191-196.

Cohen, J., 1988, *Statistical Power Analysis for the Behavioral Sciences,* 2nd ed.,

(Hillsdale, NJ: Lawrence Eribaum Associates).

Cometti, G., Maffiuletti, N.A., Pousson, M., Chatard, J.C. and Maffulli, N., 2001, Isokinetic strength and anaerobic power of elite, subelite and amateur French soccer players. *International Journal of Sports Medicine,* **22**, pp. 45-51.

Cook, J.L., Kiss, Z.S., Khan, K.M., Purdam, C.R. and Webster, K.E., 2004, Anthropometry, physical performance, and ultrasound patellar tendon abnormality in elite junior basketball players: a cross-sectional study. *British Journal of Sports Medicine,* **38**, pp. 206-209.

Dawson, B., Gow, S., Modra, S., Bishop, D. and Stewart, G., 2005, Effects of immediate post-game recovery procedures on muscle soreness, power and flexibility levels over the next 48 hours. *Journal of Science and Medicine in Sport,* **8**, pp. 210-221.

Goerres, G.W., Samuel, M., Jenkins, I.H. and Brooks, D.J., 1998, Cerebral control of unimanual and bimanual movements: An $H_2{}^{15}O$ PET study. *Neuroreport,* **9**, pp. 3631-3638.

Hoffman, J.R., Epstein, S., Einbinder, M. and Weinstein, Y., 1999, The influence of anaerobic capacity on anaerobic performance and recovery indices in basketball players. *Journal of Strength and Conditioning Research,* **13**, pp. 407-411.

Hollmann, W. and Strüder, H., 2000, Das menschliche Gehirn bei körperlicher Arbeit und als leistungsbegrenzender Faktor. *Leistungssport,* **30**, pp. 53-58.

Impellizzeri, F. M., Rampinini, E., Coutts, A.J., Sassi, A. and Marcora, S.M., 2004, Use of RPE-based training load in soccer. *Medicine and Science in Sports and Exercise,* **36**, pp. 1042–1047.

Izquierdo, M., Ibanez, J., Gorostiaga, E., Garrues, M., Zuniga, A., Anton, A., Larrion, J.L. and Hakkinen, K., 1999, Maximal strength and power characteristics in isometric and dynamic actions of the upper and lower extremities in middle-aged and older men. *Acta Physiologica Scandinavica,* **167**, pp. 57-68.

Kalinski, M.I., Norkowski, H., Kerner, M.S. and Tkaczuk, W.G., 2002, Anaerobic power characteristics of elite athletes in national level team-sport games. *European Journal of Sport Sciences,* **2**, pp. 1-14.

Kelso, J.A.S., 1994, Elementary coordination dynamics. In *Inter-limb Coordination: Neural, Dynamical and Cognitive Constraints,* edited by Swinnen, S., Heuer, H., Massion, J. and Casaer, P., (San Diego, Ca: Academic Press Inc), pp. 301-318.

Kohno, T., O'Hata, N., Shirahata, T., Hisatomi, N., Endo, Y., Onodera, S. and Sato, M., 1991, Change with age of cardiopulmonary function and muscle strength in middle and advanced-aged soccer players. In *Science and Football II,* edited by Reilly, T., Clarys, J. and Stibbe, A., (London: Routledge), pp. 53-58.

Kramer, A.F., Hahn, S., Cohen, N.J., Banich, M.T., Mcauley, E., Harrison, C.R., Chason, J., Vakil, E., Bardell, L., Boileau, R.A. and Colcombe, A., 1999, Aging, fitness and neurocognitive function. *Nature,* **400**, pp. 418-419.

Laplaud, D., Hug, F. and Menier, R., 2004, Training-Induced changes in aerobic aptitudes of professional basketball players. *International Journal of Sports Medicine*, **25**, pp. 103-108.

Lupo, C., Tessitore, A., Cortis, C., Perroni, F., Pesce, C. and Capranica, L., 2006, Correlation between strength, power and inter-limb coordination in soccer players. 11th Annual Congress of the European College of Sport Science, Lausanne, July 5-8.

Marcell, T.J., Hawkins, S.A., Tarpenning, K.M., Hyslop, D.M. and Wiswell, R.A., 2003, Longitudinal analysis of lactate threshold in male and female master athletes. *Medicine and Science in Sports and Exercise,* **35**, pp. 810-817.

Meesen, R.L., Wenderoth, N., Temprado, J.J., Summers, J.J. and Swinnen, S.P., 2006, The coalition of constraints during coordination of the ipsilateral and heterolateral limbs. *Experimental Brain Research*, **174**, pp. 367-375.

Mohr, M., Krustrup, P. and Bangsbo, J., 2003, Match performance of high-strandard soccer players with special reference to development of fatigue. *Journal of Sports Sciences*, **21**, pp. 519-528.

Ostojic, S.J., Mazic, S. and Dikic, N., 2006, Profiling in basketball: Physical and physiological characteristics of elite players. *Journal of Strength and Conditioning Research*, **20**, pp. 740-744.

Pesce, C., Cereatti, L., Casella, R., Baldari, C. and Capranica, L., 2007, Preservation of visual attention in older expert orienteers at rest and under physical effort. *Journal of Sport Exercise Psychology*, **29**, pp. 78-99.

Sadato, N., Yonekura, Y., Waki, A., Yamada, H. and Ishii, Y., 1997, Role of the supplementary motor area and the right premotor cortex in the coordination of bimanual finger movements. *Journal of Neuroscience,* **17**, pp. 9667-9674.

Sayer, A.A., Syddall, H.E., Martin, H.J., Dennison, E.M., Roberts, H.C. and Cooper, C., 2006, Is grip strength associated with health-related quality of life? Findings from the Hertfordshire Cohort Study. *Age and Ageing,* **35**, pp. 409-415.

Spirduso, W.W., Francis, K.L. and Macrae, P.G., 2005, *Physical Dimensions of Aging.* (Champaign, IL: Human Kinetics).

Stathokostas, L., Jacob-Johnson, S., Petrella, R.J. and Paterson, D.H., 2004, Longitudinal changes in aerobic power in older men and women. *Journal of Applied Physiology,* **97**, pp. 781-789.

Stephan, K.M., Binkofski, F., Halsband, U., Dohle, C., Wunderlich, G., Schnitzler, A., Tass, P., Posse, S., Herzog, H., Sturm, V., Zilles, K., Seitz, R.J. and Freund, H.J., 1999, The role of ventral medial wall motor areas in bimanual co-ordination. A combined lesion and activation study. *Brain*, **122**, pp. 351-368.

Swinnen, S.P. and Jardin, K., 1997, Egocentric and allocentric constraints in the expression of patterns of inter-limb coordination. *Journal of Cognitive Neuroscience*, **9**, pp. 348-377.

Tessitore, A., Meeusen, R., Tiberi, M., Cortis, C., Pagano, R. and Capranica, L., 2005, Aerobic and anaerobic profile, heart rate and match analysis in older soccer players. *Ergonomics*, **48**, pp. 1365-1377.

Tessitore, A., Tiberi, M., Cortis, C., Pagano, R., Meeusen, R. and Capranica, L.,

2006, Aerobic and anaerobic profiles, heart rate and match analysis in older basketball players. *Gerontology*, **52**, pp. 214-222.

Tessitore, A., Meeusen, R., Cortis, C. and Capranica, L., 2007, Effects of different recovery interventions on anaerobic performances following soccer training. *Journal of Strength and Conditioning Research,* **21**, pp. 745-750.

Tessitore, A., Meeusen, R., Pagano, R., Benvenuti, C., Tiberi, M. and Capranica, L., 2008, Effectiveness of active vs passive recovery strategies after futsal games. *Journal of Strength and Conditioning Research*, **22**, pp. 1402-1412.

Tomporowski, P.D., 2003, Effects of acute bouts of exercise on cognition. *Acta Psychologica,* **112**, pp. 297-324.

Turvey, M.T., 1990, Coordination. *The American Psychologist*, **45**, pp. 938-53.

Vetter, R.E., 2007, Effects of six warm-up protocols on sprint and jump performance. *Journal of Strength and Conditioning Research*, **21**, pp. 819–823.

Visnapuu, M. and Jurimae, T., 2007, Handgrip strength and hand dimensions in young handball and basketball players. *Journal of Strength and Conditioning Research*, **21**, pp. 923-929.

Wiswell, R.A., Jaque, S.V., Marcell, T.J., Hawkins, S.A., Tarpenning, K.M., Constantino, N. and Hyslop, D.M., 2000, Maximal aerobic power, lactate threshold, and running performance in athletes. *Medicine and Science in Sports and Exercise*, **32**, pp. 1165-1170.

INDEX